STECK-VAUGHN
GED Science

MW00710401

STECK-VAUGHN ADULT EDUCATION ADVISORY COUNCIL

Donna D. Amstutz
Assistant Professor
Adult Education
Northern Illinois
   University
DeKalb, Illinois

Sharon K. Darling
President, National Center
   for Family Literacy
Louisville, Kentucky

Roberta Pittman
Director, Project C3 Adult
   Basic Education
Detroit Public Schools
Detroit, Michigan

Elaine Shelton
President, Shelton
   Associates
Consultant, Competency-
   Based Adult Education
Austin, Texas

RSVP
RAINTREE
STECK-VAUGHN
PUBLISHERS
The Steck-Vaughn Company

STAFF CREDITS

Supervising Editor: Ron Lemay
Editor: Carolyn Hall
Design Director: D Childress
Design Manager: Scott Huber
Production Coordinator: Pam Heaney

Editorial Consultant: Donna Amstutz
Photo Editor: Margie Matejcik

Writer/Editor: Marion B. Castellucci, Kathryn R. Fox

Design/Production: Dodson Publication Services
Cover Photograph: David Mackenzie

Editorial Development: McClanahan & Company, Inc.
Project Director: Mark Moscowitz

Photograph Credits: p. 38 © Sobel/Klonsky/The Image Bank;
p. 111 (pigeons) National Audubon Society/Photo Researchers;
p. 111 (auk) © John James Audubon/National Audubon Society/Photo Researchers;
p. 132 © Jose Acosta/Gamma-Liaison;
p. 180 © Lewis Portnoy/Uniphoto;
p. 206 © E. Ilasenko/Gamma-Liaison

Illustration Credits: Merry Finley
John Harrison
Pam Heaney
Alan Klemp
Tonia Klingensmith
Dwain Osborne

ISBN 0-8114-4701-4

Copyright © 1991 Steck-Vaughn Company

Printed in the United States of America.

4 5 6 7 8 9 CP 97 96 95 94 93 92

Table of Contents

INTRODUCTION TO THE STUDENT

PRETEST

LIFE SCIENCE

EARTH SCIENCE

CHEMISTRY

PHYSICS

POSTTEST

SIMULATED GED TEST

What You Should Know About the GED Test

What is the GED Test?

You are taking a very big step toward changing your life with your decision to take the GED test. By opening this book, you are taking your second important step: to prepare for the test. You may feel nervous about what is ahead, which is only natural. Relax and read the following pages to find out the answers to your questions.

The GED, the test of General Educational Development, is given by the GED Testing Service of the American Council on Education for adults who did not graduate from high school. When you pass the GED, you will receive a certificate that is the equivalent of a high school diploma. It is regarded as being the same as a high school diploma. Employers in private industry and government, as well as admissions officers in colleges and universities, accept the GED certificate as they would a high school diploma.

The GED tests cover the same subjects people study in high school. The five subject areas are: Writing Skills, Interpreting Literature and the Arts, Social Studies, Science, and Mathematics. In Writing Skills Part II, you will be asked to write a short essay on a specific subject. You will not be required to know all the information that is usually taught in high school. You will, however, be tested on your ability to read and process information. In some states you may be required to take a test on the U.S. Constitution or on your state government. Check with your local adult education center to see if your state requires this test.

Each year hundreds of thousands of adults take and pass the GED test. The *Steck-Vaughn GED Series* will help you to develop and refine your reading and thinking skills in order to pass the GED test.

What You Should Know About GED Scores

After you complete the GED test, you will get a score for each section and a total score. The total score is an average of all the other scores. The highest score possible on a single test is 80. The scores needed to pass the GED test vary, depending on where you live. The chart on page 2 shows the minimum state score requirements. A minimum score of *40 or 45* means that no test score can be less than 40, or if one or more scores is less than 40, an average of at least 45 is required. Scores of *35 and 45* mean that no test score can be less than 35, and an average of at least 45 is required.

GED Score Requirements

Area	Minimum Score on Each Test		Minimum Average on All Five Tests
UNITED STATES			
Alabama, Alaska, Arizona, Colorado, Connecticut, District of Columbia, Georgia, Hawaii, Idaho, Illinois, Indiana, Iowa, Kansas, Kentucky, Maine, Massachusetts, Michigan, Minnesota, Missouri, Montana, Nevada, New Hampshire, North Carolina, Ohio, Pennsylvania, Rhode Island, Tennessee, Vermont, Virginia, Wyoming	35	and	45
Arkansas, California, Delaware, Florida, Maryland, New York, Oklahoma, Oregon, South Dakota, Utah, Washington, West Virginia	40	and	45
Louisiana, Mississippi, Nebraska, Texas	40	or	45
New Jersey (42 is required on Test 1; 40 is required on Tests 2, 3, and 4; 45 is required on Test 5; and 225 is required as a minimum test score.)			
New Mexico, North Dakota	40	or	50
South Carolina	–		45
Wisconsin	40	and	50
CANADA			
Alberta, British Columbia, Manitoba, Northwest Territories, New Brunswick (35 and 45 for French), Nova Scotia, Prince Edward Island, Saskatchewan, Yukon Territory	45		–
Newfoundland	40	and	45
U.S. TERRITORIES & OTHERS			
Guam, Kwajalein, Puerto Rico, Virgin Islands	35	and	45
Canal Zone, Palau	40	and	45
Mariana Islands, Marshall Islands	40	or	45
American Samoa	40		–

The chart tells you what will be on each test. When you take the simulated GED test in this book, see how well you do within the time limit. In some states you do not have to take all sections of the test on the same day. If you take the test one section at a time, this chart can help you decide how much time you will need for each test. If you want to take all the test sections in one day, you will find that the GED test will last an entire day. Check with your local adult education center for the requirements in your area.

Test	Content Areas	Number of Items	Time Limit (minutes)
Writing Skills Part I	Sentence Structure Usage Mechanics	55	75
Writing Sample Part II	Essay	1	45
Social Studies	U.S. History Geography Economics Political Science Behavioral Science	64	85
Science	Life Science Earth Science Physics Chemistry	66	95
Interpreting Literature and the Arts	Popular Literature Classical Literature Commentary	45	65
Mathematics	Arithmetic Algebra Geometry	56	90

Where Do You Go to Take the GED Test?

The GED test is offered year-round throughout the United States, its possessions, U.S. military bases worldwide, and in Canada. To find out where and when a test is being held near you, contact one of the following institutions in your area:

- An adult education center
- A continuing education center
- A private business school or technical school
- A local community college
- The public board of education
- A library

In addition, these institutions can give you information regarding necessary identification, testing fees, and writing implements. Schedules vary: some testing centers are open several days a week; others are open only on weekends.

Why Should You Take the GED Test?

A GED certificate can help you in the following ways:

Employment

Employees without high school diplomas or GED certificates are having much greater difficulty changing jobs or moving up in their present companies. In many cases, employers will not consider hiring people who do not have a high school diploma or its equivalent.

Education

If you want to enroll in a college or university, a technical or vocational school, or even an apprenticeship program, you often must have a high school diploma or its equivalent.

Personal

The most important thing is how you feel about yourself. You have the unique opportunity to turn back the clock by making something happen that did not happen in the past. You can now attain a GED certificate that not only will help you in the future, but will help you feel better about yourself now.

How to Prepare for the GED Test

Classes for GED preparation are available to anyone who wants to take the GED. The choice of whether to take the class is entirely up to you; they are not required. If you prefer to study by yourself, the *Steck-Vaughn GED Series* has been prepared to guide your study. *Steck-Vaughn Exercise* books are also available to give you practice on all of the tests, including the writing sample.

Many people are taking classes to prepare for the GED test. Most programs offer individualized instruction and tutors who can help you identify areas in which you may need help. Most adult education centers offer free day or night classes. The classes are usually informal and allow you to work at your own pace and with other adults who also are studying for the GED. Attendance is usually not taken. In addition to working on specific skills, you will be able to take practice GED tests (like those in this book) in order to check your progress. For more information about classes available near you, call one of the institutions listed on page 3.

What You Need to Know to Pass
Test Three: Science

The GED Science Test will measure your ability to understand and think about science topics. You will be tested on how well you understand, think about, and apply science information. You will not be tested on your knowledge of science facts. The GED Science Test takes 95 minutes and has 66 items. The 66 items are divided into four content areas: life science, Earth science, chemistry, and physics.

Life Science

Almost half of the test items will involve life science topics. There may be passages and illustrations about cells, genetics, ecology, health, nutrition, disease, the structure and functions of plants and animals, reproduction, and photosynthesis.

Earth Science

About one-fifth of the test items will involve Earth science topics. There may be passages and illustrations about landforms, earthquakes, volcanoes, rock formation, minerals, fossils, and the Earth's history. Also covered are the atmosphere, weather, oceanography, tides, the solar system, and the Earth's magnetic field.

Chemistry

About one-fifth of the test items will involve chemistry topics. There may be passages and illustrations about atoms, elements, molecules, chemical bonding, chemical reactions and their energy changes, radioactivity, and hydrocarbons.

Physics

Fewer than one-fifth of the test items will involve physics topics. There may be passages and illustrations about energy, motion, forces, work, machines, heat transfer, changes of state, waves, optics, electricity, magnetism, and nuclear fission and fusion.

The test questions require you to read, understand, and think about the science passages and illustrations in several different ways. To answer the questions, you will be using four basic reading and thinking skills. The *Steck-Vaughn GED Science* text will train you in applying these skills.

Comprehension

You must read carefully and examine diagrams and charts closely to understand what the author is saying. You may be asked to identify and restate specific information. You may be asked for the main idea of a passage. You may also be asked to identify an implication, or logical consequence, of what you have read. Although only one-fifth or fewer questions on the test require *only* comprehension skills, they form the basis for the other reading and thinking skills.

Application

You will be asked to take information you have read and understood and look at it another way. To answer an application question, first you must understand an idea and then you must use it, or apply it, in another situation. For example, you may read that mammals are warm-blooded animals who give birth to live young. Then you may be asked whether a fish, bumblebee, or squirrel is a mammal. You will be taking general science information and applying it to specific situations.

Analysis

To analyze something means to take it apart and see how it works. When you analyze information, you break it down into smaller parts and examine the relationships among those parts. You may be asked to tell the difference between a fact that can be proved and an opinion. Sometimes you will have to identify a conclusion and the information on which the conclusion is based. At other times, you will be asked why something happened (a cause) or what is likely to happen as a result (an effect).

Evaluation

When you evaluate information, you make judgements about its accuracy. You may be asked whether certain information is true, based on what you have read. You will be asked whether certain conclusions, or generalizations, can be supported by the information provided. You will be asked to choose the best way to solve a problem.

Sample Passage and Questions

The following is a sample passage with questions. They are similar to actual GED Science Test questions in the way they are put together and what they ask you to do. An explanation of the correct answer follows each question.

Some Amazonian Indians have learned to manage farm plots so that over the years the land changes gradually from cleared farmland back to tropical rain forest. The plots go through stages. In the first stage, the plot is cleared and regular crops are grown. In the second stage, certain wild trees and plants are allowed to return gradually. These wild species provide a variety of products, such as medicines, insecticides, and pesticides. In the third stage, the rain forest eventually reclaims the plot. Meanwhile, other plots are in other stages of use. As a result, the soil and the forest constantly renew themselves even while supporting the Indians. In contrast, other settlers in the Amazon have cleared millions of acres for crops or timber. After a few years of cultivation, the land wears out.

1. What happens during the second stage of the farm plot?

   (1) The land is cleared.
   (2) Only regular crops are grown.
   (3) A mix of farm crops and wild plants are grown.
   (4) Only wild plants are grown.
   (5) The rain forest takes over the plot.

Answer: **(3) A mix of farm crops and wild plants are grown.**

Explanation: This question tests your comprehension skill. You can find the information you need in the fourth sentence of the passage, which states that during the second stage, certain wild plants are allowed to return gradually. This suggests that both farm crops and wild plants are grown during this stage.

2. The Indians' method of managing farmland in the Amazon is most similar to which of the following?

   (1) grazing cattle in fields
   (2) planting a different crop each season in a field to renew the soil
   (3) using chemical pesticides
   (4) growing a single crop on a large farm
   (5) using fertilizer

Answer: **(2) planting a different crop each season in a field to renew the soil**

Explanation: This question tests your ability to apply the information you are given to another situation. First, you must understand that some Amazon Indians manage their farmland so that the soil and forest can renew themselves. This idea is most similar to crop rotation, in which a different crop is planted each season. The other options do not involve the natural renewal of the land.

3. If the Indians kept the wild plants from gradually returning to the farm plot, what would be the result?

   (1) More crops could be grown for years.

   (2) The wild plants would die out.

   (3) The quality of the crops would improve.

   (4) The crops would wear out the soil after a few seasons.

   (5) The rain forest would take over the plot.

Answer: **(4) The crops would wear out the soil after a few seasons.**

Explanation: The question tests your ability to analyze information. In this case, you are being asked to predict the result of an action. If the Indians continued to grow crops, it is likely that the same thing that happened to large areas cleared for crops would happen to the small farm plots also. The land would wear out.

4. Which of the following conclusions is supported by the information in the passage?

   (1) Large-scale clearing for farming is the best long-term use of the rain forest.

   (2) Areas of rain forest should not be used for growing crops.

   (3) The rain forest is being destroyed at a rate of millions of acres per year.

   (4) Dairy farming would be a better use of the rain forest.

   (5) Some farmers can use the rain forest without destroying it.

Answer: **(5) Some farmers can use the rain forest without destroying it.**

Explanation: This question requires that you evaluate information for its accuracy. First, you must understand the passage. Then you must choose the conclusion that logically follows from the ideas in the passage. In this case, the conclusion supported by the passage is that a resource such as a rain forest can be used without being destroyed. Options (1) and (2) are clearly contradicted by the passage. Option (3) is true, but nothing in the passage proves it is true. Option (4) is not mentioned in the passage.

To help you develop your reading and thinking skills, the answer key for each question in this book has an explanation of why the correct answer is right and why the incorrect answers are wrong. By studying these explanations, you will learn strategies for understanding and thinking about science.

Test-Taking Tips for the GED Science Test

- When you read to find the main idea, if you do not find a directly stated main idea, it is probably implied in the details and examples. To understand an unstated main idea, ask yourself questions such as: What is happening? Where is it happening? Why is it happening? (Ask who, what, when, where, why, and how.)

- Remember that restating an idea does not mean just repeating it. Each time the author adds another detail or example, the central idea of the passage is made clearer. Each supporting detail adds to the main idea, much as adding another piece to a puzzle makes it easier to identify the picture.

- Science passages often use examples that show how a scientific principle can be applied to everyday life. As you read each example, think of related examples that you have seen in your life. Then think about how your examples are related to the scientific principle.

- When you are distinguishing between fact and opinion, look for supporting details that prove a point. Remember that facts can be proved, but opinions cannot.

- When you are drawing a conclusion from a science passage, you will have to use reasoning skills as well as comprehension skills. Before you come to a conclusion, be sure to identify the main idea and determine the meaning of any unfamiliar terms from the context.

- As you read a science passage, look for cause and effect relationships. Sometimes the order in which events take place gives a hint of these relationships. Recognizing cause and effect relationships can help you analyze a passage.

- Remember to look at any diagrams. Although diagrams often repeat information that is presented in the passage, they also add extra details. Notice the terms that label parts of the diagram, and look for these terms as you read the passage.

And last but not least:

- Before you answer a question, be sure you have found evidence in the science passage that supports your choice. Do not rely on things you know outside the context of the passage.

Study Skills

Organize your time.

- If you can, set aside an hour every day to study. If you do not have time every day, set up a schedule of the days you can study and stick to that schedule. Be sure to pick a time when you will be the most relaxed and least likely to be bothered by outside distractions.

- Let others know your study time. Ask them to leave you alone for that period. It helps to explain to others why this is important.

- You should be relaxed when you study, so find an area that is comfortable and quiet. If this is not possible at home, go to a local library. Many libraries have areas for reading and studying. If there is a college or university near you, check the library there. Most college libraries have spaces set aside for studying.

Organize your study materials.

- Be sure to have sharp pencils, pens, and paper for any notes you might want to take.

- Keep all your books together. If you are taking an adult education class, you probably will be able to borrow some books or other study material. Keep them separate from your own books so that there is no mix-up.

- Make a separate notebook or folder for each subject you are studying. A folder with pockets is useful for storing loose papers.

- If you can study at home, keep all your material near your study area so you will not waste time looking for it each time you study.

Read!

- Read the newspaper, read magazines, read books. Read whatever appeals to you—but read! If it sounds as though this idea has been repeated a lot, you are right. Reading is the most important thing you can do to prepare for the tests.

- Go to your local library. If you are not familiar with the library, ask the librarian for help. If you are not sure what to read, ask the librarian to suggest something. Be sure to tell the librarian the kinds of things you are interested in. Ask for a library card if you do not have one.

- Try to read something new every day.

Take notes.

- Take notes on things that interest you or things that you think might be useful.

- When you take notes, do not simply copy the words directly from the book. Restate the information in your own words.

- You can take notes any way you want. You do not have to write in full sentences. Just be sure that you will be able to understand your notes later.

- Use outlines, charts, or diagrams to help you organize information and make it easier to learn.

- You may want to take notes in a question-and-answer form, such as: What is the main idea? The supporting details are . . .

Improve your vocabulary.

- As you read, do not skip over a word you do not know. Instead, try to figure out what the difficult word means. First, omit it from the sentence. Read the sentence without the word and try to put another word in its place. Is the meaning of the sentence the same?

- Make a list of unfamiliar words, look them up in the dictionary, and write down the meanings.

- Since a word may have several meanings, it is best to look up the word while you have the passage with you. You can then try out the different meanings in the sentence.

- When you read the definition of a word, restate it in your own words. Use the word in a sentence or two.

Make a list of subject areas that give you trouble.

- As you go through this book, make a note whenever you do not understand something.

- Go back and review the problem when you have time.

- If you are taking a class, ask the teacher for special help with the problem areas.

Use the glossary at the end of this book to review the meanings of the key science terms.

- All the words you see in boldface type are defined in the back of the book.

- In addition, definitions of other important words are included.

Self-Inventory of Study Skills

Ask yourself the following questions:

- Am I organized?
- Do I have a study schedule?
- Do I stick to my study schedule?
- Do I have a place to study?
- Do I have a place for my study material?
- Do I have a notebook and a pencil or pen?
- Do I rcad something at least once a day?
- Do I take notes on what I read?
- Do I review my notes later?
- Do I pay attention to words I don't know?
- Do I look up words in the dictionary and write down their meanings?
- Do I know where my local public library is?
- Do I have a library card so that I can check out books to read?
- Do I keep a list of things that give me trouble?
- Do I ask for help when I need it?

If you said <u>yes</u> to all the questions, good for you! If there were a few <u>no's</u>, why not give those areas a try?

PRETEST

Science

The Science Pretest consists of multiple-choice questions intended to measure your understanding of the general concepts in science. The questions are based on short readings that often include a graph, chart, or figure. Study the information given and then answer the question(s) following it. Refer to the information as often as necessary in answering the questions.

You should spend no more than 95 minutes answering the 66 questions on the Science Pretest. Work carefully, but do not spend too much time on any one question. Be sure you answer every question. You will not be penalized for incorrect answers.

Record your answers to the questions on the answer sheet provided on page 270. To record your answers, mark one numbered space on the answer sheet beside the number that corresponds to the question on the Science Pretest.

Example   Which of the following is the smallest unit in a living thing?

(1) tissue
(2) organ
(3) cell
(4) muscle
(5) capillary

The correct answer is "cell"; therefore, answer space 3 should be marked on the answer sheet.

Do not rest the point of your pencil on the answer sheet while you are considering your answer. Make no stray or unnecessary marks. If you change an answer, erase your first mark completely. Mark only one answer space for each question; multiple answers will be scored as incorrect. Do not fold or crease your answer sheet.

Items 1 to 6 refer to the following article.

Ants are social insects. This means that they live in a group called a colony and that each ant has a job to do to help the whole group survive. Ants must live in a colony; an ant cannot survive alone.

You will find three kinds of ants in a typical colony:

1. There is one queen who lays eggs.
2. There are winged males. The males develop from unfertilized eggs. Their role is to mate with the queen, after which they die.
3. There are workers. Most of the ants in a colony are worker ants. They are females who cannot lay eggs. Instead, they find food and sometimes fight to help the colony.

Ants are usually helpful to humans. Their underground tunneling mixes and enriches the soil. In some places ants turn more earth than earthworms. Ants spread plant seeds and feed on dead insects and other animals. Many ant species eat insects that are crop pests.

One species, the leaf-cutting ant, harms farmers' crops in Texas and Louisiana. Large leaf-cutting worker ants carry big pieces of leaves to the ant colony. There the smaller workers chop the leaves into tiny bits. The smallest worker ants transplant small pieces of fungus to the bits of leaves. The fungus then grows on the leaves, and the ants use the fungus for food.

1. All of the following activities performed by ants are helpful to humans EXCEPT

    (1) enriching the soil
    (2) feeding on dead insects
    (3) cutting leaves
    (4) eating crop pests
    (5) spreading plant seeds

2. A sociobiologist studies the way living creatures behave with one another in a group. What would a sociobiologist most likely study about ants?

    (1) the diet of the ant
    (2) the anatomy of the ant
    (3) the location of the colony
    (4) the roles of the queen, winged males, and workers
    (5) the enrichment of the soil

3. In a colony there are fewer males than worker ants. Which sentence best explains this?

    (1) Since there is only one queen to be fertilized, only a few males are required.
    (2) Winged males are always away from the colony.
    (3) The formation of wings requires a specialized environment.
    (4) Winged males develop from unfertilized eggs.
    (5) After they fertilize the queen, winged males lose their wings and die.

4. The role of the queen ant is to produce eggs so that the colony continues. What is the most important contribution other ants make to the life of the colony?

    (1) enriching the soil
    (2) feeding other members of the colony
    (3) cutting leaves
    (4) spreading plant seeds near the colony
    (5) eating crop pests

5. Which of the following is the smallest unit that can survive?

    (1) the individual queen
    (2) the winged male
    (3) the worker
    (4) the queen and her winged males
    (5) the colony

6. Which of the following human behaviors is most similar to the behavior of the leaf-cutting worker ants?

    (1) following a political system with a king and queen
    (2) performing tasks one after another on an assembly line
    (3) raising crops
    (4) raising livestock
    (5) maintaining an army

Items 7 and 8 refer to the following map.

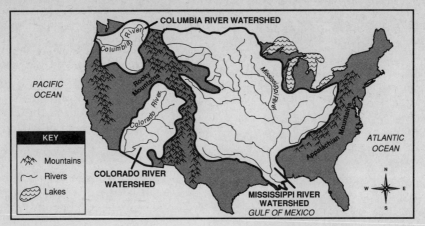

7. The area of land drained by a river is known as a watershed. Why does all the water in a watershed eventually drain into an ocean?

(1) The oceans are the largest bodies of water on Earth.
(2) Each land mass is surrounded by an ocean.
(3) Water flows toward the equator.
(4) Water flows from high elevations, such as mountains, to sea level.
(5) Too much water flows into the ocean too quickly.

8. Early explorers of the Northwest could use what they knew about watersheds to guide them. How did they know they were approaching the Pacific Ocean?

(1) They conserved water for the journey over the Rocky Mountains.
(2) They saw that the Columbia River was flowing to the west.
(3) They found the source of the Columbia River.
(4) They encountered flooding in the Columbia River watershed.
(5) They determined that the elevation of the Columbia River was too high to make it a major watershed.

Items 9 and 10 are based on the following information.

Gas is a form of matter whose molecules fill whatever space is available. The gas laws describe what happens to the gas when pressure, volume, or temperature changes.

Boyle's law tells how the volume of a gas changes when the pressure on it changes. For example, doubling the pressure on a gas will decrease its volume by one half. If all other conditions stay the same, the volume becomes smaller as the pressure becomes greater.

Charles's law tells how the volume of a gas changes when the temperature changes. Again, if all other conditions remain the same, the volume becomes greater when the temperature is higher. For example, if the absolute temperature of a gas were doubled, its volume would double.

9. Oxygen gas is often stored under pressure in metal tanks called cylinders. What is the best reason given below for storing oxygen under pressure?

(1) Oxygen would otherwise mix with air.
(2) The oxygen can be cooled.
(3) Charles's law suggests that this is the best way.
(4) More oxygen will fit in the container.
(5) The pressure turns the oxygen into a liquid.

10. To explain why hot-air balloons rise, you would need to know

A. Boyle's law
B. Charles's law
C. the number of molecules in the balloon

(1) A only
(2) B only
(3) C only
(4) A and B
(5) A and C

Items 11 to 14 refer to the following article.

Earth's crust is made of plates thousands of miles across and 30 or 40 miles thick. Scientists think that at one time all the continents formed a single land mass called Pangaea. Pangaea broke into huge pieces that drifted apart, forming the present continents.

Plate tectonics is the study of how plates form, move, and interact. Plates come together at three types of continually changing boundaries:

1. Diverging boundaries separate plates that are moving apart. These are generally located beneath the oceans. The Gulf of California is located over a diverging boundary.
2. Converging boundaries are located where plates are colliding, forming high mountains, deep ocean trenches, earthquakes, and volcanoes. The Himalayas are located at a converging boundary.
3. Transform fault boundaries occur where plates are sliding past one another. Like converging boundaries, they also cause earthquake activity. The San Andreas Fault in California is located along a transform fault boundary.

11. The formation of the Himalaya Mountains was caused by

    (1) two plates moving toward each other
    (2) two plates moving away from each other
    (3) two plates moving along each other
    (4) volcanic activity
    (5) the breakup of Pangaea

12. All of the following are related to the theory of plate tectonics EXCEPT

    (1) patterns of volcanic activity
    (2) earthquake zones
    (3) deep ocean trenches
    (4) the locations of the continents
    (5) the rise and fall in the water level of major rivers

13. According to the theory of plate tectonics, what is true of Earth's continents?

    (1) The continents are stable in their present locations.
    (2) The continents moved, causing new formations of land and water.
    (3) The separation of continents was caused by a rising ocean.
    (4) Earthquakes, volcanoes, and other dramatic geologic activity may stop someday.
    (5) Diverging boundaries usually occur on the continents.

14. What is likely to have caused the formation of the Andes Mountains along the west coast of South America, an area of earthquakes and volcanoes?

    (1) plates drifting apart
    (2) plates sliding past each other
    (3) two plates colliding
    (4) the lowering of sea level
    (5) the Ice Age

Items 15 and 16 are based on the following graph.

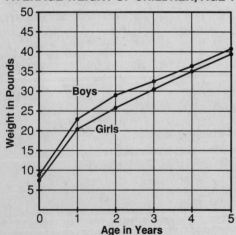

AVERAGE WEIGHT OF CHILDREN, AGE 1–5

15. During what year do children gain the most weight?

    (1) 0 – 1 year
    (2) 1 – 2 years
    (3) 2 – 3 years
    (4) 3 – 4 years
    (5) 4 – 5 years

16. At age 4, Billy weighs 39 pounds. Billy's weight is

    (1) average for his age and sex
    (2) below average for his age and sex
    (3) above average for his age and sex
    (4) above average for his sex at age 5
    (5) below average for girls at age 4

Items 17 to 20 are based on the following information.

The animal kingdom is made up of thousands of species of animals. They are divided into ten major groups, called phyla. Five of the phyla are described below:

1. Chordata. Animals have a notochord, or stiff rod of cells, as the primary skeletal support during some stage of development. Most members of this phylum are vertebrates, animals with backbones.
2. Mollusca. Animals have soft bodies and prominent shells.
3. Arthropoda. Animals have sectioned bodies covered by a jointed external skeleton.
4. Annelida. Soft-bodied, segmented animals are symmetrical, or the same on both sides.
5. Coelenterata. Unsegmented animals are symmetrical around a central point and lack a true body cavity.

Each of the following animals belongs to one of the phyla described above. For each item, choose the appropriate phylum. Each phylum above may be used more than once in items 17 to 20.

17. The earthworm, a cylindrical segmented worm, belongs to the phylum

(1) Chordata
(2) Mollusca
(3) Arthropoda
(4) Annelida
(5) Coelenterata

18. The clam, a marine animal with a shell, belongs to the phylum

(1) Chordata
(2) Mollusca
(3) Arthropoda
(4) Annelida
(5) Coelenterata

19. A cat, an animal with a backbone, belongs to the phylum

(1) Chordata
(2) Mollusca
(3) Arthropoda
(4) Annelida
(5) Coelenterata

20. A grasshopper, a long, slender-winged insect with an external jointed skeleton, belongs to the phylum

(1) Chordata
(2) Mollusca
(3) Arthropoda
(4) Annelida
(5) Coelenterata

If you drop a bar magnet into a pile of iron filings, the filings will stick to the magnet. Iron and steel and a few other metals are attracted to magnets. The force of attraction is called magnetic force. The area around a magnet where the magnetic force acts is called a magnetic field. The field is strongest at the poles, or ends, of the magnet.

Each magnet has a north and a south pole. If you hang a magnet from a string, its north pole will turn to the north. This is because the Earth has a magnetic field. The magnetic north pole of the Earth is located near the geographic north pole. A compass used for navigation points to the magnetic north pole, and not the geographic north pole.

Many substances that are attracted to a magnet can be magnetized, or made into a magnet. Inside these substances, groups of atoms act as tiny magnets. These groups, called magnetic domains, are usually pointing in different directions. If they can be made to line up in the same direction, they will cause a magnetic field. The substance will be a magnet.

21. Which of the following actions could magnetize an iron bar?

    A. leaving it on top of a magnet for a few days
    B. piling iron filings on it
    C. stroking it in one direction with a magnet

    (1) A only
    (2) B only
    (3) C only
    (4) A and B
    (5) A and C

22. Which of the following items would not be affected by a magnetic field?

    (1) iron filings
    (2) a rubber eraser
    (3) a magnetic compass
    (4) a horseshoe magnet
    (5) a steel girder in a building

23. The magnetic needle of a compass always points to the north because

    (1) compasses are used for finding direction
    (2) Earth has a magnetic north pole
    (3) the needle touches the north pole
    (4) the magnetic domains of the needle point in different directions
    (5) Earth's magnetic field is strongest at the south pole

24. Charts are available that show the difference between the geographic north pole and the magnetic north pole. These charts would be most useful for

    (1) hikers on a five-mile hike
    (2) a sailor navigating by the stars
    (3) a sailor navigating with the use of a compass
    (4) a person using a compass to find steel nails in a beam
    (5) a driver deciding which direction to go on a highway

25. Suppose you cut a bar magnet in half across the middle. What will you have?

    (1) one magnet with two north poles and one magnet with two south poles
    (2) two magnets, each with one north pole and one south pole
    (3) two magnets with only north poles
    (4) two magnets with only south poles
    (5) one magnet with a north and a south pole and one bar that is not a magnet

Items 26 and 27 are based on the following information.

Density is the relationship between the mass of a substance and its volume. For example, a substance of great mass and small volume is very dense. One substance can be distinguished from another if you know the density of each. The following chart shows the densities of some metals.

DENSITY OF SELECTED METALS	
Metal	Density in Grams per Cubic Centimeter
Aluminum	2.7
Iron	7.9
Copper	8.9
Lead	11.3
Mercury	13.6
Gold	19.3

26. An object will sink in a liquid if it is more dense than the liquid. Which of the following metals will sink in liquid mercury?

(1) iron
(2) copper
(3) lead
(4) gold
(5) all of the above

27. Pure gold is 24-karat gold. Since pure gold is too soft to hold its shape, jewelers almost always mix copper with gold to make jewelry. For example, 14K gold contains 14 parts of pure gold by weight to 10 parts of pure copper. Which of the following would have the greatest density?

(1) 24K gold
(2) 18K gold
(3) 14K gold
(4) 10K gold
(5) They would all have the same density.

Items 28 and 29 refer to the following chart.

FORMATION OF SEDIMENTARY ROCKS		
Agent of Formation	Source of Material in the Rock	Type of Rock Produced
Streams, winds, glaciers	Boulders, pebbles Sand Silt, clay	Conglomerate Sandstone Shale
Chemical reaction in seawater, evaporation	Mineral crystals	Rock salt, gypsum, some limestone
Microorganisms	Vegetation Remains of marine animals including shells	Peat and coal Most limestone

28. Which is the best conclusion that can be drawn about the formation of limestone?

(1) It is formed only by mechanical agents such as wind, streams, and glaciers.
(2) It is formed only by chemical reactions in seawater.
(3) It is formed only by microorganisms acting on the remains of marine animals, including shells.
(4) It is formed in the ocean.
(5) It consists of fine particles of silt.

29. Human beings often copy the methods and materials of nature. In which of the following activities do people copy nature's methods in the formation of sedimentary rock?

A. mixing concrete with pebbles
B. using mud to build an adobe house
C. obtaining salt from seawater by evaporation

(1) A only
(2) B only
(3) A and B
(4) B and C
(5) A, B, and C

Items 30 to 32 are based on the following article.

A gardener often builds a compost pile in a corner of a garden because the compost pile produces a mixture called humus. To create the compost pile, the gardener collects materials that were living, including kitchen waste (vegetable parings) and garden waste (lawn clippings and leaves). In the compost pile, microscopic organisms change this waste into humus. The gardener can then spread the humus on the soil to help plants grow well.

The most important microscopic organisms in compost are bacteria that specialize in breaking down organic matter. They thrive in temperatures up to 170°F and need air to survive. As the compost decomposes and cools, the bacteria give way to other microorganisms.

The new microorganisms include actinomycetes. These are a higher form of bacteria. As they become more active, they produce antibiotics, chemicals that kill the original bacteria. Protozoa, small animal organisms, are also present in compost. These are similar to bacteria, but their role in composting is less important. Other compost microorganisms are called fungi. They are primitive plants without chlorophyll. They are active during the final stages of composting.

30. Which microorganisms are active at the start of composting?

(1) bacteria
(2) antibiotics
(3) protozoa
(4) fungi
(5) All of the microorganisms are active at the start of composting.

31. The removal of which of the following would have the greatest impact on the decomposition occurring in the compost pile?

(1) bacteria
(2) antibiotics
(3) protozoa
(4) fungi
(5) The impact would be the same, regardless of which microorganism is removed.

32. All of the following material would be suitable for a compost pile EXCEPT

(1) manure
(2) straw
(3) lawn clippings
(4) sawdust
(5) plastic wrap

Items 33 and 34 refer to the following map.

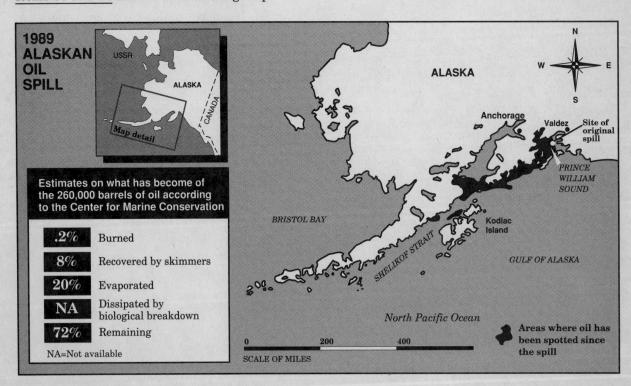

1989 ALASKAN OIL SPILL

Estimates on what has become of the 260,000 barrels of oil according to the Center for Marine Conservation

.2% Burned
8% Recovered by skimmers
20% Evaporated
NA Dissipated by biological breakdown
72% Remaining

NA=Not available

ALASKA

Anchorage
Valdez
Site of original spill

PRINCE WILLIAM SOUND

BRISTOL BAY

Kodiac Island

SHELIKOF STRAIT

GULF OF ALASKA

North Pacific Ocean

Areas where oil has been spotted since the spill

0   200   400

SCALE OF MILES

33. The map was compiled two months after the original oil spill. What is suggested by the spread of oil and the estimated amount of remaining oil?

  (1) Most of the oil was cleaned up during the first two months.
  (2) The oil could spread farther south and west.
  (3) The oil will be cleared in four months.
  (4) The ocean is able to absorb huge amounts of oil.
  (5) The remaining oil will evaporate within two months.

34. An oil spill has occurred on the east coast of the United States. Based on the map, what have scientists learned that can be applied to cleaning up the new spill?

  (1) The moving of barriers into position earlier will help.
  (2) Skimmers should not be used.
  (3) They should look into biological methods of cleaning up the oil.
  (4) They will be able to clear most of the oil quickly.
  (5) Evaporation will probably help clear some of the oil.

Items 35 and 36 refer to the following diagram.

HURRICANE IN NORTHERN HEMISPHERE

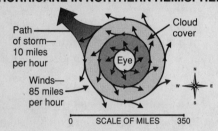

35. A family on the coast of Virginia took steps to protect their house from a strong hurricane. For more than ten hours, heavy rain fell and strong winds blew from the east. Then the storm quieted, and the sky cleared. What should the family do?

  (1) Remove the storm protection measures because the hurricane is over.
  (2) Take down the storm protection measures because the rest of the hurricane will be mild.
  (3) Leave storm protection measures in place because the rain and winds will start again—this time from the other direction.
  (4) Leave the storm protection measures in place because there may be another hurricane tomorrow.
  (5) Resume normal activities.

36. Which of the following is least likely to cause serious damage to life and property in a hurricane?

  (1) heavy rains
  (2) flooding
  (3) high winds
  (4) passage of the eye
  (5) All of these are likely to cause serious damage.

Items 37 and 38 refer to the following information.

Corrosion, or rusting, is a process in which iron combines with air and water to form rust. Rust is brittle and flakes off, so the fresh surfaces of the iron continue to be exposed to corrosion. Rusting can be prevented by keeping air and water from the iron surface.

37. A practical method of protecting cast-iron cookware from rusting is

  (1) washing it in detergent and water
  (2) not exposing it to liquids
  (3) not exposing it to air
  (4) oiling it to create a barrier against air and water
  (5) heating it slowly

38. Plating — or coating iron with chromium, tin, nickel, or zinc — prevents rust because

  (1) the rust has been permanently removed
  (2) the plating metal destroys the rust
  (3) the iron is mixed with the other materials
  (4) the rust no longer flakes off
  (5) the iron is not exposed to air and water

CELL DIVISION

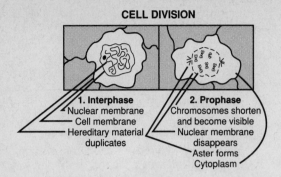

1. **Interphase**
Nuclear membrane
Cell membrane
Hereditary material
duplicates

2. **Prophase**
Chromosomes shorten
and become visible
Nuclear membrane
disappears
Aster forms
Cytoplasm

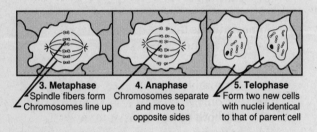

3. **Metaphase**
Spindle fibers form
Chromosomes line up

4. **Anaphase**
Chromosomes separate
and move to
opposite sides

5. **Telophase**
Form two new cells
with nuclei identical
to that of parent cell

39. In cell division, one cell becomes two. Each cell must have chromosomes exactly like those of the parent cell. What step ensures that the chromosomes are the same?

(1) duplication of hereditary material (Interphase)
(2) disappearance of the nuclear membrane (Prophase)
(3) formation of the aster (Prophase)
(4) formation of spindles (Metaphase)
(5) division of the cell material (Telophase)

40. According to the diagram, where does the material that was inside the nucleus go when the nuclear membrane disappears?

(1) It joins the cell membrane.
(2) It links to the asters.
(3) It enters the cytoplasm.
(4) It makes up the spindles.
(5) It is part of the nuclear membrane.

Yeasts are one-celled fungi that are mixed with flour and other ingredients to make bread dough. The yeast changes some of the starches and sugars in the dough and produces alcohol and the gas, carbon dioxide. When dough becomes warm, the carbon dioxide expands and causes the bread to rise. During the actual baking, the bread continues to rise for a while, and the alcohol evaporates.

41. Before baking the bread, the dough is usually allowed to rise. To ensure that the dough rises, which is the best location to let it rest?

(1) in the freezer
(2) in the refrigerator
(3) in a cool spot in the kitchen
(4) in a warm spot in the kitchen
(5) in the heated oven

42. You have probably noticed the pores, or small holes, in a slice of bread. These holes are the result of

(1) starches in the dough
(2) sugars in the dough
(3) flour in the dough
(4) salt in the dough
(5) carbon-dioxide bubbles in the dough

Items 43 to 46 refer to the following article.

Energy passes through ecosystems in the form of food. In the process of photosynthesis, plants use the Sun's energy to make organic, or living, material. This organic material is used as food for the plants as well as for the animals that eat plants. The feeding relationships in an ecosystem make up a food chain.

Organisms in a food chain operate at one or more levels in the chain. Producers are plants that use the Sun's energy to manufacture food. Other creatures consume, or eat, the food. Primary (first-level) consumers are animals that eat plants. Secondary (second-level) consumers are animals that eat primary consumers. Tertiary (third-level) consumers are animals that eat secondary consumers. Some animals called omnivores eat plants and animals. Omnivores are at more than one consumer level. The final stage in the food chain is made up of decomposers. These microorganisms break down the remains of dead plants and animals and return the components to the soil, to be used again by plants.

Energy passes from plant to animal to animal to decomposer. Although decomposers return the remains of dead organisms to the soil, they do not return energy to the soil. Each consumer in the food chain uses up some of the energy. As a result, less energy is available to the next consumer level.

43. On an African prairie, antelopes eat grasses, and lions and cheetahs eat antelopes. Vultures eat what the cats leave. When a lion dies, vultures feed on it as well. Last, bacteria and fungi decompose what remains. In this food chain, when the vultures eat the antelopes, the vultures are functioning as

(1) producers
(2) primary consumers
(3) secondary consumers
(4) tertiary consumers
(5) decomposers

44. Which of the following is true of energy in a food chain?

(1) All the energy at one level is passed to the next level.
(2) Animals get their energy directly from the Sun.
(3) There is more energy available to secondary consumers than to tertiary consumers.
(4) There is more energy available to secondary consumers than to primary consumers.
(5) Plants get their energy from the soil.

45. Some animals are classified at more than one level in the food chain because they

(1) eat many varieties of plants
(2) eat many types of insects
(3) eat only plants
(4) eat both plants and animals
(5) have different levels of energy

46. What is the first source of energy in a food chain?

(1) nutrients in the soil
(2) the Sun
(3) plants
(4) omnivores
(5) decomposers

Items 47 and 48 refer to the following diagram.

BUOYANCY

 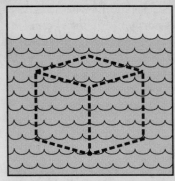

Weight of object = weight of displaced liquid (liquid pushed aside)

Weight of object is greater than weight of displaced liquid

47. An object will float if

(1) it displaces or pushes aside liquid
(2) its weight equals the weight of the liquid it displaces
(3) its weight is greater than the weight of the liquid it displaces
(4) its weight equals the weight of the liquid in which it is placed
(5) its weight is greater than the weight of the liquid in which it is placed

48. A solid steel cube will sink when placed in water, but a steel ship will float. Which of the following is the best explanation of why a steel ship will float?

(1) Steel is heavier than the water it displaces.
(2) Steel is lighter than the water it displaces.
(3) Steel and air are heavier than the water they displace.
(4) Steel and air are equal in weight to the water they displace.
(5) Steel and air are lighter than the water they displace.

Items 49 and 50 refer to the following diagram.

AN ALLUVIAL FAN

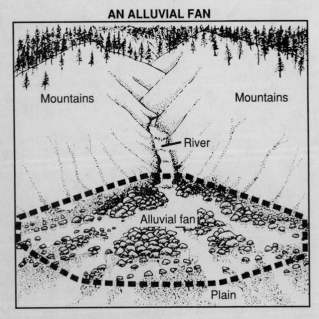

49. From the diagram you can conclude that alluvial fans

(1) are ancient land formations
(2) are formed by the action of glaciers
(3) are formed by sediment or soil carried by rivers from mountains to the plain
(4) contain rich soil that is good for agriculture
(5) occur in regions of volcanic activity

50. When the river comes out of the mountains and into the plain, it

(1) speeds up and becomes colder
(2) becomes narrower and slows down
(3) becomes narrower and speeds up
(4) spreads out and slows down
(5) goes underground

In order to be healthy, plants must take in certain minerals from the soil. These minerals contain the chemical elements that plants need to grow. Nitrogen is used for leaf and stem growth. Phosphorus helps strengthen roots and stems. Potassium helps prevent disease. Other elements are also needed but to a lesser extent. As plants grow, they use up the minerals in the soil. When plants are harvested, these chemicals are not returned to the soil. For this reason, fertilizers are added to the soil to restore its ability to support growing plants.

Artificial fertilizers consist of mined or manufactured minerals that contain the missing chemicals. These are mixed in different proportions. When the composition of the soil is analyzed, it is easy to see which chemical is missing and to select the proper mix. Fertilizer labels show the percentages of the three major ingredients. For example, 8-12-6 means that the fertilizer contains 8 percent nitrogen, 12 percent phosphorus, and 6 percent potassium. Artificial fertilizers contain these elements in compounds that are easy for plants to absorb.

Some people prefer to use organic or living matter, such as compost or manure, as fertilizer to replace the missing chemicals. These natural fertilizers contain minerals in complex compounds. Plants cannot absorb these compounds until they have been broken down by the decay process. In addition, the exact mineral content of the organic matter cannot be controlled.

51. Fertilizers are added to the soil because they

(1) replace soil worn away by erosion
(2) restore chemicals lost in the composting process
(3) restore chemicals used up as plants grow
(4) help control the microorganisms in the soil
(5) are better than manure

52. The addition of fertilizer to the soil is most similar to adding

(1) vegetables to your diet
(2) vitamins and minerals to processed flour
(3) artificial sweetener to a soft drink
(4) fluoride to the water supply
(5) sugar to cereal

53. On what basis could a farmer who uses organic fertilizers claim that these vegetables are better than vegetables grown using artificial fertilizers?

(1) Organically grown vegetables use organic forms of minerals.
(2) Organically grown vegetables use simpler forms of minerals.
(3) Organic fertilizers act on the soil in a manner more similar to nature than do chemical fertilizers.
(4) Organic fertilizers can be more easily adjusted to achieve the correct balance of minerals than chemical fertilizers.
(5) Organic fertilizers do not contain pesticides.

Items 54 to 56 are based on the following information.

Mammals are a class of animals whose females produce milk to feed their young after birth. Mammals are divided into several groups, or orders. Five of the orders are described below.

1. <u>Carnivores</u> are mammals that are meat-eaters.
2. <u>Cetaceans</u> are mammals with very large heads and tapering bodies that live entirely in the water, usually the ocean.
3. <u>Marsupials</u> are mammals whose females have pouches in which the young develop after birth.
4. <u>Rodents</u> are small mammals with a single pair of incisors, upper teeth that are used for gnawing.
5. <u>Primates</u> are mammals with complex brains, specialized limbs used for grasping, and eyes that can perceive depth.

54. Which of the following mammals would be of most interest to a physical anthropologist, a person who studies the evolution of humans?

(1) carnivores
(2) cetaceans
(3) marsupials
(4) rodents
(5) primates

55. The domestic cat is often valued as a mouse catcher. The domestic cat is a member of which group of mammals?

(1) carnivores
(2) cetaceans
(3) marsupials
(4) rodents
(5) primates

56. Kangaroos are mammals with powerful hind legs, long feet, and front paws used for grasping. The young kangaroo is born in an undeveloped state. It spends about six months in the mother's pouch, where it feeds on milk she produces. The key characteristic that classifies kangaroos as marsupials is

(1) powerful hind legs
(2) production of milk by the mother
(3) handlike forepaws used for grasping
(4) a pouch in which the young stay after birth
(5) two upper incisor teeth used for gnawing

Items 57 to 60 refer to the following article.

Solar energy is energy from the Sun. Because sunlight shines daily, solar energy is a renewable resource, unlike fossil fuels such as oil and coal. Fossil fuels will someday be used up. Solar energy does not produce smoke or ash and is a clean source of energy.

One way to gather solar energy for use in a building is through a solar collector. A solar collector looks like a box with a glass or plastic top. Inside the box are tubes filled with water. The Sun heats the water in the tubes, and the heated water is pumped into a storage tank. From there the hot water is piped through the building.

Another way to gather solar energy is to build houses that act as solar collectors. Sunlight comes in through the windows that face southward. Materials inside the house absorb and release the heat. To make sure that enough heat is stored during sunny days to warm the house on cloudy days and at night, a dense material, such as stone, cement, or brick, is used to collect the energy. The dense material is called a storage mass. At night and on sunless days, the storage mass continues to release energy into the air around it.

To convert a building so that it can use solar heat may require the redesign of windows, the creation of storage masses, and the addition of collectors to the roof or surrounding areas.

57. What is the major reason that solar heating of homes is not widespread?

(1) Solar heating always requires large windows.
(2) Solar heating always requires a storage mass.
(3) Most houses do not receive direct sunlight.
(4) It is often difficult and expensive to convert existing houses to solar energy.
(5) Solar heating pollutes the environment.

58. The purpose of a storage mass is to

(1) serve as the foundation of the house
(2) provide a decorative surface for the southern wall
(3) heat water
(4) store cement, stone, and brick
(5) store reserves of energy to be used on cloudy days and at night

59. How does a storage mass provide heat to areas of the house that are not near it?

(1) Heated air is piped into radiators throughout the house.
(2) Heated water is piped into radiators throughout the house.
(3) Steam is piped into radiators throughout the house.
(4) Heated air near the storage mass creates air currents that circulate warm air throughout the house.
(5) The storage mass generates electricity, which is used to heat the house.

60. Which of the statements listed below are included in the article as evidence of the importance of developing solar energy?

A. The supply of fossil fuels will run out.
B. Fossil fuels do not produce much energy.
C. Fossil fuels are sources of pollution.
D. Fossil fuels are imported into the United States.

(1) A and B
(2) B and C
(3) A and C
(4) A and D
(5) A, B, and C

Items 61 to 63 are based on the following information.

The temperature at which a liquid begins to boil is called the boiling point. At the boiling point, the vapor pressure in the liquid is equal to the pressure of the atmosphere. In higher places, the pressure of the atmosphere decreases, and the boiling point is lowered. No matter how much more heat is applied, a liquid never gets hotter than its boiling point. It simply boils until there is no liquid left.

BOILING POINTS OF SOME LIQUIDS	
Liquid	Boiling Point at Sea Level (°C)
Chloroform ($CHCl_3$)	61.7
Ethanol (C_2H_5OH)	78.5
Water (H_2O)	100.0
Octane (C_8H_{18})	126.0

61. In New York City, which has an elevation of 0 meters, water boils at 100°C. In Denver, Colorado, at an elevation of 1,609 meters, water boils at

(1) 126°C
(2) 100°C
(3) 95°C
(4) 0°C
(5) −5°C

62. At sea level, which of the following substances would boil first if they were all heated at the same rate?

(1) chloroform
(2) ethanol
(3) water
(4) octane
(5) They would all reach the boiling point at the same time.

63. Which of the following conclusions can be supported by the information provided?

(1) Once a liquid reaches its boiling point, its temperature increases until all the liquid evaporates.
(2) Water has a higher boiling point than octane.
(3) Adding salt to water raises its boiling point.
(4) The boiling point of a liquid increases as the air pressure increases.
(5) Chloroform is used as an anesthetic because its boiling point is lower than that of water.

Items 64 to 66 refer to the following article.

Jellyfish are invertebrate animals that live in the sea. Some of these invertebrates have a two-stage life cycle. The free-swimming jellyfish mate and produce young called polyps. Polyps are unable to move, but when they mature they become free-swimming jellyfish.

The body of a jellyfish is shaped like a bell or umbrella. A clear material that looks like jelly fills the space between the top and the bottom of the bell. The jellyfish's mouth is located in the middle of the bottom surface of the bell. Tentacles grow around the edge of the bell. Jellyfish move up and down in the water by contracting and relaxing muscles around the edge of the bell. They depend on ocean currents and waves to move sideways.

Jellyfish usually catch their prey by using stinging cells located on the tentacles. The sting of some jellyfish can be irritating or even dangerous to humans. The sea wasp, for example, is a well-known stinger. It has a tall, rigid bell and four two-part tentacles. It lives in tropical oceans but sometimes strays as far north as Long Island Sound. The Australian variety of the sea wasp is dangerous to humans. A sting from the Australian sea wasp can be fatal.

64. The jellyfish has muscles around the edge of the bell. What kind of movement do these muscles permit?

(1) movement in any direction
(2) vertical movement only
(3) horizontal movement only
(4) movement at a 45° angle
(5) movement of tentacles

65. Which of the following statements is supported by the information provided?

(1) Jellyfish stings are usually fatal to humans.
(2) Jellyfish are closely related to mussels and clams.
(3) Jellyfish are concentrated in tropical oceans.
(4) Jellyfish reproduce sexually.
(5) Polyps look like jellyfish.

66. Why might a person who is standing still in the ocean be stung by a jellyfish?

(1) The waves or currents cause the jellyfish to bump into the person.
(2) The jellyfish uses its muscles to move up to the surface.
(3) The jellyfish uses its muscles to move down to the bottom.
(4) The jellyfish cannot move in shallow water.
(5) Jellyfish sting only moving creatures.

Answers and Explanations

1. (Comprehension) **(3) cutting leaves** According to the article, leaf-cutting ants harm crops in Texas and Louisiana. The other activities of ants in options (1), (2), (4), and (5) help make food for humans. These are described in the third paragraph of the article.

2. (Evaluation) **(4) the roles of the queen, winged males, and workers** Each type of ant has a specific role to play in the colony. This would most interest a sociobiologist. Options (1), (2), (3), and (5) describe only physical aspects of ant life, not the way ants act in a group.

3. (Evaluation) **(1) Since there is only one queen to be fertilized, only a few males are required.** Since each colony has only one queen, only a few winged males are needed to mate with her, whereas many workers are needed to care for the offspring of the queen. Options (2) and (3) are not related to the number of males that belong to a colony. Options (4) and (5) are true, but they are not the reason the colony has more workers than males.

4. (Evaluation) **(2) feeding other members of the colony** This is the primary job of worker ants. Options (1), (4), and (5) describe effects of ant activities that might benefit farmers. Option (3) describes an activity, cutting leaves, that can benefit one kind of ant, the leaf-cutter. Cutting leaves is viewed in the article mainly as a problem for farmers.

5. (Evaluation) **(5) the colony** Only a colony can survive for any length of time. Options (1), (2), and (3) are individual types of ants that cannot survive on their own. Option (4) is the reproductive unit of the colony, but it cannot survive without the help of the worker ants.

6. (Application) **(2) performing tasks one after another on an assembly line** In assembly-line work, each worker performs a task in the manufacture of a product. This is most like the behavior of the leaf-cutting worker ants that divide the labor according to size. Option (1) is not correct because ants do not have political systems. Options (3) and (4) are similar to the function of one kind of leaf-cutting worker, not to all leaf-cutting ants. Option (5) is incorrect because the main job of the leaf-cutting workers is to feed the colony, not fight for it.

7. (Analysis) **(4) Water flows from high elevations, such as mountains, to sea level.** The map shows that the watershed areas are bounded by mountainous areas, which are higher than sea level. Water flows downward. Options (1) and (2) are true but do not tell why rivers flow in a certain direction. Option (3) is untrue. Option (5) describes the reason for a flood.

8. (Application) **(2) They saw that the Columbia River was flowing to the west.** After they crossed the Rocky Mountains, explorers saw that water was flowing toward the west, not the southeast. This change suggested that water was flowing to a western ocean, the Pacific. Option (1) would not tell the explorers anything. Option (5) is not true. Options (3) and (4) would not tell the explorers about the location of the nearest ocean.

9. (Analysis) **(4) More oxygen will fit in the container.** According to Boyle's law, the volume of a gas becomes smaller as the pressure increases. This means that more oxygen can be stored in a cylinder if the oxygen is under pressure. Option (1) is not true. If the oxygen is contained in the cylinder, it will not mix with the air. Option (2) may be true, but it does not explain why oxygen is pressurized. Option (3) is incorrect because it mentions Charles's law, which discusses only questions of temperature, not pressure. Option (5) is incorrect because there is also no reason given in the passage to suggest that liquifying the gas would be helpful in storing it.

10. (Analysis) **(2) B only** Because changes in temperature involve Charles's law and not Boyle's law, options (1), (4), and (5) are incorrect. Option (3) is incorrect because knowing the number of molecules inside the balloon will not explain why the balloon rises, nor will this number change as the balloon rises.

11. (Comprehension) **(1) two plates moving toward each other** The article states that the Himalayas were formed as a result of two plates converging, or colliding. Options (2) and (3) describe plate move-

ments other than collision. Option (4), volcanic activity, may have accompanied the formation but did not cause it. Option (5) involves the separation of plates, not the formation of mountains.

12. (Comprehension) **(5) the rise and fall in the water level of major rivers** Option (5) has to do with changes on the surface of the land. Plate tectonics is concerned with the deeper movements of the Earth's crust, not the surface. Options (1), (2), (3), and (4) involve other events associated with plate tectonics.

13. (Comprehension) **(2) The continents moved, causing new formations of land and water.** Option (2) indicates that the movement of plates is changing the location of continents and oceans. Option (1) is incorrect because plate movement will continue, thus moving land masses. Option (3) happened when the land mass, Pangaea, broke into huge pieces. Option (4) implies that plate movement has stopped, which is incorrect. Option (5) is false because diverging boundaries are usually beneath the oceans.

14. (Analysis) **(3) two plates colliding** The collision of plates causes earthquakes, volcanic activity, and the formation of high mountains. These are activities not found in option (1). Mountain formation does not happen in option (2). Options (4) and (5) do not fit the study of the Earth's plates.

15. (Comprehension) **(1) 0–1 year** Children gain about 14 pounds from birth to their first birthday. The largest gain is indicated by the steepest slope on the graph. Between 1 and 2

years, they gain about 6 pounds, option (2). Between 2 and 3 years, they gain about 4 pounds, option (3). Between 3 and 4 years, they gain about 4 pounds, option (4). Between 4 and 5 years, they gain about 4 pounds, option (5).

16. (Analysis) **(3) above average for his age and sex** At 39 pounds, Billy weighs about 2 1/2 pounds more than the average boy his age, who weighs about 36 1/2 pounds. Options (1) and (2) are incorrect because Billy weighs more than the average four-year-old boy. Option (4) is incorrect because Billy weighs less than the average five-year-old boy, who weighs about 41 pounds. Option (5) is incorrect because Billy weighs more than the average four-year-old girl, who weighs about 35 1/2 pounds.

17. (Application) **(4) Annelida** The earthworm belongs to the phylum Annelida, which consists of soft-bodied, segmented animals that are symmetrical, or the same on both sides. Option (1) is incorrect because the earthworm does not have a shell. Option (3) is incorrect because earthworms are not covered by an external skeleton. Option (5) is incorrect because Coelenterates are not segmented.

18. (Application) **(2) Mollusca** A clam has a soft body and a hard shell, which is characteristic of animals of the phylum Mollusca. Option (1) is incorrect because the clam does not have a notochord. Option (3) is incorrect because the clam does not have a jointed external skeleton. Option (4) is incorrect because a clam has a shell, which members of the phylum Annelida do not have. Option (5) is incorrect because

clams are not symmetrical around a central point.

19. (Application) **(1) Chordata** The phylum Chordata is the only one that has animals with backbones as members. All the other options are phyla whose members are invertebrates, animals without backbones.

20. (Application) **(3) Arthropoda** The grasshopper, an insect, belongs to the phylum Arthropoda, along with other animals having external jointed skeletons. None of the other phyla contain animals with external jointed skeletons.

21. (Analysis) **(5) A and C** Leaving the iron bar in contact with a magnet and stroking the bar with a magnet would have the same effect. The magnetic domains would line up and produce a magnet. Options (1) and (3) are only partly correct. Options (2) and (4) are incorrect because iron filings are attracted to a magnet, but they cannot make an object into a magnet.

22. (Application) **(2) a rubber eraser** The article mentions that objects containing iron and steel are affected by magnets. Options (1), (3), (4), and (5) all contain iron or steel. Options (3) and (4) describe two kinds of magnets; in paragraph 2, there is also a description of the way magnetic fields affect magnets. Option (5), the steel girder, is large and may not visibly shift but would be affected by magnetism. A magnet would certainly stick to it!

23. (Analysis) **(2) Earth has a magnetic north pole** The Earth, with a magnetic north pole, causes the suspended compass needle to point north. Option (1) is a result of the

compass's behavior, not a cause. Option (3) is not true. Option (4) is incorrect because if the domains pointed in different directions, the needle would not be a magnet. Option (5) is not true because one pole is not stronger than another.

24. (Application) **(3) a sailor navigating with the use of a compass** This person might be directed off course by the difference between the two poles. The hikers in option (1) are traveling only a short distance and the driver in option (5) has only two directions to choose from. These people would not be led off course by the difference. Option (2) is incorrect because this sailor is not affected by the magnetic poles. Option (4) is incorrect because the compass will point to the nails, and not to the north pole.

25. (Analysis) **(2) two magnets, each with one north pole and one south pole** The article states that each magnet has a north and a south pole. This statement makes options (1), (3), and (4) untrue. Option (5) makes no sense; if the domains are lined up in one half of the magnet, they will also be lined up in the other half.

26. (Analysis) **(4) gold** Gold is the only metal listed with a density greater than that of liquid mercury. Therefore, it is the only one that will sink in liquid mercury. Options (1), (2), and (3) have densities less than that of liquid mercury, and they will float in it. Option (5) is incorrect since only gold will sink.

27. (Application) **(1) 24K gold** Pure gold has a density of 19.3. This is greater than the densities of the mixtures of gold and copper represented by options (2), (3), and (4). These densities are in the range 8.9 to 19.3, depending on the amount of copper. Option (5) is incorrect, since the densities of different materials differ.

28. (Evaluation) **(4) It is formed in the ocean.** The chart shows that limestone is formed by chemical reaction in seawater and from the remains of marine animals. Both of these methods of formation occur in the ocean. Option (1) is the method of formation for conglomerates, sandstones, and shales, not limestone. Options (2) and (3) are each only partially correct. Option (5) is incorrect because silt makes up shale, not limestone.

29. (Application) **(5) A, B, and C** People use methods and materials, such as those that form sedimentary rocks, in all three cases described. Concrete, with its pebbles, is similar to conglomerate. Constructing an adobe house of mud is similar to the formation of rock (shale) out of clay. Evaporation of seawater is similar to the process found in nature. Options (1), (2), (3), and (4) do not include all the methods and are therefore incorrect.

30. (Comprehension) **(1) bacteria** Bacteria start the decomposition process, later to be replaced by other microorganisms. Options (2), (3), and (4) are microorganisms active in the later stages of composting. Option (4) is a substance produced by the actinomycetes that contributes to the death of bacteria in the pile. Option (5) is incorrect because each organism is active at a different time.

31. (Analysis) **(1) bacteria** The bacteria do most of the initial decomposition of the organic matter; their removal would, in effect, stop the composting process. Options (2), (3), (4), and (5) would have a lesser impact; therefore, option (1) is correct.

32. (Analysis) **(5) plastic wrap** Plastic wrap is the only material listed that is not living. Options (1), (2), (3), and (4) are all from living plants or animals and so are incorrect.

33. (Evaluation) **(2) The oil could spread farther south and west.** Option (3) is incorrect because the map shows a southwestward movement of oil, and there is not enough information to conclude that this spread has stopped. The spread of the oil over a vast area of ocean and coast suggests that clean-up efforts will take a long time. Option (1) is incorrect because the oil spread during the first two months and was not contained. Option (4) is incorrect because the oil is not easily absorbed by the ocean, but floats on top. Option (5) is incorrect because the amount estimated to have evaporated during the first two months is only 20 percent. If an equal amount were to evaporate during the next two months, half the oil would still remain.

34. (Application) **(5) Evaporation will probably help clear some of the oil.** According to the map, 20 percent of the oil in the Alaskan spill evaporated. Option (1) mentions barriers, but the map does not show how barriers were or were not used. Option (2) suggests that skimmers should not be used, but the map shows that they were useful in the Alaskan accident. Accord-

ing to the map, information about biological dissipation is not available, so you cannot use the map to argue for option (3). Option (4) is contradicted by the map; in Alaska, most of the oil was not cleared in two months.

35. (Application) **(3) Leave storm protection measures in place because the rain and winds will start again—this time from the other direction.** The family is experiencing the lull in the storm caused by the passage of the eye over them. When the eye passes, the rain and winds resume, only from the opposite direction, or in this case west. Option (1) is incorrect because the hurricane has not passed; if the winds had been from the west, they could have concluded that the hurricane was over when the storm ended. Option (2) is based on an assumption that may not be true, so removing the storm protection measures does not make sense. Option (4) is incorrect because hurricanes are storms of such vast size that two of them would probably not occur within a space of two days. Option (5) is incorrect because the hurricane is not over.

36. (Evaluation) **(4) passage of the eye** The eye is a calm area in the middle of the hurricane. The eye is not likely to cause damage. Options (1), (2), (3), and (5) are incorrect because rain, flooding, and wind are all likely to cause serious damage.

37. (Application) **(4) oiling it to create a barrier against air and water** Oiling the cookware provides an air- and water-resistant barrier to protect the iron from rusting. Option (1) is incorrect because water will cause the iron to

rust. Options (2) and (3) are incorrect because they offer only partial protection, and they are impractical when considering cooking utensils. Option (5) does not suggest a way to keep air and water away from the iron.

38. (Analysis) **(5) the iron is not exposed to air and water** Plating iron with a rust-resistant metal protects the surface from contact with air and water, and thus protects it from corrosion. Options (1) and (2) are incorrect because plating is described as a coating process, not a removal process. Option (3) is incorrect because the iron is plated with a layer of metal; it is not mixed with another metal. Option (4) is incorrect because rust is prevented from forming.

39. (Analysis) **(1) duplication of hereditary material (Interphase)** The division of hereditary material into two sets of chromosomes ensures that each new cell has the same chromosomes as the parent cell. (See Interphase in the cell division diagram.) Options (2), (3), (4), and (5) are aspects of the cell-division process, but they are not the direct cause of each new cell receiving a complete set of chromosomes from the parent.

40. (Analysis) **(3) It enters the cytoplasm.** The nuclear material goes into the cytoplasm, the cell material outside the nucleus. This is shown in the prophase stage. Option (1) is incorrect because the cell membrane functions as a protective surface for the cell contents. Options (2) and (4) are cell structures with other functions. Option (5) is incorrect because the nuclear membrane disappears at this stage.

41. (Application) **(4) in a warm spot in the kitchen** Warmth is needed for yeast action and expansion of the carbon dioxide gas. Options (1), (2), and (3) are all cool or cold places, so they would stop or slow the action of the yeast. Option (5) is incorrect because the dough needs to rise before it is placed in the oven.

42. (Comprehension) **(5) carbon-dioxide bubbles in the dough** The pores, or small holes, in bread are formed by trapped bubbles of carbon dioxide gas. Options (1), (2), (3), and (4) are all solids that do not create pores.

43. (Application) **(3) secondary consumers** The vultures are eating a primary consumer when they eat the antelope, so they are functioning as secondary consumers. Option (1) is incorrect because only plants are producers. Option (2) is incorrect because the vultures are not eating plants. Although vultures are acting as tertiary consumers, option (4), when they are eating the lion, they are acting as secondary consumers when they are eating the antelope. Option (5) is incorrect because it applies to microorganisms that decompose organic matter.

44. (Analysis) **(3) There is more energy available to secondary consumers than to tertiary consumers.** The article states that each consumer uses some energy, so less energy is available to the next level consumer. Thus, there is more energy available to secondary consumers than to tertiary consumers. This same information shows options (1) and (4) to be incorrect. Options (2) and (5) are not true. Animals get their

energy from plants, either directly or indirectly, and plants get their energy from the Sun.

45. (Comprehension) **(4) eat both plants and animals** Such animals can be classified at various levels in the food chain because they eat a variety of foods at different levels. Options (1) and (3) are true, but such animals would be classified only as primary consumers. Option (2) is also true, but such animals would be classified as secondary consumers. Option (5) has nothing to do with food chains.

46. (Comprehension) **(2) the Sun** This is the basic source of energy for the food chain because the Sun is used by plants to manufacture food. Option (1) is incorrect because nutrients in the soil are produced by decomposers, which get their energy from plants and animals. Options (3), (4), and (5) are incorrect because they all get their energy, directly or indirectly, from the Sun.

47. (Comprehension) **(2) its weight equals the weight of the liquid it displaces** According to the diagram, the cube floats when its weight is the same as the weight of the liquid it pushes aside, or displaces. Option (1) is incorrect because objects that sink also displace water. Option (3) is incorrect because, if the weight of the object is greater than that of the displaced liquid, the object sinks as shown in the second diagram. Options (4) and (5) are incorrect because the total weight of the liquid does not matter. What is important is the weight of the object in relation to the weight of the liquid pushed aside.

48. (Evaluation) **(4) Steel and air are equal in weight to the water they displace.** Steel ships float because they are not made of solid steel, but of steel and air (hollow parts of the ship). Thus, they weigh less than an equal volume of solid steel, which would sink. Although option (1) is true, it is not correct because ships contain air as well as steel. Option (2) is not true since steel sinks. Option (3) is not true since steel ships float. Option (5) is not true since floating objects are equal in weight to the liquid they displace, not lighter.

49. (Analysis) **(3) are formed by sediment or soil carried by rivers from mountains to the plain** The diagram shows the river emerging from a mountain range and dividing into many little channels. The slowdown of the river's flow causes it to deposit the sediment it carries. Option (1) is incorrect because nothing in the diagram indicates whether the formation is new or old. Option (2) is incorrect because the diagram indicates the presence of a river, not a glacier. Options (4) and (5) cannot be assumed from the information given.

50. (Analysis) **(4) spreads out and slows down** The picture shows the river spreading out and dividing into channels. The action of the river suddenly flowing into a flat area from the mountains causes the flow to spread out and slow. None of the other options is supported by the picture. Option (1) is incorrect because the speed of the water slows when the river spreads out. The temperature is not shown. Options (2) and (3) are incorrect because the river becomes wider, not narrower. Option (5)

is incorrect because nothing in the picture indicates that the stream goes underground.

51. (Comprehension) **(3) restore chemicals used up as plants grow** The first paragraph of the article indicates that this is the purpose of using fertilizer. Option (1) is incorrect because fertilizer does not replace soil. Option (2) is incorrect and untrue. Minerals are not lost in the composting process. Option (4) is incorrect because the article is not about microorganisms in the soil. Option (5) is incorrect because manure is a form of fertilizer.

52. (Application) **(2) vitamins and minerals to processed flour** When flour is processed, vitamins and minerals are lost. They are added again at a later stage of processing. This is similar to restoring minerals lost through agriculture. Option (1) is too general to be correct. Option (3) is similar to adding fertilizer to soil, but the sweetener has no nutritional value. Fertilizer adds nutritional value to the soil. Option (4) is also similar to adding fertilizer to soil, except that the fluoride does not replace something lost in the water. The water is just a convenient place to put fluoride for human consumption. Option (5) is simply a matter of taste. Most cereals contain natural sugars.

53. (Evaluation) **(3) Organic fertilizers act on the soil in a manner more similar to nature than do chemical fertilizers.** Since the action of compost and manure imitates the process of decay in nature, the farmer can claim that this crop is produced naturally and is therefore better. Option (1) is untrue since the article states that minerals are not used by the plants until they have decayed into a simpler form. Option (2) is true but is incorrect because the article states that plants use simpler forms from both natural and artificial fertilizers. Option (4) is not true. Option (5) is also untrue; pesticides are not discussed in the article.

54. (Application) **(5) primates** Humans are members of the group of mammals called primates. Thus, knowing something about primates would be important to a physical anthropologist. Option (1) is incorrect because it includes animals such as lions and wolves. Option (2) includes marine mammals such as whales and dolphins. Option (3) includes the opossum and the kangaroo. Option (4) includes mice, squirrels, and rats. Options (1), (2), (3), and (4) are incorrect because these groups are not closely related to humans.

55. (Application) **(1) carnivores** As meat-eaters, cats are members of the carnivore group, as are their relatives, tigers and lions. They are not cetaceans, option (2), because they are not ocean-dwellers. They are not marsupials, option (3), because the females do not have pouches. They are not rodents, option (4), because they do not have specialized upper teeth for gnawing. And they are not primates, option (5), because they do not have specialized limbs for grasping (such as thumbs) or complex brains.

56. (Analysis) **(4) a pouch in which the young stay after birth** The pouch is listed as the key characteristic that separates marsupials from other types of mammals. Option (1) is a characteristic of mammals, not just of marsupi-

als. Option (2) is the characteristic that makes kangaroos members of the larger group, mammals. Option (3) would suggest a relationship to the primates, who also have the ability to grasp. However, the kangaroo does not have a highly complex brain or eyes that can perceive depth, so the kangaroo is not a primate. Option (5) is incorrect because kangaroos do not have the two specialized upper incisor teeth of the rodent.

57. (Evaluation) **(4) It is often difficult and expensive to convert existing houses to solar energy.** The redesigning of windows and adding solar collectors can be very expensive. Options (1) and (2) are incorrect because solar heating does not always involve these designs. Option (3) is not true in most areas. Option (5) is not true since solar energy is a very clean form of energy.

58. (Comprehension) **(5) store reserves of energy to be used on cloudy days and at night** The density of the material used allows for the absorption and the gradual release of stored energy. Options (1) and (2) are not purposes of a storage mass. Solar collectors—not storage masses—are generally used to heat water, option (3). Option (4) is not true.

59. (Analysis) **(4) Heated air near the storage mass creates air currents that circulate warm air throughout the house.** Since warm air rises, a pattern of air flow is established in a well-designed solar house. Options (1), (2), and (3) are incorrect because houses that receive all their heat by means of solar energy do not have radiators

or heating pipes. Option (5) is incorrect because the storage mass generates radiant energy, or heat, not electrical energy.

60. (Evaluation) **(3) A and C** The article supports the idea that fossil fuels will run out and that they are sources of pollution. Statement B is not true. Although statement D is true, it is not mentioned in the article. Any option— (1), (2), (4), and (5)—that includes B or D is incorrect.

61. (Application) **(3) 95°C** Since the boiling point of a liquid decreases with a drop in air pressure, the boiling point of water in Denver must be less than the boiling point of water in New York. This is so because Denver, at a higher elevation than New York, has a lower atmospheric pressure. Option (1) is incorrect because it is a higher temperature than the boiling point of water in New York, which is 100°C. Option (2) is incorrect because the boiling point of water will differ at different elevations. The boiling point of water at sea level is 100°C. Options (4) and (5) are incorrect because they are the freezing and below-freezing temperatures of water.

62. (Analysis) **(1) chloroform** According to the chart, chloroform would boil first because it has the lowest boiling point, 61.7°C. Options (2), (3), and (4) are incorrect because it would take longer to heat these liquids to their boiling points than it would take to heat chloroform. Option (5) is not true, since the liquids have different boiling points. Therefore, the liquids will start to boil at different times.

63. (Evaluation) **(4) The boiling point of a liquid increases as the air pressure increases.** The information provided indicates a direct relationship between air pressure and the boiling points of liquids. In other words, as one increases, so does the other. Option (1) is incorrect because the passage indicates that once the boiling point is reached, the liquid does not get any hotter. It just boils until it evaporates. Option (2) is incorrect because the boiling point of water, 100°C, is lower than that of octane, 126°C, as indicated in the chart. Option (3) is true but is not supported by the information provided. The passage and chart do not deal with the effect on the liquid's boiling point of adding other substances. It is therefore incorrect. Option (5) is incorrect. Although the boiling point of chloroform is lower than that of water, this fact does not necessarily have any bearing on the use of chloroform in anesthesia. No information is given to support or disprove option (5).

64. (Analysis) **(2) vertical movement only** The article states that jellyfish contract and relax muscles to move up and down in the water. This is vertical movement. The article states that jellyfish depend on currents and waves to move sideways, implying that the jellyfish cannot cause motion in this direction. Options (1), (3), and (4) are not correct because the jellyfish cannot use its muscles to move in these directions. Option (5) is incorrect because the question asks about muscles in the bell, not in the tentacles.

65. (Evaluation) **(4) Jellyfish reproduce sexually.** The article states in the first paragraph that jellyfish mate and produce polyps. Sexual reproduction is another term for mating. Option (1) is not supported by the information provided, which is that jellyfish stings can be irritating and sometimes fatal. Option (2) is not true since mussels and clams are not mentioned. Option (3) is incorrect because the article

states only that sea wasps, a type of jellyfish, are usually found in tropical oceans. Option (5) may or may not be true; however, there is not enough information to determine this.

66. (Analysis) **(1) The waves or currents cause the jellyfish to bump into the person.** Since a jellyfish cannot move sideways on its own, another force must move it toward a person standing still. Options (2) and (3) are incorrect because upward or downward movement in the water would not cause the jellyfish to bump into a person standing still. Option (4) is incorrect because the jellyfish can move in shallow water, both vertically and by wave action. Option (5) is not supported by the article and is not true.

PRETEST Correlation Chart

Science

The chart below will help you determine your strengths and weaknesses in thinking skills and in the content areas of life science, Earth science, chemistry, and physics.

Directions

Circle the number of each item that you answered correctly on the Pretest. Count the number of items you answered correctly in each column. Write the amount in the Total Correct space for each column. (For example, if you answered 8 comprehension items correctly, place the number 8 in the blank above *out of 12*.) Complete this process for the remaining columns.

Count the number of items you answered correctly in each row. Write that amount in the Total Correct space for each row. (For example, in the Life Science row, write the number correct in the blank before *out of 30*.) Complete this process for the remaining rows.

Cognitive Skills / Content	Comprehension	Application	Analysis	Evaluation	Total Correct
Life Science (pages 38–131)	1, **15**, 30, 45, 46, 51	6, 17, 18, 19, 20, 43, 52, 54, 55	**16**, 31, 32, **39**, **40**, 44, 56, 64, 66	2, 3, 4, 5, 53, 65	_____ out of 30
Earth Science (pages 132–179)	11, 12, 13, 58	**8**, **29**, **34**, **35**	7, 14, **49**, **50**, 59	**28**, **33**, **36**, 57, 60	_____ out of 18
Chemistry (pages 180–205)	42	**27**, 37, 41, **61**	9, 10, **26**, 38, **62**	**63**	_____ out of 11
Physics (pages 206–241)	**47**	22, 24	21, 23, 25	**48**	_____ out of 7
Total Correct	_____ out of 12	_____ out of 19	_____ out of 22	_____ out of 13	Total correct: ____ out of 66 1–54 → Need more review 55–66 → Congratulations! You are ready!

Boldfaced items are based on diagrams, maps, charts, or graphs.

If you answered fewer than 66 questions correctly, determine which areas are hardest for you. Go back to the *Steck-Vaughn GED Science* book and review the content in those areas. In the parentheses under the item type heading, the page numbers tell you where you can find specific instruction about that area of science in the *Steck-Vaughn GED Science* book.

LIFE SCIENCE

Planting new trees to replace fallen trees

◆ **organism**
a living thing

◆ **environment**
all the surround-ings in which an organism lives

◆ **hereditary**
capable of being passed from a parent to a new organism

◆ **photosynthesis**
the process by which plants use light energy to make food

Life science is the field of science that studies **organisms** and how they interact with one another and their **environment**. Under-standing life science helps us understand ourselves and the world around us. It also helps us learn to improve the quality of life and to protect our environment.

In the first lesson of this section you will learn about the structure of plant and animal cells. Cells are the microscopic units of which all living things are made. They take in food, carry out hundreds of chemical reactions, and reproduce. You will learn how, through the process of reproduction, cells pass on **hereditary** information from one generation to the next.

☞ *See Also: GED Exercise Book Science, pages 4–14*

- **respiration**
  the chemical process by which living things obtain energy for their use

- **biosphere**
  the thin layer of the Earth where life exists

- **species**
  a group of similar organisms that can mate and produce fertile offspring

- **organ system**
  several organs working together to perform a function

- **infectious**
  capable of being spread from one organism to another

- **offspring**
  the direct descendants of an animal or plant

- **genetics**
  the study of inherited characteristics

- **ecosystem**
  a selected area where living and nonliving things interact

In the second lesson you will learn how plants make their own food through the process of **photosynthesis**. You will learn how all living organisms use the process of **respiration** to obtain the energy they need. Then you will see how these two processes are related. You will also learn how nitrogen is cycled through the **biosphere**.

The third lesson discusses theories about the origin of life and how living things evolved from small, simple organisms to the complex plants and animals we see around us today. You will learn how the environment affects the development of new **species** through the process of natural selection.

The fourth lesson explains how plants grow from a single seed and develop such features as stems, leaves, roots, and flowers. Then you are introduced to the system scientists use to classify all organisms. You will learn how this system of classification is based on the idea that all living things are related to one another.

In the fifth lesson you will learn about some of the **organ systems** of the human body. For example, you will learn how the digestive system works and how the heart pumps blood through the body. You will also learn about the glands of the body and how they regulate many of the body's functions.

The sixth lesson begins with a discussion of how scientists discovered what causes **infectious** diseases. Then you are introduced to two types of disease-causing organisms, bacteria and viruses. You will learn about the structure of bacteria and viruses and how they cause diseases. At the end of the lesson you will learn about some scientific developments used to fight infectious diseases.

The seventh lesson discusses how parent organisms pass on characteristics to their **offspring**. You will learn about the experiments that opened the field of modern **genetics**. You will also see how all genetic information is coded in each cell.

In the last lesson you will learn how organisms interact with one another and their environment in **ecosystems**. The lesson closes with a discussion of how environmental pollution has affected all living things.

LESSON 1 Comprehension Skill: Identifying the Main Idea

When you are studying or taking a test, you are reading to understand the material. This means you must read slowly and carefully, looking for main ideas and details.

How can you find the main ideas of a passage? First look over the passage quickly, noting how many paragraphs there are. If there are three paragraphs, you should find three main ideas.

Each paragraph is a group of sentences about a single topic, or **main idea**. The main idea of a paragraph is usually in the **topic sentence.** Often the topic sentence is the first or last sentence of the paragraph, but sometimes the topic sentence is in the middle. Wherever it is, the sentence with the main idea has a meaning general enough to cover all the points made in the paragraph. Sometimes the main idea is not stated clearly in one sentence. In that case, read the whole paragraph to find the main idea.

When you think you have found the main idea of a paragraph, look for **supporting details.** These may be reasons, proofs, examples, or details that add to the main idea. Sometimes the supporting details are in the form of a list.

Read the following paragraph to find the main idea and the supporting details.

A **reflex** is an automatic response to a condition in the environment or surroundings. For example, when you try to pick up a hot pan from the stove, the heat from the pan causes you to pull your hand away. Or, when a light is suddenly turned on in a dark room, you blink and squint against the glare. Reflexes prevent you from further injuring yourself. Because reflexes occur without your having to think about them, they are said to be automatic.

You were correct if you said the first sentence of the paragraph tells the main idea. This sentence gives a broad definition of a reflex, which is the topic of the paragraph. The second and third sentences offer examples of reflexes. The fourth sentence tells why reflexes are important. The last sentence tells why reflexes are said to be automatic.

Remember, the main idea of a paragraph is the general idea. The supporting details are more specific than the main idea.

Practicing Comprehension

Read the following paragraphs.

The skin of our bodies forms a protective surface. Skin is made up of epithelial tissue, which is flat and broad. This tissue also forms a protective lining for organs, such as the stomach. The cells that make up epithelial tissue are close together. As a result, this tissue can control which substances pass through it.

Another type of human tissue is connective tissue. This tissue supports and holds together parts of the body. The cells in connective tissue are not close together. Nonliving material, such as calcium, fills the spaces between cells. Such nonliving substances give connective tissues strength. Bone and cartilage are the most familiar types of connective tissue.

Questions 1 to 3 refer to the paragraphs. Circle the best answer for each question.

1. What is the main idea of the first paragraph?

(1) Epithelial tissue controls the passage of substances through it.
(2) The stomach contains epithelial tissue.
(3) Skin is a type of epithelial tissue.
(4) The cells of epithelial tissue are close together.
(5) Epithelial tissue forms a protective surface for parts of the body.

2. What is the main idea of the second paragraph?

(1) Human connective tissue supports and holds together parts of the body.
(2) Connective tissue contains nonliving material.
(3) Bone is a familiar form of connective tissue.
(4) Cartilage is a type of connective tissue.
(5) Calcium is found in some connective tissue.

3. In both paragraphs, the author gives examples as supporting details for the main idea. Which of the following is in the form of an example?

(1) Connective tissue contains nonliving material.
(2) The cells of epithelial tissue are close together.
(3) Skin is a type of epithelial tissue.
(4) Connective tissue supports and holds together parts of the body.
(5) Epithelial tissue controls the passage of substances through it.

To check your answers, turn to page 48.

Topic 1: The Biology of Cells

Read the following passage and look at the diagram.

All living organisms are made up of microscopic units called **cells**. Some organisms consist of only a single cell. Other organisms are made up of many cells. Almost all cells have certain things in common, and their basic structure is similar. Most cells take in food. They break down food to get energy, and then give off waste. Cells also grow, reproduce, and die.

Most cells are divided into two general parts, the **nucleus** and the **cytoplasm**. The nucleus controls the cell's activities, and the cytoplasm carries out these activities.

The nucleus of a cell is a round body inside the cell. The three parts of the nucleus include:

1. **Nucleolus**. The nucleolus plays a role in making protein.
2. **Chromatin**. The chromatin are thin, dark strands of matter. When the cell divides, the chromatin contract to form chromosomes, which carry the hereditary information for the cell.
3. **Nuclear membrane**. The nuclear membrane divides the nucleus from the cytoplasm.

The cytoplasm contains the other cell structures called **organelles**. Some organelles found in the cytoplasm include:

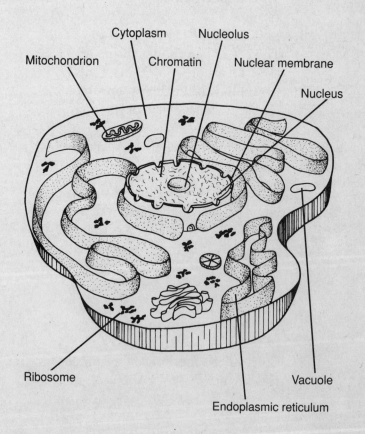

1. **Mitochondria.** Mitochondria are rod-shaped organelles that produce almost all the energy a cell needs. Some cells have hundreds of mitochondria.
2. **Endoplasmic reticulum**. The endoplasmic reticulum (ER) is a system of tubes running through the cytoplasm. It is thought that the ER carries materials throughout the cell.
3. **Ribosomes.** Ribosomes are attached to the ER. They help make proteins.
4. **Vacuoles.** Vacuoles are storage spaces in the cytoplasm. They may contain food or water, or they may collect and excrete waste.

Exercises

Questions 1 to 6 refer to the previous passage and diagram. Circle the best answer for each question.

1. Which of the following sums up the main idea of the first paragraph?

   (1) All cells need energy to live.
   (2) Most cells have things in common and have a similar structure.
   (3) Food gives the cell energy.
   (4) The nucleus is the center of a cell.
   (5) The nucleus is an organelle.

2. Which detail most completely supports the main idea?

   (1) All cells grow, reproduce, and die.
   (2) Most cells have things in common and have a similar structure.
   (3) Some organisms consist of a single cell.
   (4) The nucleus is the center of a cell.
   (5) The nucleus is an organelle.

3. What is the main idea of the second paragraph?

   (1) All cells have a nucleus.
   (2) The nucleus controls the activities of the cell.
   (3) All cells are divided into two basic parts.
   (4) All cells have cytoplasm.
   (5) The activities of the cell are carried out by the cytoplasm.

4. What is the main idea of the third paragraph?

   (1) The nucleus is round.
   (2) The nucleolus is part of the nucleus.
   (3) The nucleus contains a nucleolus, chromatin, and a nuclear membrane.
   (4) The nuclear membrane divides the nucleus from the cytoplasm.
   (5) Chromatin contract to form chromosomes.

5. Which of the following sums up the main idea of the fourth paragraph?

   (1) Cytoplasm contains all the structures inside the cell membrane except the nucleus.
   (2) Organelles are cell structures.
   (3) Mitochondria produce the energy a cell needs for its activities.
   (4) Vacuoles are storage spaces in the cytoplasm.
   (5) Cytoplasm includes the nucleus.

6. Which detail most completely supports the main idea of the fourth paragraph?

   (1) Organelles are cell structures.
   (2) Mitochondria are smaller than vacuoles.
   (3) The ER is used to transport material through the cell.
   (4) A cell may contain hundreds of mitochondria.
   (5) The cytoplasm contains mitochondria, the ER, ribosomes, and vacuoles.

To check your answers, turn to page 48.

Reviewing Lesson 1

Read the following passage and look at the diagram.

The process by which a cell divides is called **mitosis**. During mitosis, the chromosomes in the original cell, or **parent cell**, duplicate and divide into two identical sets. One set will go to each of two new cells, or **daughter cells**. The process of cell division (mitosis) is divided into five phases.

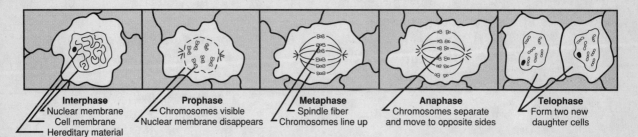

Interphase
Nuclear membrane
Cell membrane
Hereditary material

Prophase
Chromosomes visible
Nuclear membrane disappears

Metaphase
Spindle fiber
Chromosomes line up

Anaphase
Chromosomes separate
and move to opposite sides

Telophase
Form two new
daughter cells

During the first phase, called **interphase**, the hereditary material in the nucleus duplicates itself.

During the second phase, called **prophase**, the hereditary material shortens and thickens to form the **chromosomes**. Each chromosome is made of two identical parts that are attached at their centers. Protein fibers come from opposite poles, or ends, of the cell to form a **spindle**. The membrane around the nucleus disappears.

During the third phase, or **metaphase**, the chromosomes line up across the middle of the cell. A spindle fiber attaches to each chromosome.

During the fourth phase, called **anaphase**, the identical parts of chromosomes separate. They move toward opposite sides of the cell, pulled by the spindle fibers.

During the final phase, or **telophase**, the chromosomes again become threads of hereditary material and the spindle fibers disappear. A new membrane forms around each nucleus. The cytoplasm of the cell divides, producing two daughter cells.

Questions 1 to 6 refer to the passage and diagram. Circle the best answer for each question.

1. According to the diagram, when do the spindles first attach to the chromosomes?

    (1) interphase
    (2) prophase
    (3) metaphase
    (4) anaphase
    (5) telophase

2. What appears to be pulling the chromosomes apart during anaphase?

    (1) cell membrane
    (2) cytoplasm
    (3) nuclear membrane
    (4) nucleus
    (5) spindle

3. What is the subject of the passage?

  (1) In telophase, the parent cell finally forms two new cells.
  (2) Mitosis is the process by which cells divide.
  (3) Spindle fibers are important to the process of the division.
  (4) Chromosomes divide in half during mitosis.
  (5) The cytoplasm divides during mitosis.

4. Which phase makes sure that both daughter cells will receive all hereditary material?

  (1) interphase
  (2) prophase
  (3) metaphase
  (4) anaphase
  (5) telophase

5. According to the diagram, what do the spindle fibers seem to do in mitosis?

  (1) guide the chromosomes
  (2) produce the nucleus
  (3) form the nuclear membrane
  (4) form the hereditary material
  (5) form a cell plate

6. If the number of chromosomes in the parent cell is 46, how many chromosomes will there be in each daughter cell?

  (1) 12
  (2) 18
  (3) 23
  (4) 46
  (5) 60

Read the following passage.

Sexual reproductive cells are formed by a process called **meiosis**. This process differs from mitosis in that the resulting cells, called **gametes**, have half the number of chromosomes as the parent cell. Human gametes are sperm and eggs. During sexual reproduction, the two gametes combine to form a new cell called a **zygote**. Because the zygote contains the chromosomes from both gametes, the original number of chromosomes is restored.

Chromosomes occur in pairs. A parent cell contains a complete set of paired chromosomes. A gamete receives only one chromosome from each pair in the parent cell. Thus a gamete has half the number of chromosomes found in the parent cell.

Questions 7 and 8 refer to the passage. Circle the best answer for each question.

7. The chromosome number for human body cells is 46. What is the chromosome number for human gametes?

  (1) 12
  (2) 23
  (3) 32
  (4) 46
  (5) 48

To check your answers, turn to pages 48–49.

8. The chromosome number of a cell produced by meiosis differs from the chromosome number of a cell produced by mitosis in that the cell produced by meiosis has

  (1) half the number
  (2) no chromosomes
  (3) the same number
  (4) one-fourth the number
  (5) twice the number

GED Mini-Test

Directions: Choose the <u>best answer</u> to each item.

<u>Items 1 to 4</u> refer to the following paragraphs.

Plant cells have some special structures that animal cells do not have. For example, most plant cells have a cell wall surrounding the cell membrane. A cell wall is composed of several layers. The middle layer, called the middle lamella, contains the jellylike substance pectin. Primary walls on either side of the middle lamella are made of cellulose and pectin. Secondary walls form on the outside of the primary walls. The stiff walls are made of cellulose. They remain long after the cells have died.

Plant cells contain organelles called plastids. Chloroplasts are one kind of plastid. Chloroplasts contain the green pigment chlorophyll. Chlorophyll makes it possible for the plant to carry out photosynthesis. In photosynthesis, the plant uses energy from the Sun to combine water and carbon dioxide. This process produces oxygen and sugar, which the plant uses for food. Another kind of plastid is called a leucoplast. Leucoplasts contain enzymes that change sugar into starch. The starch is then stored in the leucoplasts.

1. What structures do plant cells have that animal cells do not?

   (1) cell wall, chloroplast, and leucoplast
   (2) cell membrane, chloroplast, and leucoplast
   (3) cell membrane, chromatin, and nucleolus
   (4) cell wall, chromatin, and nucleolus
   (5) middle lamella, pectin, and cytoplasm

2. Which structure must a plant contain for the production of food?

   (1) sugar
   (2) nucleus
   (3) cell wall
   (4) chloroplast
   (5) leucoplast

3. The process of photosynthesis is responsible for which of the following?

   (1) cell division
   (2) transportation of water
   (3) absorption of nitrogen
   (4) stiffness of stems
   (5) production of sugar

4. Which cell structure is most responsible for wood retaining its stiffness after a tree dies?

   (1) the secondary cell wall
   (2) the middle lamella
   (3) the leucoplasts
   (4) the primary cell wall
   (5) the plastids

To check your answers, turn to page 49.

Items 5 to 8 refer to the following paragraphs.

Like all living things, trees are subject to disease, decay, and death. When a tree is wounded, fungus spores get into the wound, germinate, and send out creeping threads that attack the cell tissues. In time the tree dies unless a tree surgeon saves it.

A tree surgeon removes injured or rotted branches close to the trunk, or parent branch, so as not to leave a stub with the fungus in it. To saw off a large branch, the surgeon first undercuts it to keep the bark from tearing as the limb falls. The surgeon makes the first cut on the underside 12 to 18 inches from the trunk and continues until the saw begins to bend from the pressure of the limb.

The next cut is made from above and about two inches outward from the first cut. After the limb falls, the stub is removed near the trunk and the wound is treated.

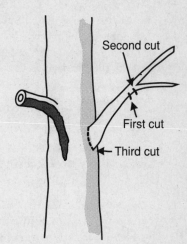

5. Why are injured or rotted branches removed from wounded trees?

(1) to prevent the stub from showing
(2) to prevent the bark from tearing
(3) to prevent fungus from killing the tree
(4) to make a clean cut
(5) to leave a stub so that a new branch will grow

6. In the diagram, the limb on the left was not properly cut. What was the result of the improper cut?

(1) The third cut was not necessary.
(2) The natural swaying of the tree was prevented.
(3) The bark was torn.
(4) The bark grew inward and healed.
(5) A fungus continued to grow.

7. Why is it important to remove the stub of the injured branch?

(1) to make sure that all the fungus is removed
(2) to prevent the bark from peeling off
(3) to protect the underside of the branch
(4) to make sure that another branch will grow
(5) to prevent injury to the tree surgeon

8. Why is the first cut done on the underside of the injured limb?

(1) The underside is easier to reach.
(2) The saw will get stuck, causing the branch to fall.
(3) The limb will fall without tearing the bark and wood.
(4) More fungus grows on the underside.
(5) The wound needs to be treated.

To check your answers, turn to page 49.

Answers and Explanations

1. (Comprehension) **(5) Epithelial tissue forms a protective surface for parts of the body.** Option (5) describes, in general, the function of epithelial tissue. The other statements are more specific. Option (1) tells something epithelial tissue does. Options (2) and (3) are examples. Option (4) describes a detail about epithelial tissue.

2. (Comprehension) **(1) Human connective tissue supports and holds together parts of the body.** This is a general statement about the purpose of connective tissue, the topic of the paragraph. Option (2) describes a detail about connective tissue. Options (3) and (4) are examples of connective tissue. Option (5) contains an example of the nonliving materials found in connective tissue.

3. (Comprehension) **(3) Skin is a type of epithelial tissue.** This is an example. The other options are incorrect. Options (1), (2), and (5) provide details describing tissues. Option (4) is the main idea of the second paragraph.

Exercises (page 43)

1. (Comprehension) **(2) Most cells have things in common and have a similar structure.** This is stated in the third sentence of the first paragraph. Options (1) and (3) are true but do not state the main idea. Options (4) and (5) name a structure that most cells have in common.

2. (Comprehension) **(1) All cells grow, reproduce, and die.** These are things all cells have in common. Option (2) is incorrect because it is the main idea, not a detail. Option (3) refers to only some cells. Options (4) and (5) are structures that cells have in common, so each only partly supports the main idea.

3. (Comprehension) **(3) All cells are divided into two basic parts.** Although options (1), (2), (4), and (5) are all true, they give specific information about the two basic parts of a cell. Therefore, option (3), the most general statement, is correct.

4. (Comprehension) **(3) The nucleus contains a nucleolus, chromatin, and a nuclear membrane.** Note that this is an implicit main idea; the paragraph does not contain a topic sentence. Options (1), (2), (4), and (5) provide details about the nucleus and its parts. Option (3), which simply names the parts of the nucleus, is the main idea.

5. (Comprehension) **(1) Cytoplasm contains all the structures inside the cell membrane except the nucleus.** Options (2), (3), and (4) are incorrect because they give specific information about the parts of the cytoplasm. Option (5) is not true.

6. (Comprehension) **(5) The cytoplasm contains mitochondria, the ER, ribosomes, and vacuoles.** In naming several parts of the cytoplasm, option (5) supports the main idea, that the cytoplasm contains all the cell

structures except the nucleus. Option (1) is a definition of organelles. Options (2), (3), and (4) are true, but they are specific details about various parts of the cytoplasm.

**Reviewing Lesson 1
(pages 44–45)**

1. (Comprehension) **(3) metaphase** During interphase, option (1), the spindles have not yet appeared. In prophase, option (2), the spindles are forming. During the last two phases, options (4) and (5), the spindles are already contracting and disappearing.

2. (Comprehension) **(5) spindle** By observing the diagram, you can see that the spindle fibers are pulling the chromosomes apart. Option (1) remains in one place and does not change its appearance. Option (2) has no way to pull the chromosome apart. Options (3) and (4) are not seen during anaphase.

3. (Comprehension) **(2) Mitosis is the process by which cells divide.** This option is correct because the paragraph and diagram describe the process of mitosis. Options (1), (3), (4), and (5) are all true, but they describe only specific aspects of mitosis.

4. (Comprehension) **(1) interphase** According to the passage, the hereditary material that makes up the chromosomes is duplicated during the first phase, or interphase. The already duplicated chromosomes are present during each of the other phases, options (2), (3), (4), and (5).

5. (Comprehension) **(1) guide the chromosomes** By observing the diagram and reading the passage, you can see that the spindle fibers guide the chromosomes toward the opposite sides of the cell. They could not produce the nucleus, option (2), or form the nuclear membrane, option (3), since they have disappeared already. The hereditary material, option (4), has already become chromosomes. There is no cell plate discussed, option (5).

6. (Comprehension) **(4) 46** According to the passage, each daughter cell will have the same number of chromosomes as the parent cell. Since the parent cell has 46 chromosomes, each daughter cell will also have 46.

7. (Application) **(2) 23** According to the passage, a gamete has half the number of chromosomes found in a parent cell. Since the cells have a chromosome number of 46, the gametes have half that number, or 23.

8. (Comprehension) **(1) half the number** The number of chromosomes resulting from meiosis is half that of the parent cell. A cell produced by mitosis has the same number as the parent cell.

GED Mini-Test (pages 46–47)

1. (Comprehension) **(1) cell wall, chloroplast, and leucoplast** The first paragraph states that the cell wall is a structure in plant cells, not animal cells; the second paragraph describes chloroplasts and leucoplasts, two kinds of plastids. Option (2) is incorrect because the cell membrane is found in both plants and animals. Option (3) is incorrect because the cell membrane

and nucleolus are found in both plants and animals. Option (4) is incorrect because the chromatin and nucleolus are found in both plants and animals. Option (5) is incorrect because cytoplasm is found in both plants and animals.

2. (Comprehension) **(4) chloroplast** Photosynthesis cannot take place without chlorophyll, which is stored in the chloroplast. Option (1) is the food manufactured in the chloroplast. Option (2) is the part of the cell that controls the cell's activities. Option (3) provides stiffness to the cell. Option (5) stores the sugar after it is produced by photosynthesis in the chloroplast.

3. (Comprehension) **(5) production of sugar** According to the passage, the products of photosynthesis are sugar and oxygen. Option (1) is controlled by the nucleus of the cell. Options (2), (3), and (4) have nothing to do with photosynthesis.

4. (Comprehension) **(1) the secondary cell wall** This layer is made stiff by cellulose and often remains after the cell has died. The middle lamella, option (2), is made up of pectin, a jellylike substance, which would not make wood stiff. Options (3) and (5) are organelles that have nothing to do with cell stiffness. The primary cell wall, option (4), is made up of cellulose and pectin and would not be stiff enough to maintain wood's hard structure, especially after the tree has died.

5. (Comprehension) **(3) to prevent fungus from killing the tree** If the injured branch is not removed, the fungus will spread, eventually killing the tree. Option (1) has nothing to do with injured branches.

Options (2) and (4) are incorrect because they are not the reason the branch is being removed. Option (5) is not true.

6. (Comprehension) **(3) The bark was torn.** The diagram shows that the bark was torn, so option (3) is the only possible answer. The other options cannot be seen in the diagram.

7. (Comprehension) **(1) to make sure that all the fungus is removed** The stub is removed so that the fungus will not spread to other branches or the trunk. Options (2) and (3) are not the reasons for removing the stub. Options (4) and (5) are not true.

8. (Comprehension) **(3) The limb will fall without tearing the bark and wood.** When the limb breaks away, the cut in the limb keeps the bark from being torn from the branch or trunk. Option (1) does not matter. Options (2) and (4) are not true. Option (5) is true, but it is not the reason for undercutting the branch.

LESSON 2 Comprehension Skill: Restating Information

When you restate information, you say it in another way. Sometimes you simply use different words. At other times you may restate information with a diagram or formula. Restating information is one way to make sure you understand what you read.

There are three basic ways to restate information. The most common way is called **paraphrasing**. This means that you rewrite an idea, sentence, or paragraph in your own words. When you take notes in class or while reading a textbook, you are often paraphrasing the information.

Read the following paragraph and then see how the information can be restated.

> The average woman gains about 24 pounds during pregnancy. About two-thirds of this weight is taken up by the uterus and its contents. The fetus averages 7.7 pounds, the amniotic fluid 1.8 pounds, the placenta 1.4 pounds, and the uterus 2 pounds. The remaining weight is distributed in other parts of the body. The breasts gain 0.9 pound, the blood weighs 4 pounds, and other body fluids weigh 2.7 pounds. The last 3.5 pounds is added in body fat.

Now restate what you have read in different words.

> By the time her baby is born, the average woman has gained 24 pounds. The baby itself weighs only 7.7 pounds. What is the other weight? The placenta accounts for 1.4 pounds and the uterus accounts for 2 pounds. About 0.9 pound is added to the breasts and 3.5 pounds is body fat. Additional body fluids make up another 8.5 pounds: the weight of the blood increases by 4 pounds, the amniotic fluid makes up 1.8 pounds, and other body fluids average 2.7 pounds.

You will notice that all the facts remain the same in the restated paragraph. The way the facts are presented, however, has changed. In addition to changes in wording, the information has been rearranged a bit. Instead of grouping the uterus and its contents, the writer has made body fluids a group. The wording and order of the information has changed, but its meaning has not.

Now look at another way to restate information. Instead of using words, you can present information in a diagram, chart, or graph.

This bar graph has the same information as the two paragraphs. The graph makes it easy to tell at a glance where the extra weight is during pregnancy. The weight of the baby and the weight gain of each part of the mother's body are represented by bars. The higher the bar, the more weight.

There is a third way to restate information. Instead of using words or diagrams, you can use formulas or math expressions. For example, you could show the various weights as a list of numbers to be added.

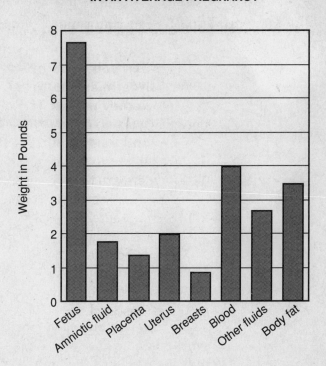

DISTRIBUTION OF WEIGHT GAIN IN AN AVERAGE PREGNANCY

Practicing Comprehension

Read the following paragraph.

> **Malnutrition** is a poor state of health that results from an unbalanced diet. Malnutrition is often thought of as undernutrition. This problem can be caused by eating too little food or by eating too little of a particular nutrient. However, malnutrition can also be caused by overnutrition. People who eat too much food or too much of certain nutrients such as fats or vitamins also suffer from malnutrition.

Questions 1 and 2 refer to the paragraph. Circle the best answer for each question.

1. Malnutrition is a condition in which

   (1) not enough food is eaten
   (2) too little of a specific nutrient is eaten
   (3) a person's health is affected by poor diet
   (4) a person eats too much
   (5) too much fat is eaten

2. What causes undernutrition?

   (1) not eating enough food or not eating enough of a specific nutrient
   (2) eating foods rich in fat
   (3) eating foods with very large amounts of particular vitamins
   (4) poor health
   (5) malnutrition

To check your answers, turn to page 58.

Topic 2: Photosynthesis

Read the following passage and look at the diagram.

Almost all animal life depends on plants for food. Plants, however, produce their own food through a process called photosynthesis. In **photosynthesis**, green plants take water, carbon dioxide, and energy from sunlight and use these to make sugar, oxygen, and water. The sugar is used as food and to build other substances that the plant needs, such as starches and protein. Oxygen is a **byproduct** of photosynthesis. Plants release oxygen into the atmosphere where animals breathe it.

Photosynthesis can be shown as a chemical equation. In the equation shown below, the arrow means <u>yields</u>.

$$\text{water} + \text{carbon dioxide} \xrightarrow[\text{chlorophyll}]{\text{light}} \text{sugar} + \text{oxygen} + \text{water}$$

This equation is a summary of all the reactions that make up photosynthesis. The diagram below shows photosynthesis in more detail. Notice that the reactions are grouped into two phases — the light reactions and the dark reactions.

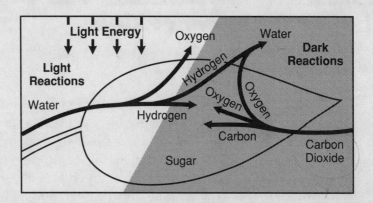

The first phase of photosynthesis is called the light reactions. The light reactions take place in sunlight. Light energy from the Sun is trapped by the plant's **chlorophyll**, or green coloring. The chlorophyll changes light energy to chemical energy, which is used to split particles of water into hydrogen and oxygen.

The second phase of photosynthesis is called the dark reactions. Light is not needed for the dark reactions to take place. During this phase, the hydrogen produced in the light reactions combines with carbon dioxide from the atmosphere to form sugar. The rest of the hydrogen combines with oxygen to form water.

Exercises

Questions 1 to 6 refer to the previous passage and diagram. Circle the best answer for each question.

1. Which substance produced in the light reactions of photosynthesis combines with carbon dioxide to produce sugar in the dark reactions?

   (1) oxygen
   (2) carbon
   (3) hydrogen
   (4) chlorophyll
   (5) water

2. What substance produced during photosynthesis does the plant use for food?

   (1) sugar
   (2) oxygen
   (3) water
   (4) carbon dioxide
   (5) chlorophyll

3. Which of the following restates the equation for photosynthesis?

   (1) Water plus carbon dioxide, in the presence of light and chlorophyll, yields sugar, oxygen, and water.
   (2) Water plus carbon dioxide, in the presence of chlorophyll, yields sugar, oxygen, and water.
   (3) Water plus carbon dioxide, in the presence of light and chlorophyll, yields sugar, hydrogen, and water.
   (4) Water plus oxygen, in the presence of light and chlorophyll, yields sugar, carbon dioxide, and water.
   (5) None of the above restates the photosynthesis equation.

4. Which of the following restates the main idea of the passage?

   (1) Animals eat plants for food.
   (2) Plants make their own food through photosynthesis.
   (3) Photosynthesis takes place in green plants.
   (4) Photosynthesis replaces the oxygen in the atmosphere that is used by animals.
   (5) Sunlight is needed for photosynthesis to occur.

5. Which of the following supports the statement, "The first phase of photosynthesis is called the light reactions"?

   (1) Chemical energy splits water into hydrogen and oxygen.
   (2) The plant's green coloring is called chlorophyll.
   (3) Oxygen is a by-product of photosynthesis.
   (4) Carbon dioxide is absorbed from the atmosphere.
   (5) The light reactions take place in sunlight.

6. Which of the following restates the main idea of the third paragraph?

   (1) Light energy from the Sun is absorbed by chlorophyll.
   (2) Chlorophyll makes chemical energy from light energy.
   (3) Chemical energy splits water into hydrogen and oxygen.
   (4) The light reactions make up the first phase of photosynthesis.
   (5) Sunlight is needed for the light reactions.

To check your answers, turn to page 58.

Reviewing Lesson 2

Read the following passage.

> **Respiration** is the process by which cells obtain the energy they need in order to function. During respiration, sugar is broken down to release chemical energy. As a result, carbon dioxide and water are released. Many of the reactions in respiration are the opposite of those in photosynthesis, where sugar is formed from water and carbon dioxide. The equation for respiration is:

$$\text{sugar} + \text{oxygen} \longrightarrow \text{carbon dioxide} + \text{water} + \text{energy}$$

> There are two kinds of respiration — direct and indirect respiration. **Direct respiration** occurs in single-celled organisms, like the amoeba. These organisms are able to directly exchange gases with the environment. This exchange takes place through cell membranes. **Indirect respiration** occurs in many-celled organisms where most of the body's cells are not in direct contact with the environment.
>
> Indirect respiration can be divided into two phases. **External respiration** exchanges gases between the environment and the blood. Gills and lungs perform this function. **Internal respiration** exchanges gases between the blood and cells. The circulatory system provides this function.

Questions 1 to 4 refer to the passage. Circle the best answer for each question.

1. Which organism functions with direct respiration?

  (1) human
  (2) bird
  (3) lizard
  (4) amoeba
  (5) dog

2. Which two substances would you expect to be exhaled by the lungs?

  (1) oxygen and carbon dioxide
  (2) hydrogen and oxygen
  (3) glucose and carbon dioxide
  (4) hydrogen and carbon
  (5) water vapor and carbon dioxide

3. Which two substances are the direct result of respiration?

  (1) oxygen and hydrogen
  (2) carbon dioxide and water
  (3) glucose and water
  (4) glucose and hydrogen
  (5) water and oxygen

4. How can the equation for respiration be stated in a sentence? Sugar plus oxygen

  (1) plus carbon dioxide yields water plus energy.
  (2) yields carbon dioxide plus water plus energy.
  (3) yields carbon dioxide plus water.
  (4) yields carbon dioxide plus water plus sugar.
  (5) yields carbon dioxide plus gases plus energy.

To check your answers, turn to page 58.

Study the following diagram and read the passage.

RESPIRATION

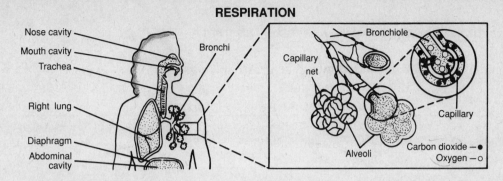

We breathe in air through the nose or mouth. The air passes through a tube called the **trachea**, which runs from the back of the mouth to the lungs. The trachea divides into two **bronchi**, one for each lung. Each of these tubes divides, forming branches called **bronchioles**. The bronchioles end in small air sacs called **alveoli**.

The exchange of oxygen and carbon dioxide takes place at the alveoli. They are covered by a network of tiny blood vessels called **capillaries**. Oxygen from the air in the alveoli passes into the bloodstream. Carbon dioxide in the blood passes from the capillaries into the air in the lungs. The carbon dioxide is carried out of the body as we exhale.

Questions 5 to 8 refer to the diagram and passage. Circle the best answer for each question.

5. After inhaled air passes through the trachea, it goes through the

(1) alveoli, bronchi, bronchioles
(2) bronchi, bronchioles, alveoli
(3) bronchioles, alveoli, bronchi
(4) alveoli, bronchioles, bronchi
(5) bronchioles, bronchi, alveoli

6. Which of the following accurately describes the air we breathe?

(1) Inhaled air contains more carbon dioxide than exhaled air.
(2) Inhaled air contains more oxygen than exhaled air.
(3) Inhaled air passes directly from the trachea to the alveoli.
(4) Exhaled air contains more oxygen than inhaled air.
(5) Exhaled air passes through the diaphragm.

7. What is the main function of the lungs?

(1) supporting the bronchi
(2) warming the air we breathe
(3) absorbing oxygen and releasing carbon dioxide
(4) absorbing carbon dioxide and releasing water vapor
(5) splitting water into oxygen and hydrogen

8. Most of the information in the second paragraph can also be found in the
  A. first paragraph of the passage
  B. left side of the diagram
  C. right side of the diagram

(1) A only
(2) B only
(3) C only
(4) A and B
(5) A and C

To check your answers, turn to page 59.

GED Mini-Test

Directions: Choose the best answer to each item below.

Items 1 to 4 refer to the following diagram and passage.

THE NITROGEN CYCLE

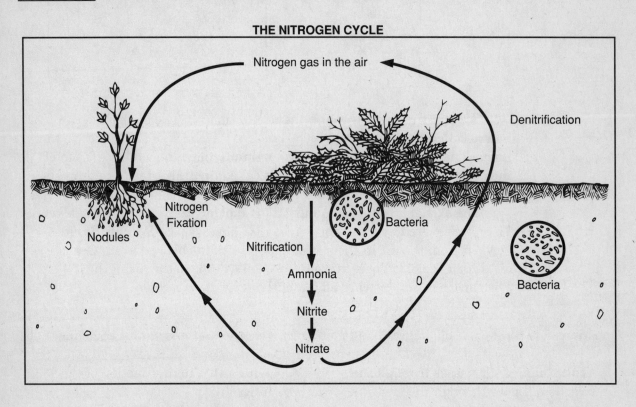

Plants and animals must have nitrogen to make proteins. Air is mostly nitrogen, but plants and animals cannot use nitrogen in gas form. Animals get nitrogen by eating plants or other animals. Plants get nitrogen from the soil. They rely on bacteria to convert nitrogen into forms that they can use.

Through a process called nitrogen fixation, certain kinds of bacteria can take nitrogen from the air and combine it with other substances into a form that plants can use. These bacteria are found only in bumps, or nodules, on the roots of plants called legumes.

Other soil bacteria take the nitrogen found in decomposing plants and animals. This nitrogen is converted to ammonia, nitrite, and nitrate in a process called nitrification. This process provides another source of nitrogen for plants.

Other bacteria return nitrogen to the atmosphere by changing nitrates into gaseous nitrogen, a process called denitrification.

This nitrogen is useless to living things until nitrogen-fixing bacteria take nitrogen from the air and convert it into a form plants can use.

1. Where do plants get their nitrogen?

    (1) from other plants
    (2) directly from the atmosphere
    (3) from animals
    (4) from the soil
    (5) from protein

2. According to the diagram, the process of nitrogen fixation takes place

    (1) in decaying organisms
    (2) in the leaves of plants
    (3) in the roots of all plants
    (4) in the atmosphere
    (5) in nodules on the roots of legumes

3. By what process do bacteria break down the protein of decaying organisms into nitrite and nitrate compounds?

    (1) nitrogen fixation
    (2) denitrification
    (3) erosion
    (4) nitrification
    (5) ionization

4. Which process captures nitrogen from the atmosphere and turns it into forms that plants can use?

    (1) nitrification
    (2) denitrification
    (3) decomposition
    (4) nitrogen fixation
    (5) photosynthesis

Items 5 and 6 refer to the following passage and diagram.

The two main processes involved in the carbon-oxygen cycle are respiration and photosynthesis. Each of the processes produces substances used in the other process. Respiration takes place in both plants and animals. Only plants, however, are able to carry on photosynthesis. During respiration, sugar is broken down and carbon dioxide is released. During photosynthesis, water, carbon dioxide, and light energy from the Sun produce oxygen, sugar, and water.

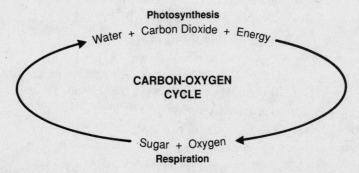

5. According to the diagram, what reacts with sugar during respiration to produce water, carbon dioxide, and energy?

    (1) carbon
    (2) hydrogen
    (3) oxygen
    (4) water
    (5) carbon dioxide

6. If there were no photosynthesis, the amount of oxygen in the air would

    (1) remain unchanged
    (2) increase
    (3) go into the atmosphere
    (4) decrease
    (5) be produced by animals

To check your answers, turn to page 59.

Answers and Explanations

1. (Comprehension) **(3) a person's health is affected by poor diet** The paragraph states that malnutrition is a poor state of health that can result from overeating as well as undereating. Options (1), (2), (4), and (5) define malnutrition as <u>either</u> overeating or undereating. Therefore, they are not complete definitions of malnutrition.

2. (Comprehension) **(1) not eating enough food or not eating enough of a specific nutrient** Undernutrition can be a matter of not getting enough total calories, or it can be a matter of a vitamin deficiency. Options (2) and (3) are incorrect because they are examples of overnutrition. Option (4) is incorrect because poor health is described as a result of undernutrition, not a cause. Option (5) is incorrect because undernutrition is a form of malnutrition; malnutrition does not cause undernutrition.

Exercises (page 53)

1. (Comprehension) **(3) hydrogen** The fourth paragraph states, and the diagram shows, that hydrogen produced in the light reactions combines with carbon dioxide to form sugar in the dark reactions. The other options are therefore incorrect.

2. (Comprehension) **(1) sugar** The first paragraph states that sugar is used as food and to build other substances such as starches and proteins that the plant needs. The other products of photosynthesis are waste materials, options (2)

and (3). Option (4), carbon dioxide, is not a product of photosynthesis. Neither is chlorophyll, option (5).

3. (Comprehension) **(1) Water plus carbon dioxide, in the presence of light and chlorophyll, yields sugar, oxygen, and water.** Option (2) is incorrect because it leaves out light. Option (3) is incorrect because hydrogen is not a product of photosynthesis. Option (4) is incorrect because oxygen is not used in photosynthesis; it is a product of photosynthesis. Option (5) is incorrect because option (1) is correct.

4. (Comprehension) **(2) Plants make their own food through photosynthesis.** The passage describes how photosynthesis works. Although options (1), (3), (4), and (5) are true statements, they are not the main idea of the passage. They are all supporting details.

5. (Comprehension) **(5) The light reactions take place in sunlight.** This statement explains why the light reactions are called light reactions. Options (1), (2), (3), and (4) are all true statements, but they do not support the idea in the question.

6. (Comprehension) **(4) The light reactions make up the first phase of photosynthesis.** This restates the topic sentence of the third paragraph, the sentence that gives the general idea of the paragraph. Options (1), (2), (3), and (5) are details supporting the general topic, the light reactions.

1. (Comprehension) **(4) amoeba** Direct respiration occurs in single-celled organisms. Indirect respiration occurs in most many-celled organisms. Option (4) is the only single-celled organism listed. Options (1), (2), (3), and (5) are incorrect because they all are many-celled organisms.

2. (Comprehension) **(5) water vapor and carbon dioxide** Since water and carbon dioxide are the result of respiration, it follows that carbon dioxide and water vapor would be exhaled. In option (1), oxygen is not a result of respiration. In option (2), neither hydrogen nor oxygen is the result of respiration. In option (3), sugar is not a result of respiration. In option (4), neither hydrogen nor carbon is the result of respiration.

3. (Comprehension) **(2) carbon dioxide and water** According to the equation, carbon dioxide and water are the only two substances that result from respiration. Options (1), (3), (4), and (5) are incorrect because neither hydrogen, sugar, nor oxygen is a direct result of respiration.

4. (Comprehension) **(2) yields carbon dioxide plus water plus energy.** This restates the respiration equation. Option (1) is incorrect because carbon dioxide is a product of respiration. Option (3) is incorrect because it leaves out energy. Option (4) is incorrect because sugar is not a result of respiration. Option (5) is incorrect because respiration yields carbon dioxide and water, not "gases."

5. (Comprehension) **(2) bronchi, bronchioles, alveoli** According to the passage and the diagram, the air we inhale passes through the trachea, bronchi, and bronchioles before arriving at the alveoli. Options (1), (3), (4), and (5) are incorrect because the structures are listed in the wrong order.

6. (Comprehension) **(2) Inhaled air contains more oxygen than exhaled air.** Since oxygen in the air is absorbed into the bloodstream in the alveoli, the air that is left to exhale contains less oxygen. Option (1) is incorrect because there is more carbon dioxide in exhaled air. Option (3) is incorrect because inhaled air must go through the bronchi and bronchioles before reaching the alveoli. Option (4) is incorrect because inhaled air contains more oxygen. Option (5) is incorrect because no air passes through the diaphragm.

7. (Comprehension) **(3) absorbing oxygen and releasing carbon dioxide** The main function of the lungs, as described by the passage, is to exchange oxygen in inhaled air and replace it with carbon dioxide in exhaled air. Its main function is not to support the bronchi, option (1), or to warm the air, option (2). It does not absorb carbon dioxide, option (4), nor does it split water into oxygen and hydrogen, option (5).

8. (Comprehension) **(3) C only** The right side of the diagram shows a close-up of the exchange in the lungs of oxygen and carbon dioxide, which is the topic of the second paragraph. The first paragraph (A) describes how air gets into the alveoli, not what happens there. The left side of the diagram (B) shows the path of air into the lungs, not the exchange of oxygen and carbon dioxide.

GED Mini-Test (pages 56–57)

1. (Comprehension) **(4) from the soil** This is stated in the first paragraph. Options (1), (2), (3), and (5) are therefore incorrect.

2. (Comprehension) **(5) in nodules on the roots of legumes** The diagram shows that nitrogen fixation takes place in the nodules on roots. The second paragraph also states this fact. Options (1), (2), (3), and (4) are therefore incorrect.

3. (Comprehension) **(4) nitrification** Nitrification is the process by which nitrogen is made available to plants by the breakdown of decaying organisms. Options (1) and (2) name other nitrogen-producing processes. Options (3) and (5) have nothing to do with the nitrogen cycle.

4. (Comprehension) **(4) nitrogen fixation** This is clearly shown on the diagram. Option (1), nitrification, is how nitrogen from decomposing matter is put back into the soil. Option (2), denitrification, is the release of nitrogen back into the air. Option (3) is the breakdown of animal and plant matter. Option (5) is not part of the nitrogen cycle.

5. (Comprehension) **(3) oxygen** Oxygen reacts with glucose to produce water, carbon dioxide, and energy. The other items are all incorrect because they name the wrong substance.

6. (Analysis) **(4) decrease** Photosynthesis provides most of the oxygen for the oxygen-carbon cycle. Without it, the amount of oxygen in the air would decrease. This is why Options (1), (2), and (3) are incorrect. Option (5) is incorrect because animals do not produce oxygen.

LESSON 3 Analysis Skill:
Cause and Effect

When you feel hungry, you eat. In this situation, being hungry is a cause; eating is an effect. Situations in which one thing makes another happen are called *cause and effect relationships*.

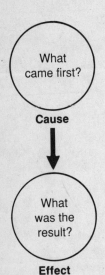

What came first?

Cause

What was the result?

Effect

Causes and effects can be many things. They can be facts, conditions, things, events, or actions. For example, a baseball player who hits the winning home run is the cause of his or her team's victory. Too much sunlight causes sunburn. Your body's fight against an infection causes fever.

Science is full of cause and effect relationships. Scientists try to discover general laws that can predict what will result from specific causes.

Read this passage and identify the cause and effect relationships.

> Tobacco smoke contains a poisonous chemical called nicotine. Inhaling nicotine immediately causes changes in the body. It increases the heart rate and blood pressure, narrows blood vessels, and slows movements of the digestive system. Nicotine also has long-term effects. It damages the lining of the air passages in the throat and lungs.

What is the cause in this passage? Nicotine is the cause. What are the effects of nicotine? Nicotine has immediate effects, such as increased heart rate and blood pressure, narrowed blood vessels, and slowed digestive movements. It also has a long-term effect—it damages the throat and lungs.

When you read, look for cause and effect relationships. Watch for words or phrases such as <u>cause</u>, <u>effect</u>, <u>because</u>, <u>as a result of</u>, <u>leads to</u>, <u>develops from</u>, <u>due to</u>, <u>thus</u>, <u>therefore</u>, <u>so</u>, and <u>the reason is</u>.

Cause and effect relationships are stated very clearly in the passage about nicotine. Sometimes, however, you have to figure out a cause or effect. For example, what might be a result of damaged air passages in the lungs? The author does not say. Perhaps damaged air passages might result in breathing difficulties. Another effect might be diseases of the lungs such as cancer or emphysema.

Practicing Analysis

Read the following paragraph.

Injuries, heavy bleeding, disease, and some drugs can change the way the circulatory system works. Injuries can cause a large amount of blood to be lost. Some drugs make blood vessels expand, reducing the flow of blood through the body. When either condition happens, a person is said to be "in shock." The person has low blood pressure and a weak pulse. He or she also becomes pale and may sweat heavily.

Questions 1 and 2 refer to the paragraph. Circle the best answer for each question.

1. What causes shock?

(1) injuries, heavy bleeding, disease, and some drugs
(2) a weak pulse
(3) heavy sweating
(4) blood clots
(5) oxygen starvation

2. What first-aid measure would help stop shock?

(1) artificial respiration
(2) setting a broken bone
(3) stopping bleeding
(4) giving a drink of water
(5) inducing vomiting

Study the following diagram.

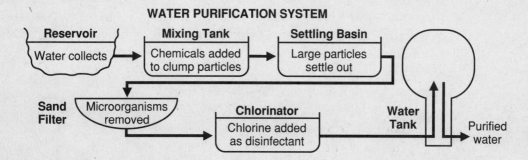

WATER PURIFICATION SYSTEM

Reservoir — Water collects
Mixing Tank — Chemicals added to clump particles
Settling Basin — Large particles settle out
Sand Filter — Microorganisms removed
Chlorinator — Chlorine added as disinfectant
Water Tank — Purified water

Questions 3 and 4 refer to the diagram. Circle the best answer for each question.

3. What would happen if the sand filter were removed from the water purification system?

(1) The water would contain more large particles.
(2) The water would contain more microorganisms.
(3) The water would contain more chlorine.
(4) The water would contain less chlorine.
(5) There would be no effect on the water.

4. Why is purification of drinking water necessary?

(1) to pump water from reservoirs
(2) to move water to water storage tanks
(3) to conserve water supplies
(4) to prevent the spread of harmful substances through drinking water
(5) to prevent tooth decay

To check your answers, turn to page 68.

Topic 3: Evolution

Read the following passage and look at the diagram.

Evolution means an orderly change. In science, the word <u>evolution</u> usually refers to the theory of evolution. The theory of evolution holds that all organisms living today have a common ancestor that evolved from the first living cells 3.5 billion years ago.

There is much evidence to support the theory of evolution. One piece of evidence is the presence of similar structures in different organisms. Called **homologous structures,** these body parts from different organisms have similar structures but perform different functions. For example, a whale's flipper, a human's arm, a dog's front leg, and a bird's wing are homologous structures. Each of these limbs is used in a different way. When you first look at these limbs, they appear different. But if you look closely, you will see that the bones of each limb are very similar. They have even been given the same names. The similarity of these four different limbs suggests that these four organisms evolved from a common ancestor. The differences are the result of adaptation to different environments.

Another piece of evidence that supports the theory of evolution is that the embryos of different organisms are similar. An **embryo** is the early stage of development from a fertilized egg. The embryos of a fish, a bird, and a human are similar at first. For example, at one stage all of these embryos have gill slits and tail buds. As the embryos develop, the differences among the organisms become more clear. However, the similarities in the early embryos suggest that fish, birds, and humans evolved from a common ancestor.

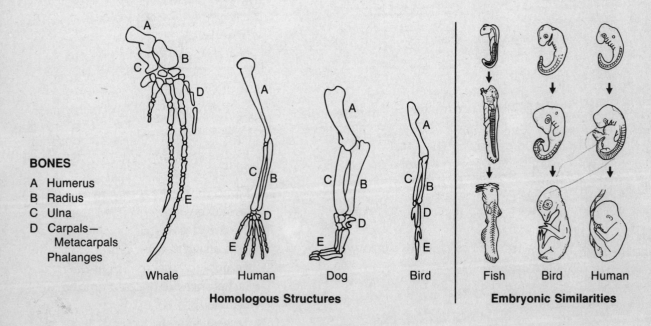

BONES

A Humerus
B Radius
C Ulna
D Carpals—
    Metacarpals
    Phalanges

Whale   Human   Dog   Bird

Homologous Structures

Fish   Bird   Human

Embryonic Similarities

Exercises

Questions 1 to 6 refer to the previous passage and diagram. Circle the best answer for each question.

1. According to the diagram, what is similar about the radius and ulna bones of whales, humans, dogs, and birds?

    (1) They form fingerlike structures.
    (2) They form footlike structures.
    (3) They are positioned side by side below the humerus.
    (4) They are part of an arm, leg, and wing.
    (5) They are evidence of similarity between embryos.

2. What is one reason that scientists support the theory of evolution?

    (1) Humans are the most highly developed organisms on the Earth.
    (2) Different types of organisms have homologous structures.
    (3) All organisms developed 3.5 billion years ago.
    (4) All embryos have gill slits and tail buds.
    (5) All organisms function in the same way.

3. What caused the ancestors of whales and birds to develop into different types of animals?

    (1) homologous structures
    (2) similarity between embryos
    (3) Whales and birds had a common ancestor.
    (4) Whales and birds are not related.
    (5) Whales and birds evolved in different environments.

4. Why does a human embryo have a tail bud?

    (1) Human ancestors had tails.
    (2) Fish and bird embryos have tail buds.
    (3) Tails are homologous structures.
    (4) Humans have evolved.
    (5) The early stages of the human embryo are similar to the early stages of the fish embryo.

5. What is the main idea of the passage?

    (1) Homologous structures are evidence supporting the theory of evolution.
    (2) Similarity among embryos exists in organisms with a common ancestor.
    (3) Birds, whales, humans, and dogs had a common ancestor.
    (4) Homologous structures and similarity among embryos are evidence supporting the theory of evolution.
    (5) Humans are related to birds.

6. How will the passage of a million years affect today's plant and animal species?

    (1) Plants and animals will continue to evolve.
    (2) Fertilization of the egg will no longer be required.
    (3) Plants and animals will develop into a single life form.
    (4) Animals will continue to change, but plants will remain the same.
    (5) Birds and whales will become land animals.

To check your answers, turn to page 68.

Reviewing Lesson 3

Read the following passage.

One of the key points in Charles Darwin's theory of evolution is the idea of **natural selection.** According to natural selection, the individual organisms that have traits, or characteristics, that help them live in their environment are more likely to survive and reproduce. Natural selection can be summarized in five steps.

1. Overproduction. Most organisms have many offspring. But because the resources of the environment are limited, the environment cannot support as many organisms as are born.
2. Competition. Too many offspring and limited resources cause organisms to compete for food, water, and other needs.
3. Variation. Individual organisms vary a great deal in the traits they inherit.
4. Survival. Some individuals have inherited traits that help them use the resources of the environment. These individuals are more likely to survive.
5. Reproduction. Individuals that survive pass these traits or adaptations on to their offspring.

Questions 1 to 4 describe an event that is an example of one of the steps in the theory of natural selection. Circle the answer that best matches the event described.

1. In a herd of giraffes, some have longer necks than others.

    (1) overproduction
    (2) competition
    (3) variation
    (4) survival
    (5) reproduction

2. Birds that blend well with their surroundings had parents that also blended with their surroundings.

    (1) overproduction
    (2) competition
    (3) variation
    (4) survival
    (5) reproduction

3. In one season, a pine tree produces over 100 cones, each of which contains about 35 seeds.

    (1) overproduction
    (2) competition
    (3) variation
    (4) survival
    (5) reproduction

4. Only cactus plants that can store large amounts of rainwater will still be alive when the desert's next rainy season begins.

    (1) overproduction
    (2) competition
    (3) variation
    (4) survival
    (5) reproduction

To check your answers, turn to pages 68–69.

Study the following diagram and read the passage.

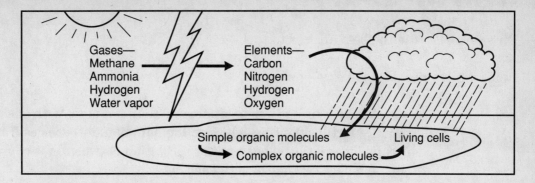

Life is thought to have first appeared about 3.5 billion years ago. Many scientists believe that the Earth's atmosphere was then made up of ammonia, hydrogen, methane, and water vapor. These gases contain the elements carbon, nitrogen, hydrogen, and oxygen, which are found in organic molecules. **Organic molecules** are the building blocks of all living things. Organic molecules all contain carbon in combination with one or more of the other elements.

According to one theory of the origin of life, energy from lightning or sunlight split some of the gas molecules in the atmosphere. As a result, unattached particles of the elements carbon, nitrogen, hydrogen, and oxygen were present. Some of these combined to form simple organic molecules. Rain washed some of the organic molecules down from the atmosphere to the Earth's surface. Here the organic molecules combined with one another. As time passed they became more complex and formed the first living cells.

Questions 5 to 8 refer to the passage and diagram. Circle the best answer for each question.

5. What caused some of the gas molecules of the early atmosphere to break up?

(1) water vapor
(2) organic molecules
(3) methane
(4) rain
(5) lightning

6. Organic molecules must have existed before life could form because organic molecules

(1) make up all living things
(2) give rise to carbon
(3) provide energy
(4) make up the atmosphere
(5) produce rain

To check your answers, turn to page 69.

7. A current theory concerning the origin of life assumes that the Earth's early atmosphere contained the gases

(1) ammonia, nitrogen, and hydrogen
(2) methane, ammonia, and water vapor
(3) carbon, methane, and ammonia
(4) silicon, hydrogen, and water
(5) nitrogen, hydrogen, and water

8. What element is present in all organic molecules?

(1) carbon
(2) nitrogen
(3) hydrogen
(4) oxygen
(5) all of the above

GED Mini-Test

Directions: Choose the best answer to each item.

Items 1 to 4 refer to the following passage.

A species is a group of organisms that can mate with one another and produce fertile offspring. The development of a new species from an old one is called speciation. The result of speciation is two groups of organisms that can no longer reproduce with each other. 47

As long as the environment remains more or less the same and the entire population of a species remains isolated and together, speciation is unlikely to occur. However, environments do change and individuals do move from one place to another. In these situations, speciation is more likely to occur.

Sometimes many species evolve from one species. This kind of speciation is called adaptive radiation. Adaptive radiation occurs when small groups of individuals become separated from the rest of the population. This often happens on islands or in areas bounded by mountains. The separated populations become adapted to their separate environments. Adaptations, or traits that help individuals survive, are passed down to their offspring. Over a long period of time, each isolated group develops into a new species.

The results of adaptive radiation can be seen in Australia, where about 200 species of marsupials, mammals with pouches, have developed. One common ancestor produced many different species. For example, the kangaroo is adapted to life on the plains. The koala is adapted to life in the trees.

1. Which of the following is the most likely to contribute to speciation?

   (1) environmental changes
   (2) environmental stability
   (3) isolation of an entire species
   (4) decreased reproduction
   (5) increased population size

2. Which of the following best describes the pattern of changes that occur in species during adaptive radiation?

   (1) becoming more alike
   (2) random
   (3) becoming less alike
   (4) equal
   (5) none of the above

To check your answers, turn to page 69.

3. The marsupials of Australia are an example of

   (1) extinction
   (2) fossilization
   (3) convergent evolution
   (4) adaptive radiation
   (5) acquired traits

4. How has speciation probably affected the variety of animal and plant life?

   (1) created less variety
   (2) not affected variety
   (3) stopped the formation of new varieties
   (4) slowed the formation of new varieties
   (5) created more variety

A gene is a part of a cell that determines a particular trait. A particular gene may occur in most individuals or in only a few individuals. How often a gene occurs in a population is called gene frequency. When the frequency of a particular gene increases or decreases by chance, the change in frequency is called genetic drift.

For example, suppose that a population of squirrels consists mostly of white squirrels with a few gray individuals (a). Then suppose that a waterway is constructed, and the squirrel population is divided into two isolated groups (b). The larger group still has several gray squirrels. The smaller group has only one gray squirrel. The frequency of the gene for grayness is lower in the smaller population.

(a)　　　　　　　　　　　　　　　　(b)

5. Genetic drift is largely the result of

(1) isolation of species
(2) speciation
(3) adaptive radiation
(4) chance
(5) mating between species

6. What sort of genetic drift can be expected among the larger population of squirrels after the construction of the waterway?

(1) increase the gray gene's frequency
(2) decrease the gray gene's frequency
(3) increase the white gene's frequency
(4) make the frequency of both genes equal
(5) increase the frequency of both genes

To check your answers, turn to page 69.

Answers and Explanations

1. (Comprehension) **(1) injuries, heavy bleeding, disease, and some drugs** In the first sentence of the passage, these conditions are said to interfere with the circulation of the blood, which is what happens in shock. Options (2) and (3) are symptoms of shock, not causes. Options (4) and (5) are effects of shock, not causes.

2. (Analysis) **(3) stopping bleeding** Since one of the causes of shock is a loss of blood, it makes sense that stopping blood loss would help stop shock. The other first-aid measures do not affect the circulatory system and therefore would not be helpful.

3. (Analysis) **(2) The water would contain more microorganisms.** According to the diagram, the purpose of the sand filter is to remove microorganisms from the water. If the sand filter were removed, the result would be more microorganisms in the water.

4. (Analysis) **(4) to prevent the spread of harmful substances through drinking water** The diagram shows that water purification removes unwanted material, such as microorganisms, from the water supply. The effect of purification is that people do not drink such materials along with the water. Options (1) and (2) involve the movement of water, not water purification. Option (3) is incorrect because conservation has nothing to do with water purification. Option (5) involves adding a chemical, fluoride, to the water. This is not a step in purifying water.

1. (Comprehension) **(3) They are positioned side by side below the humerus.** The diagram shows that in all four animals the radius and ulna are next to each other just below the humerus. They are not part of the fingers, option (1), or the feet, option (2). Option (4) is incorrect because it lists different functions of the bones in the four different animals. Option (5) is not true.

2. (Analysis) **(2) Different types of organisms have homologous structures.** This fact is one piece of evidence that causes scientists to believe the theory of evolution. Option (1) may be true, but it does not offer support for the theory of evolution. Option (3) is not true; organisms have evolved gradually over time. Option (4) is incorrect because the passage states that fish, bird, and human embryos have gill slits and tail buds, but it does not say all embryos would have these structures. Option (5) is not true.

3. (Analysis) **(5) Whales and birds evolved in different environments.** Differences in structure between whales and birds are the result of one species surviving in the ocean and another surviving in the air. Options (1) and (2) are incorrect because homologous structures and similarity between embryos are a result of evolution, not a cause. Option (3) is true but does not explain why whales and birds developed differently. Option (4) is not true.

4. (Analysis) **(1) Human ancestors had tails.** The presence of a tail bud indicates that at one time the ancestors of humans had tails. Options (2), (4), and (5) are true, but they are incorrect because they do not explain why human embryos have tail buds. Option (3) may be true of some organisms, but not humans, since humans have no tails.

5. (Comprehension) **(4) Homologous structures and similarity among embryos are evidence supporting the theory of evolution.** The passage is about facts that support the idea that organisms have developed over time from a common ancestor. Options (1) and (2) are each only partly correct. Options (3) and (5) are true, but they are supporting details, not the main idea.

6. (Analysis) **(1) Plants and animals will continue to evolve.** Evolution is an ongoing process; it will not stop with today's organisms. Options (2), (3), and (5) are specific events that are unlikely and cannot be predicted based on the passage. Option (4) is not true; both plants and animals will continue to change.

1. (Application) **(3) variation** The differences in the necks of giraffes are variations within a species. Options (1) and (5) are incorrect because the statement about giraffes says nothing about reproduction. Option (2) is incorrect because the statement does not describe any competition among giraffes. Option (4) is incorrect. Although the long necks help giraffes feed on trees, this survival value is not included in the statement.

2. (Application) **(5) reproduction** The parent birds passed on their coloration trait to their offspring. Option (1) is incorrect; the statement does not mention how many baby birds there were. Option (2) is incorrect; the statement does not refer to any competition between these birds and any other birds. Option (3) is incorrect; the statement does not describe any differences among the birds. While it can be assumed that their coloration helped the birds survive, it is not part of the statement, so option (4) is incorrect.

3. (Application) **(1) overproduction** In one season a single pine tree produces enough seeds to start a forest. Since the statement does not say what happens to the seeds after they are formed, the other options are incorrect.

4. (Application) **(4) survival** Only plants that can store water will survive between rainy periods. Options (1) and (5) are incorrect; the statement does not mention reproduction. Option (2) is incorrect because although competition for water exists in the desert, it is not described here. Option (3) is incorrect because although there is variation in the ability of plants to store water, the point being made in the statement is that only certain cactus plants survive.

5. (Analysis) **(5) lightning** Energy from lightning was a cause of splitting the gas molecules in the early atmosphere. Options (1) and (3) are incorrect because these gases were present in the atmosphere, not causes of splitting molecules. Option (2) was a result of the recombining of elements, not a cause. Option (4) did not split the gas molecules.

6. (Analysis) **(1) make up all living things** According to the passage, organic molecules are the building blocks of all living things. Therefore, organic molecules had to exist before living things could evolve. Organic molecules do not give rise to carbon, option (2); carbon is part of any organic molecule. Energy, option (3), was obtained from other sources before life emerged. Organic molecules are not part of the atmosphere, option (4); gases are. Option (5) is not true.

7. (Comprehension) **(2) methane, ammonia, and water vapor** According to the passage, the Earth's atmosphere was composed of ammonia, hydrogen, methane, and water vapor. Option (2) contains three of the gases. Options (1) and (5) are incorrect because they include nitrogen. Option (3) is incorrect because it includes carbon. Option (4) is incorrect because it includes silicon.

8. (Comprehension) **(1) carbon** According to the passage, carbon is present in combination with one or more of the other elements listed. Carbon is therefore found in all organic molecules. Option (5) is incorrect because not all organic molecules contain all these elements.

GED Mini-Test (pages 66–67)

1. (Analysis) **(1) environmental changes** According to the passage, environmental changes are one of the things that contribute to speciation. Options (2) and (3) do not contribute to speciation. Options (4) and (5) have little effect on speciation.

2. (Analysis) **(3) becoming less alike** During adaptive radiation, species change and become less alike. Option (1) is the opposite of what happens. Random differences, option (2), are not a factor in adaptive radiation. Option (4) is incorrect because equal changes in population would not create different species. Option (5) is incorrect because there is a good description.

3. (Comprehension) **(4) adaptive radiation** The passage uses the development of marsupials in Australia as a classic example of adaptive radiation. With option (1) there would be no marsupials. Option (2) is incorrect because the passage says nothing about fossils. Option (3) is the opposite result of adaptive radiation. Option (5) does not affect adaptive radiation.

4. (Analysis) **(5) created more variety** Speciation allows new species of organisms to evolve, creating more variety. Option (1) would be the result if no speciation occurred. Options (2) and (3) are incorrect since more variety occurred. Option (4) is incorrect because speciation adds more variety.

5. (Analysis) **(4) chance** Chance variation of the gene frequency within a population is genetic drift. Options (1), (2), and (3) describe ways that new species develop. Option (5) is incorrect because mating between species is rare and not related to genetic drift.

6. (Analysis) **(1) increase the gray gene's frequency** Since the larger group of squirrels now has more gray squirrels in proportion to the number of white squirrels than before, the frequency of the gray gene is increased. Options (2), (3), (4), and (5) are incorrect because they do not describe the new proportion of gray to white squirrels.

LESSON 4 Comprehension Skill: Identifying Implications

You have learned to look for main ideas and supporting details when you read. This helps you understand what the author is saying. Sometimes, however, the author leaves out some statements. If you think about what you read, you can figure out what is missing.

For example, suppose you read, "It was a beautiful day, and we went swimming at the beach." What can you figure out about this event that the author does not tell you? You might figure that the weather was sunny and warm. You might decide that they brought swimsuits. Can you think of anything else that is likely to be true about that trip to the beach?

When you take what is written and then figure out other things that are probably true, you are **identifying implications.** Implications are not expressed in words by the author. Rather, they are **implied** by what is said. Implications are facts that you can be reasonably sure are true because they follow from what the author wrote. They are the logical outcomes of what has actually been said.

Read this paragraph and identify some of its implications.

> The thyroid gland produces a substance that helps regulate growth. One of the ingredients of this substance is iodine. When the body does not contain enough iodine, the thyroid gland becomes enlarged, a condition called goiter. Since seafood contains relatively large amounts of iodine, goiter is not common in coastal areas where people eat a lot of seafood. Goiter is more common in mountainous and inland areas where the soil contains small amounts of iodine. In these areas, locally grown food and drinking water do not provide people sufficient supplies of iodine.

The passage explains the relationship between a lack of iodine and the occurrence of goiter. What can we figure out from what we have read about this? First, the author says that people who eat a lot of seafood containing iodine usually don't get goiter. And people who live in areas where the local food and water supplies are low in iodine often do get goiter. Therefore we can figure out that iodine must be provided in the diet to prevent goiter.

Another implication of the passage is that iodine is not evenly distributed in the foods people eat. It is easier for some groups to develop goiter because not enough iodine is present in their diets. What does this suggest about preventing goiter? It implies that iodine could be added to some common substance that most people eat or drink in order to prevent goiter. (Indeed, in the United States iodine is added to table salt for this reason.)

Another implication is that goiter is not contagious—you can't catch it from someone else. As long as you have enough iodine in your diet, you will not develop goiter. Can you think of any other implications of this passage?

Remember, when you identify an implication you are figuring out something that is likely to be true based on what has actually been said.

Practicing Comprehension

Read the following passage.

High mountains, deserts, oceans, lakes, rivers, and soil conditions are all geographic barriers across which many plants and animals cannot pass. These barriers prevent many species from spreading into new areas.

Questions 1 and 2 refer to the passage. Circle the best answer for each question.

1. Which of the following statements is implied by the passage?

   (1) Frogs cannot cross lakes on floating logs.
   (2) An African zebra can easily cross the Arabian desert into Eurasia to find other grasslands.
   (3) No animal can cross the Rocky Mountains.
   (4) No animal can travel from one side of an ocean to the other.
   (5) Continents can be barriers to ocean plants and animals.

2. Which animal's spread is least affected by geographic barriers?

   (1) fish
   (2) humans
   (3) mountain goats
   (4) deer
   (5) camels

To check your answers, turn to page 78.

Topic 4: Plant Growth

Read the following passage and look at the diagram.

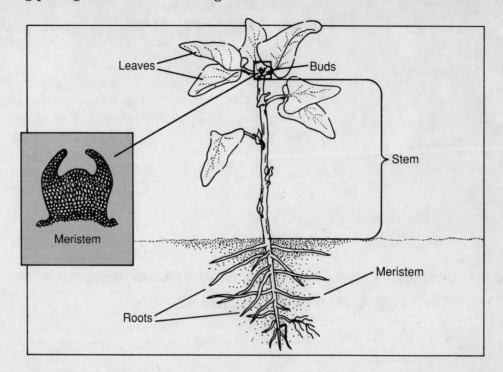

All living things grow when their cells divide, making the organism larger. In seed plants, growth occurs in specific places called **meristems**. During the growing season, cells in these areas divide rapidly, causing the plant to increase in size. Meristem cells are found at the tips of stems and roots.

The meristem cells of stems are found in the buds. The top bud of the stem is where the plant increases its height. Other buds on the side of the stem cause the plant to grow outward, or branch. Stem buds produce branches, flowers, and leaves.

Like stems, roots grow longer by adding cells in the meristem. The meristem of a root is covered by a root cap, a layer of thick dead cells. This protects the meristem cells as the root pushes through the soil.

In woody plants, another kind of meristem cell adds to the thickness of the plant. These cells form the **cambium**, a one-cell-thick layer in stems, branches, and roots. As the cells in the cambium layer divide, they add to the diameter of the plant.

Exercises

Questions 1 to 8 refer to the passage and diagram. Circle the best answer for each question.

1. Which of the following is implied by the passage?

    (1) Growth of stems does not take place between buds.
    (2) Leaves contain meristem cells.
    (3) Meristem cells in the roots are tough.
    (4) All plants have cambium.
    (5) Branches do not have cambium.

2. The fact that meristems in stems produce branches, flowers, and leaves suggests that

    (1) all meristem cells develop into the same type of structure
    (2) meristem cells become specialized as they develop
    (3) all plants have flowers
    (4) roots do not have meristem tissue
    (5) stems grow only in length

3. The immediate short-term effect of cutting off the top bud of a stem is that the

    (1) plant would stop getting taller
    (2) lower buds on the stem would stop growing
    (3) cambium cells would stop dividing
    (4) roots would branch
    (5) root caps would stop developing

4. What is the main idea of the passage?

    (1) Only living things grow by cell division.
    (2) Growth of seed plants takes place in special areas called meristems.
    (3) Roots have meristems.
    (4) Cambium is a layer of meristem cells in woody plants.
    (5) Animals and seed plants grow in similar ways.

5. The pattern of plant growth suggests that after thirty years

    (1) the bottom branch of a tree will be considerably higher than it was when it first formed
    (2) the bottom branch of a tree will be the same height above the ground that it was when it first formed
    (3) most new growth will take place between buds
    (4) the tree trunk will have the same diameter after two years
    (5) the tree trunk will have a smaller diameter after two years

6. Where can meristem cells be found in woody plants but not in most nonwoody plants?

    (1) root cap
    (2) buds at the top of the stem
    (3) buds at the sides of stems
    (4) leaves
    (5) cambium

7. If all the meristem cells in a plant were destroyed, what would most likely happen to the plant?

    (1) It would stop growing.
    (2) It would become bushy.
    (3) It would grow faster.
    (4) It would produce more leaves.
    (5) Its roots would start to branch.

8. The word apical describes the apex, or tip of a structure. Where would you expect to find apical meristems?

    (1) along stems between buds
    (2) at the ends of stems and roots
    (3) in buds on the sides of stems
    (4) in the cambium
    (5) in the leaves

To check your answers, turn to page 78.

Reviewing Lesson 4

Read the following information.

The classification of organisms into groups is called **taxonomy**. Taxonomy is based on the relationships between different groups of organisms as they evolved. Biologists divide most living things into four main groups call **kingdoms**.

Kingdom	Characteristics	Examples
Monera	One-celled organisms Nucleus of cell has no membrane Reproduce by cell division	Bacteria Blue-green algae
Protista	Mostly one-celled organisms Many-celled organisms are not organized into tissues or organs Nucleus of cell has membrane	Algae Amoeba Paramecium Slime molds Fungi
Plantae	Many-celled organisms with specialized tissues and organs Cells contain chlorophyll Cell walls contain cellulose Produce food through photosynthesis	Mosses Trees Ferns Flowering Plants
Animalia	Many-celled organisms with complex tissues and organs Most can move and respond to stimuli No chlorophyll and no cell walls Food is taken into body in some way	Insects Worms Mollusks Fish Birds Mammals

Questions 1 to 4 refer to the information. Circle the best answer for each question.

1. Which organisms would be the simplest in structure and function?

   (1) bacteria
   (2) slime molds
   (3) mosses
   (4) ferns
   (5) insects

2. What characteristic is only in plants?

   (1) one-celled
   (2) can move
   (3) cell walls contain cellulose
   (4) food is taken into the plant
   (5) reproduce by cell division

3. A goldfish would have which of the following characteristics?

   (1) nucleus of cells lack membrane
   (2) presence of chlorophyll
   (3) absence of cell walls
   (4) produce food through photosynthesis
   (5) reproduce through cell division

4. A complex organism that can move and respond to stimuli is likely to be a member of which kingdom?

   (1) Monera
   (2) Protista
   (3) Plantae
   (4) Animalia
   (5) Paramecium

To check your answers, turn to pages 78–79.

Read the following passage and look at the diagram.

Water is needed in the leaves of plants for photosynthesis to occur. But how does water from the soil reach the leaves, which may be many feet above the ground? Once water is absorbed by the roots of the plant, it passes into a type of tissue called **xylem**. Xylem is present in roots, stems, and leaves. In woody stems, xylem makes up most of the stem. Xylem has many long, dead cells that stand vertically next to one another, much like straws in a glass. Water passes upward through tubes made of xylem.

Phloem is the other type of tissue used for transport in plants. Food is moved downward from the leaves through phloem tissue. In woody stems, phloem is located under the outermost layer of plant tissue.

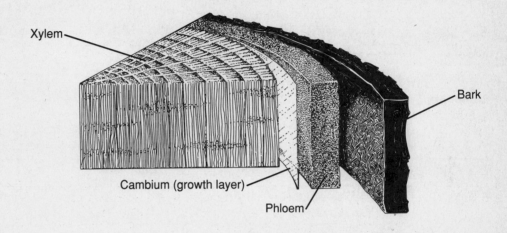

Questions 5 to 8 refer to the passage and diagram. Circle the best answer for each question.

5. What is the main source of water for most plants?

(1) well water
(2) rain and surface water soaking into the soil
(3) water vapor in the atmosphere
(4) water in the xylem tissue
(5) irrigation systems

6. What would be the immediate effect on a plant if all its xylem tissue became clogged?

(1) Photosynthesis would slow and stop because of lack of water.
(2) The stem would collapse.
(3) The roots would start carrying water to the leaves.
(4) The phloem would carry water to the leaves.
(5) The cambium would stop growing.

7. Which type of plant tissue makes up most of the wood we use?

(1) bark
(2) phloem
(3) cambium
(4) xylem
(5) meristem

8. Which type of tissue carries the food produced during photosynthesis to the rest of the plant?

(1) bark
(2) phloem
(3) cambium
(4) xylem
(5) meristem

To check your answers, turn to page 79.

GED Mini-Test

Directions: Choose the best answer to each item below.

Items 1 to 4 refer to the following diagram.

```
                         ┌──────────────────┐
                         │    SEED PLANTS   │
                         └──────────────────┘
```

Gymnosperms—Nonflowering	Angiosperms—Flowering
Exposed seeds without protective covering	Seeds protected by pod or fleshy fruit Seeds contain stored food for seedling
Examples: Pines, junipers, yews; 750 species	**Examples:** All flowering plants; 240,000 species

Monocotyledons	Dicotyledons
Seedling has 1 leaf Leaves have parallel veins Flower parts in sets of 3	Seedling has 2 leaves Leaves have netlike veins Flower parts in sets of 4 or 5
Examples: Palms, grasses, lilies, irises, corn, grain; 40,000 species	**Examples:** Woody plants, shrubs, herbs, tomatoes, potatoes, most beans; 200,000 species

1. The number of species of gymnosperms compared to the number of species of angiosperms implies that

    (1) gymnosperms outnumber angiosperms
    (2) angiosperms have spread because they have advantages that gymnosperms lack
    (3) angiosperms can be found only in tropical climates
    (4) the seeds of angiosperms are less protected than the seeds of gymnosperms
    (5) gymnosperms can be found only in deserts

2. Monocotyledons differ from dicotyledons in that monocotyledons have

    (1) seedlings with 1 leaf
    (2) seedlings with 2 leaves
    (3) leaves with netlike veins
    (4) flower parts in sets of 4
    (5) flower parts in sets of 5

3. Since wheat belongs to the family of grasses, you would expect wheat to have

    (1) flower parts in sets of 4 or 5
    (2) flower parts in sets of 3
    (3) netlike veins
    (4) 2 leaves on its seedling
    (5) unprotected seeds

4. Which of the following probably best explains the remarkable success of angiosperms?

    (1) They produce unprotected seeds.
    (2) They have netlike veins.
    (3) They produce protected seeds.
    (4) They produce seedlings with 1 leaf.
    (5) They produce seedlings with 2 leaves.

In order to germinate, or sprout, most seeds need at least three conditions: moisture, oxygen, and the correct temperature. The exact requirements vary greatly. For example, the seeds of many water plants germinate underwater where there is plenty of moisture, oxygen dissolved in the water, and an even temperature. Under these conditions, however, most land plants will not germinate.

Before a seed germinates, it absorbs a lot of water, causing the seed to swell and become soft. Too much moisture will cause the growth of fungi, which may cause the seed to rot. Most seeds germinate best at a temperature between 60°F and 80°F.

While seeds are germinating, they need plenty of oxygen because of increased activity. Seeds do best in loose soil and when planted near the surface. These conditions provide a good supply of oxygen.

5. Which of the following conditions is not important to germination?

(1) correct temperature
(2) moisture
(3) sunlight
(4) oxygen
(5) water supply

6. What is likely to be the result of planting seeds deep in the soil?

(1) Conditions for germination will improve.
(2) The seeds will not get enough water to germinate.
(3) The seeds will not get enough oxygen to germinate.
(4) The seeds will not get enough sunlight to germinate.
(5) The seeds will be too warm to germinate.

7. A maple seed can germinate on a block of ice. What is likely to happen to the newly sprouted plant?

(1) It will take root and grow into a maple tree.
(2) It will be transported to another location.
(3) It will rot.
(4) It will grow slowly and die.
(5) It will melt.

8. Some seeds can be inactive for years and still be able to germinate. What causes them to germinate after a long period of time?

(1) Water conditions improve.
(2) Oxygen is present.
(3) Light is present.
(4) The temperature is correct.
(5) There is the correct amount of warmth, moisture, and oxygen.

To check your answers, turn to page 79.

Answers and Explanations

1. (Comprehension) **(5) Continents can be barriers to ocean plants and animals.** The passage mentions various types of geographic barriers and indicates that they prevent many types of plants and animals from spreading. The example of continents acting as a barrier to ocean life is the same type of situation. Options (1), (2), (3), and (4) are not true.

2. (Comprehension) **(2) humans** Humans have spread into all types of areas because they have created ways to overcome geographic barriers. The other animals are all limited by the extent of their normal habitats.

Exercises (page 73)

1. (Comprehension) **(1) Growth of stems does not take place between buds**. The passage states that growth takes place at the buds; it implies that growth does not take place between buds. Option (2) is incorrect; the passage says nothing about growth of leaves. Option (3) is not true; meristem cells are delicate, so plants form root caps to protect them. Option (4) is not true; the passage states that woody plants have cambium. Option (5) is not true; the passage states that cambium is present in stems, branches, and roots.

2. (Comprehension) **(2) meristem cells become specialized as they develop** Meristems develop into all three structures, so they must take on the characteristics of each during growth. Option (1) is contradicted by the passage. Options (3) and (4) are not true. Option (5) is not true of woody stems.

3. (Analysis) **(1) plant would stop getting taller** Since growth from the top bud is upward growth, the effect of cutting off the top bud would be to stop upward growth, at least for a while. Cutting off the top bud would not have the effects listed in options (2), (3), (4), and (5).

4. (Comprehension) **(2) Growth of seed plants takes place in special areas called meristems.** This is the general topic, or main idea, of the passage. Options (1), (3), and (4) are details in the passage. Option (5) is incorrect because animals are not mentioned in the passage.

5. (Comprehension) **(2) the bottom branch of a tree will be the same height above the ground that it was when it first formed** Since upward growth of a seed plant is from the top of the stem only, once branches grow along the stem, they will stay at the same height above the ground. Option (1) is not true since the tree grows from the top of the stem. Option (3) is not true since there is no growth between buds. Options (4) and (5) are not true since after thirty years the cambium cells will have added to the thickness of the trunk.

6. (Comprehension) **(5) cambium** The passage states that cambium is found in woody plants and implies that it is not found in most nonwoody plants. Options (1) and (4) are incorrect because these structures do not contain meristem cells. Options (2) and (3) are incorrect because these structures are found in both woody and nonwoody plants.

7. (Analysis) **(1) It would stop growing.** Since meristem cells form the growth regions of a plant, their destruction would stop all growth. Options (2), (3), (4), and (5) all could not happen without meristem cells.

8. (Analysis) **(2) at the ends of stems and roots** The word apical describes the tip, or end, of stems and roots. Option (1) is incorrect because there is no meristem between buds. Options (3) and (4) refer to meristem cells that are not at the tips of structures. Option (5) is incorrect because there is no meristem in leaves.

Reviewing Lesson 4 (pages 74–75)

1. (Comprehension) **(1) bacteria** Bacteria, members of the kingdom Monera, are the simplest since they are one-celled and lack a membrane around the nucleus. Option (2) is incorrect because slime molds, members of the kingdom Protista, are a little more complex; their nuclei have membranes and they may have more than one cell. Options (3), (4), and (5) are members of more complex kingdoms; their cells are organized into specialized tissues and organs.

2. (Comprehension) **(3) cell walls contain cellulose** According to the chart, of the five options, this is the only characteristic that is found only in plants.

3. (Comprehension) **(3) absence of cell walls** Since a goldfish is a type of fish, it can be classified as a member of the kingdom Animalia. The only characteristic of animals among the five options is the absence of cell walls. Options (1), (2),

(4), and (5) are characteristics of other kingdoms.

4. (Comprehension) **(4) Animalia** According to the chart, most animals can move and respond to stimuli. Options (1), (2), and (3) are kingdoms whose members are either not complex or are not able to move and respond. Option (5) is not a kingdom.

5. (Comprehension) **(2) rain and surface water soaking into the soil** Since the passage states that the plant absorbs water from the soil into the roots, it follows that most of the water a plant gets is from rain and surface water that has soaked into the soil. Options (1) and (5) are incorrect because very few plants depend on wells or irrigation systems for water. Option (3) is incorrect because the passage says that water is absorbed by the roots. Option (4) is incorrect because xylem only carries the water; it is not a source of water.

6. (Analysis) **(1) Photosynthesis would slow and stop because of lack of water.** Clogging the xylem would have the effect of stopping the water flow to the leaves, where photosynthesis takes place. Since water is one of the ingredients of photosynthesis, it would slow and stop. Options (2) and (5) are incorrect because they would not happen right away. Options (3) and (4) are not true.

7. (Comprehension) **(4) xylem** By studying the diagram, you can see that the main part of the trunk of the tree is xylem tissue. Option (1) is incorrect because bark is trimmed from wood. When the bark is trimmed, it is likely that some phloem, option (2), and cambium, options (3) and (5) are also trimmed.

8. (Comprehension) **(2) phloem** The last paragraph of the passage describes the function of phloem, which is to carry food from the leaves to the rest of the plant. Therefore the other options are incorrect.

GED Mini-Test (pages 76–77)

1. (Comprehension) **(2) angiosperms have spread because they have advantages that gymnosperms lack** The dramatic difference in the number of species of gymnosperms and angiosperms implies that the more numerous type of plant has characteristics that allow adaptation to many environments. These advantages have caused the spread of angiosperms. Option (1) is not true. Options (3) and (5) are not true; both types of plants are found in a variety of environments. Option (4) is not true because angiosperms have protective coverings for their seeds.

2. (Comprehension) **(1) seedlings with 1 leaf** The diagram indicates that seedlings with 1 leaf is a distinguishing characteristic of monocotyledons. Options (2), (3), (4), and (5) are incorrect because they are characteristics of dicotyledons.

3. (Comprehension) **(2) flower parts in sets of 3** Since wheat is a grass, and grasses are monocotyledons, it follows that wheat flowers have parts in sets of 3. Options (1), (3), and (4) are characteristics of dicotyledons. Option (5) is a characteristic of gymnosperms.

4. (Analysis) **(3) They produce protected seeds.** The fact that the seeds are protected means that more of them survive to become seedlings and mature plants. This contributes to the success of the angiosperms. Option (1) is not a characteristic of angiosperms. Options (2),

(4), and (5) are characteristics of angiosperms, but they do not contribute directly to the success of these plants.

5. (Comprehension) **(3) sunlight** The passage states that the correct temperature, moisture, and oxygen are necessary for germination. This implies that other conditions, such as the amount of sunlight, are not important. Option (5) is another way to say moisture.

6. (Analysis) **(3) The seeds will not get enough oxygen to germinate.** The passage indicates that seeds are planted near the surface to ensure a good oxygen supply. Therefore the result of deep planting would be to decrease the supply of oxygen. Option (1) is incorrect because conditions do not improve with deep planting. Temperature and water are less affected by the depth of the seed, options (2) and (5). Sunlight is not necessary for germination, option (4).

7. (Comprehension) **(4) It will grow slowly and die.** Although a maple seed can germinate under poor conditions, this does not imply it can also grow successfully to maturity. Option (1) is incorrect because a maple tree cannot grow in ice. Option (2) may happen, but it is not the most likely outcome. Option (3) is unlikely since rotting usually happens under warm conditions. Option (5) is not true.

8. (Analysis) **(5) There is the correct amount of warmth, moisture, and oxygen.** That seeds stay inactive suggests that correct germination conditions do not exist. Since the passage states that three conditions must be met, option (5) is correct. Options (1), (2), and (4) name only one of the three. Option (3) is not necessary for germination.

LESSON 5 Analysis Skill: Recognizing Assumptions

People take many facts for granted when they communicate. When you read, you often must identify such facts.

Facts or ideas that are taken for granted without being explained are called **unstated assumptions**. You make unstated assumptions very often when you speak and write. For example, suppose you tell a friend that you got caught in a thunderstorm. You assume that your friend knows what a thunderstorm is. Therefore, you do not describe the thunder, lightning, wind, and rain. You take for granted that your friend will understand your experience.

When you read about science, you will find that there are many unstated assumptions. Writers take for granted that you know many common facts. In order to understand a passage, you must be able to identify the assumptions the writer makes.

Read this paragraph and identify some unstated assumptions.

Despite the barrier of the skin, microorganisms still manage to get inside the body. Some of them can affect the body's normal functioning and cause disease. When you get the flu or a cold, your body's functions are upset. However, this condition is temporary. You are not sick for the rest of your life. Your body has ways to get rid of the microorganisms in order for you to return to normal.

What does the writer of this paragraph take for granted and not explain?

- that microorganisms are tiny living things

- that they get into the body through the air we breathe, through the food and water we take in, and by direct contact

- that some microorganisms cause disease

Were you able to identify these unstated assumptions? If so, the writer did a good job of deciding what needs to be explained and what does not.

Remember, an unstated assumption is something the writer takes for granted and does not explain. It is up to you to figure out what the unstated assumptions are.

Practicing Analysis

Study the following diagram.

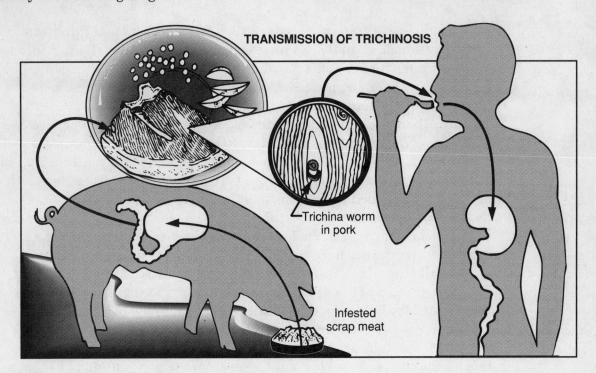

TRANSMISSION OF TRICHINOSIS

Trichina worm in pork

Infested scrap meat

Questions 1 and 2 refer to the diagram. Circle the best answer for each question.

1. Which of the following is an unstated assumption of the diagram?

(1) Trichinosis is a disease.
(2) Trichinosis is spread by eating meat infested with the trichina worm.
(3) Trichinosis is generally passed from humans to pigs.
(4) Trichinosis can be spread from one human to another through sneezing or coughing.
(5) Trichinosis can be prevented by washing your hands before eating.

2. Which of the following is an unstated assumption of the diagram?

(1) Trichinosis can be spread by eating beef, pork, or lamb.
(2) The trichina worm is a parasite, an organism that obtains its food from another living thing.
(3) The trichina worm can live outside the bodies of pigs and humans.
(4) Trichinosis can be spread through contaminated water.
(5) Trichinosis affects only adult men.

To check your answers, turn to page 88.

Topic 5: The Human Digestive System

Read the following passage and study the diagram.

The digestive system consists of a long tube called the **alimentary canal**. It is made up of the mouth, esophagus, stomach, small intestine, and large intestine. In a human adult, the alimentary canal is about 30 feet long. The liver and pancreas, which also have a role in digestion, are connected to the alimentary canal by small tubes.

Mouth and Esophagus. In the mouth, food is ground and moistened with saliva. Saliva contains a substance that begins to break down the starches in food. Saliva also contains mucus, which makes the food slippery enough to pass easily through the body. The tube that connects the mouth and stomach is called the esophagus.

Stomach. In the stomach, the breakdown of fats begins. Minerals are dissolved, and bacteria in the food are killed by acid produced by the stomach lining. The stomach makes mucus to protect the lining from the acid.

Small Intestine. The small intestine is about 1 1/4 inches wide and 23 feet long. Most of the digestive process takes place here. Nutrients from the digested food pass from the small intestine into the bloodstream.

Pancreas and Liver. The pancreas secretes substances that help break down proteins, starches, and fats. Bile from the liver breaks up fats into smaller droplets. All the substances from the pancreas and liver enter the upper part of the small intestine by means of tubes called ducts.

Large Intestine. The undigested material from the small intestine contains a lot of water. One of the main functions of the large intestine is to absorb this water. The digested material becomes more solid, and it eventually passes out of the body.

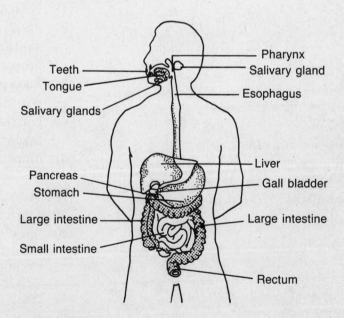

Exercises

Questions 1 to 8 refer to the previous passage and diagram. Circle the best answer for each question.

1. What is the purpose of the digestive system?

    (1) to move food from the mouth to the stomach
    (2) to break down food into substances the body can absorb and to get rid of wastes
    (3) to protect the lining of the organs from harmful substances swallowed with food
    (4) to protect the body against disease
    (5) to absorb oxygen from the air and release carbon dioxide

2. How is food ground up and mixed with saliva?

    (1) by contractions in the esophagus
    (2) by chemical action in the stomach
    (3) by swallowing
    (4) by absorption of water
    (5) by chewing

3. What is the main function of the large intestine?

    (1) to break down proteins
    (2) to break down fats
    (3) to absorb water
    (4) to produce bile
    (5) to kill bacteria swallowed with food

4. Which organ makes up most of the length of the alimentary canal?

    (1) mouth
    (2) esophagus
    (3) stomach
    (4) small intestine
    (5) large intestine

5. Nutrients from the small intestine enter the bloodstream and go to the

    (1) large intestine
    (2) pancreas
    (3) liver
    (4) stomach
    (5) rest of the body

6. How would damage to the liver affect digestion?

    (1) Foods would not be ground up.
    (2) Proteins would not be broken down.
    (3) Starches would be changed to sugars.
    (4) Fats would not be digested properly.
    (5) Too much water would be absorbed by the large intestine.

7. What is the result of chewing and swallowing too quickly?

    (1) Food absorbs too much saliva.
    (2) Food is not ground up enough and so starches are not properly processed.
    (3) Food passes through the esophagus into the stomach.
    (4) Nutrients are not absorbed through the small intestine.
    (5) More minerals are dissolved in the stomach.

8. Through what organ does solid waste pass out of the body?

    (1) pancreas
    (2) liver
    (3) small intestine
    (4) stomach
    (5) rectum

To check your answers, turn to page 88.

Reviewing Lesson 5

Study the following passage and diagram.

The **heart** is the major organ of the circulatory system. It beats an average of 72 times a minute, pumping blood throughout the body. The right and left sides of the heart are divided by a wall called the **septum**. Each side is divided into two chambers. The upper chambers are called the **atria**, and the lower chambers are called the **ventricles**.

Oxygen-poor blood from the body enters the right atrium through large veins called the **venae cavae**. When the atrium contracts, it forces the blood into the right ventricle. Next, the right ventricle contracts, forcing the blood into the pulmonary arteries, which carry it to the lungs. In the lungs the blood picks up oxygen. The oxygen-rich blood then flows through the **pulmonary veins** and back to the heart where it enters the left atrium. The left atrium contracts and forces the blood into the left ventricle. As the left ventricle contracts, it forces the blood into a large artery called the **aorta**. From the aorta, blood flows into a system of blood vessels that carry it throughout the body.

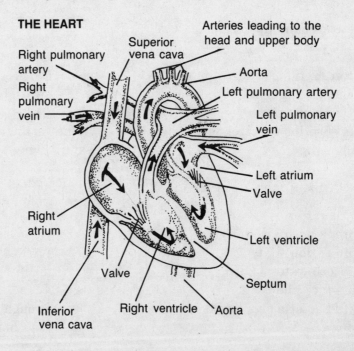

THE HEART

Questions 1 to 4 refer to the passage and diagram. Circle the best answer for each question.

1. What controls the beating of the heart?

   (1) the aorta
   (2) the venae cavae
   (3) the lungs
   (4) the brain
   (5) the intestines

2. Which chamber sends blood from the heart through the body?

   (1) the right ventricle
   (2) the right atrium
   (3) the left ventricle
   (4) the left atrium
   (5) the venae cavae

3. What would probably happen if the venae cavae became clogged?

    (1) Too much blood would enter the right atrium.
    (2) Too little blood would enter the right atrium.
    (3) Too much blood would enter the left atrium.
    (4) Too much blood would enter the lungs.
    (5) Too much blood would enter the aorta.

4. What is most likely to happen if the aorta were pinched off?

    (1) Too much blood would enter the left ventricle.
    (2) Not enough oxygen-poor blood would enter the heart.
    (3) Oxygen-poor blood would not be able to reach the lungs.
    (4) The body would not get enough oxygen-rich blood.
    (5) Oxygen-rich blood would enter the venae cavae.

Read the following passage.

    Arteries carry blood away from the heart. They branch to form smaller vessels called **arterioles**, which enter the body's tissues and branch into **capillaries**. The walls of capillaries are only one cell thick. Dissolved nutrients diffuse, or spread, through the thin capillary walls into the body's cells. Waste products diffuse out of the cells and into the capillaries to be carried away. Capillaries form **venules**, which combine to form the large blood vessels called **veins**. Veins carry the blood back to the heart.

Questions 5 to 8 refer to the passage. Circle the best answer for each question.

5. What is a blood vessel shaped like?

    (1) a fist
    (2) a pear
    (3) a sphere
    (4) a pyramid
    (5) a tube

6. When blood vessels contract, or become narrower, what is the effect on the circulatory system?

    (1) More blood circulates.
    (2) Less blood circulates.
    (3) Arteries become capillaries.
    (4) Bleeding occurs.
    (5) There is no effect.

7. In which of the following blood vessels would you expect waste products to be carried?

    (1) veins
    (2) arterioles
    (3) arteries
    (4) aorta
    (5) heart

8. Which blood vessel carries blood from the heart to the lungs?

    (1) vein
    (2) venule
    (3) artery
    (4) capillary
    (5) lymph

To check your answers, turn to pages 88–89.

GED Mini-Test

Directions: Choose the best answer to each item.

Items 1 to 6 refer to the following information.

The endocrine system is made up of glands that secrete substances called hormones. Hormones travel throughout the body but affect the functions of only some parts of the body.

Endocrine Gland	Hormone	Effect
Thyroid	Thyroxin	Controls how fast food is converted to energy in cells
Parathyroid	Parathormone	Regulates body's use of calcium and phosphorus
Thymus	Thymosin	May affect the formation of antibodies in children
Adrenal	Adrenaline	Prepares the body to meet emergencies
	Cortisone	Maintains salt balance
Pancreas	Insulin	Decreases level of sugar in the blood
Ovaries (female gonads)	Estrogen	Controls the development of secondary sex characteristics
Testes (male gonads)	Testosterone	Controls the development of secondary sex characteristics
Pituitary	Growth hormone	Controls the growth of bones and muscles
	Oxytocin	Causes uterine contractions in labor
	ACTH, TSH, FSH, LH, LTH	Regulate the secretions of the other endocrine glands

1. A person with high levels of sugar in the blood is said to have diabetes. Which hormone is lacking in a person with this condition?

   (1) thyroxin
   (2) adrenaline
   (3) insulin
   (4) oxytocin
   (5) thymosin

2. In an emergency, your heart rate and breathing quicken and you have a sudden burst of energy. Which endocrine gland causes this reaction?

   (1) adrenal
   (2) thyroid
   (3) thymus
   (4) parathyroid
   (5) pancreas

3. The body normally responds to high levels of sugar in the blood by secreting

   (1) glucagon
   (2) insulin
   (3) estrogen
   (4) testosterone
   (5) adrenaline

4. Which endocrine gland is different in males and females?

   (1) pituitary
   (2) gonads
   (3) adrenal
   (4) thymus
   (5) pancreas

5. A condition called acromegaly causes certain bones in the adult to grow larger and thicker. Which endocrine gland causes this condition?

(1) adrenal
(2) pancreas
(3) thymus
(4) thyroid
(5) pituitary

6. How are hormones carried throughout the body?

(1) by the digestive system
(2) by the nerves
(3) by the blood
(4) by the saliva
(5) by the skin

Items 7 to 10 refer to the following passage and diagram.

The semicircular canals in the ear help the body to keep its balance. They contain nerve endings and a liquid. When the head moves, the liquid sloshes against the nerve endings. These nerves send a message to the brain, which interprets the position of the head. Because the three canals in the ear are positioned at right angles to one another, any change in position moves the fluid in at least one of the canals.

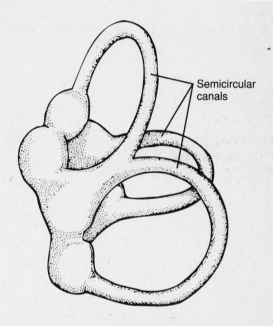

Semicircular canals

7. Where are the semicircular canals located?

(1) in the left ear
(2) in the right ear
(3) in both ears
(4) at the base of the neck
(5) below the jaw

8. When you spin around rapidly, the fluid moves to one end of the canals. It rushes to the other side if you

(1) slow down gradually
(2) stop suddenly
(3) start running
(4) jump
(5) keep spinning

9. What is likely to be the result of disease of the semicircular canals?

(1) loss of hearing
(2) damaged vision
(3) dizziness
(4) paralysis
(5) poor sense of touch

10. Which condition is associated with the normal functioning of the semicircular canals?

(1) deafness
(2) motion sickness
(3) measles
(4) common cold
(5) low blood pressure

To check your answers, turn to page 89.

Answers and Explanations

1. (Analysis) **(1) Trichinosis is a disease.** The infection of humans by trichina worms causes a disease called trichinosis. The diagram shows how this disease is transmitted, but does not state that it is a disease. Option (2) is true, but since it is clearly shown in the diagram, it cannot be called an unstated assumption. Option (3) is the opposite of what the diagram shows. Option (4) is not true; trichinosis does not affect the respiratory system and is not spread by coughing and sneezing. Option (5) is not true; since the trichina worms are inside the meat, washing your hands will not prevent the spread of trichinosis.

2. (Analysis) **(2) The trichina worm is a parasite, an organism that obtains its food from another living thing.** The diagram shows that the trichina worm lives in other animals. Such an organism is called a parasite. Option (1) is not true; the diagram does not indicate that trichinosis can be caught from beef or lamb. Options (3), (4), and (5) are not true.

Exercises (page 83)

1. (Analysis) **(2) to break down food into substances the body can absorb and to get rid of wastes** Although the passage discusses how the digestive system works, the general purpose is never stated. The purpose of the digestive system is an unstated assumption. Option (1) is true but is only part of the digestive system's function. Option (3) is not true. Option (4) is one

function of the stomach. Option (5) is the function of the respiratory system, not the digestive system.

2. (Analysis) **(5) by chewing** Although the passage does not state this, we can assume that chewing is the process by which food is ground up and mixed with saliva. Since saliva is present in the mouth, options (1) and (2) are incorrect. Swallowing, option (3), is the way food is passed through the esophagus and into the stomach. Option (4) has nothing to do with grinding up food.

3. (Comprehension) **(3) to absorb water** In the paragraph about the large intestine, it is stated that the main purpose of the large intestine is to absorb water. Options (1), (2), (4), and (5) are functions of other parts of the digestive system.

4. (Comprehension) **(4) small intestine** The passage states that the alimentary canal is about 30 feet long. Since the small intestine is about 23 feet long, it makes up most of the length of the alimentary canal. If you look at the diagram, you will see that the other organs are much shorter than the small intestine.

5. (Comprehension) **(5) rest of the body** Since nutrients pass into the blood and the blood travels throughout the body, the nutrients are carried throughout the body. Options (1), (2), (3), and (4) are therefore only partly correct.

6. (Analysis) **(4) Fats would not be digested properly.** The function of bile is to break down fats into small droplets. This process would be disrupted if anything were wrong with the

liver. Options (1), (2), (3), and (5) are not functions of the liver.

7. (Analysis) **(2) Food is not ground up enough and so starches are not properly processed.** The mouth's function is to grind and moisten the food and to begin the breakdown of starch. If food does not spend enough time in the mouth, the result is that it is not properly processed. Option (1) is the opposite of what happens when you swallow too quickly. Option (3) is true whether or not food is chewed properly. Options (4) and (5) are not part of the action of the mouth.

8. (Comprehension) **(5) rectum** The diagram shows that the large intestine empties into the rectum, which is the passage out of the body. The other options are all internal organs that are not close to a body opening.

Reviewing Lesson 5 (pages 84–85)

1. (Analysis) **(4) the brain** That the brain controls the beating of the heart, as it controls all body activities, is an unstated assumption of the passage. Options (1), (2), (3), and (5) are body parts with functions not associated with controlling activities of other organs.

2. (Comprehension) **(3) the left ventricle** The left ventricle forces blood into the aorta, which leads to blood vessels throughout the body. Option (1) sends oxygen-poor blood to the lungs. Option (2) collects oxygen-poor blood from the body. Option (4) receives oxygen-rich blood from the lungs. Option (5) is not one of

the chambers of the heart but is a blood vessel.

3. (Analysis) **(2) Too little blood would enter the right atrium.** Since the venae cavae bring blood from the body to the right atrium, a clogged vena cava would slow the rate of blood entering the right atrium. Option (1) states the opposite. The venae cavae do not connect with options (3) and (4). The slowed flow of blood to the right atrium would also slow, not increase, the flow of blood to the aorta, option (5).

4. (Analysis) **(4) The body would not get enough oxygen-rich blood.** Since the aorta sends oxygen-rich blood throughout the body, if it were pinched off this flow of blood would be reduced. Option (1) is not connected to the aorta. Oxygen-poor blood, option (2), enters the heart through the venae cavae, not the aorta. Oxygen-poor blood, option (3), passes to the lungs through the pulmonary arteries, not the aorta. Oxygen-rich blood does not enter the venae cavae, option (5).

5. (Analysis) **(5) a tube** Blood vessels carry liquid blood throughout the body and are shaped like tubes of various thicknesses. All the other shapes mentioned are closed and not suitable for carrying liquids around the body.

6. (Analysis) **(2) Less blood circulates.** If the vessels become narrower, they carry less blood. Option (1) is the opposite. Option (3) is not true. Option (4) occurs when there are breaks in the blood vessels. Option (5) is incorrect because there is an effect.

7. (Comprehension) **(1) veins** According to the paragraph, waste products are carried away from the body's tissues,

and veins carry blood from the tissues back to the heart. Options (2), (3), and (4) all refer to the arteries. Option (5) is not a blood vessel.

8. (Comprehension) **(3) artery** According to the paragraph, arteries carry blood away from the heart. Options (1) and (2) refer to veins. Option (4) connects arteries and veins. Option (5) refers to a different part of the circulatory system, which does not carry blood.

GED Mini-Test (pages 86–87)

1. (Comprehension) **(3) insulin** According to the chart, insulin secreted by the adrenal glands lowers the level of sugar in the blood. Therefore, if the blood has a high level of sugar, insulin must be lacking in adequate amounts. Options (1), (2), (4), and (5) name other hormones, which have different functions.

2. (Analysis) **(1) adrenal** According to the chart, the adrenal glands prepare the body for emergencies. The reactions stated in the question are reactions to emergency situations and are caused by the adrenal glands. Options (2), (3), (4), and (5) name other endocrine glands.

3. (Comprehension) **(2) insulin** According to the chart, insulin is the hormone that decreases the blood-sugar level. Option (1) is not mentioned in the chart. Options (3) and (4) are sex hormones. Option (5) is the hormone secreted in emergencies.

4. (Comprehension) **(2) gonads** According to the chart, the gonads are different in males and females. Options (1), (3), (4), and (5) name other endocrine glands that are the same in both males and females.

5. (Analysis) **(5) pituitary** According to the chart, the

growth hormone produced by the pituitary gland controls the growth and development of bones and muscles. Extra growth in some bones, therefore, is probably the effect of the growth hormone produced by the pituitary gland. Options (1), (2), (3), and (4) name other endocrine glands that have different functions from the one described in the question.

6. (Analysis) **(3) by the blood** The blood is the only option mentioned that can carry substances throughout the body. The other options have other functions.

7. (Analysis) **(3) in both ears** Although the author says the canals are in "the ear," you can assume that the canals are present in both ears because the human body is the same on both sides. Options (1) and (2) are only partly correct. Options (4) and (5) are not true.

8. (Analysis) **(2) stop suddenly** Stopping suddenly causes the fluid to slosh quickly to the other side. Option (1) is incorrect because the fluid would move gradually. Option (3) is incorrect because the fluid would move back if you moved forward. Option (4) is incorrect because jumping would cause it to move up and down. Option (5) would keep the fluid at one end of the canals.

9. (Analysis) **(3) dizziness** The semicircular canals help keep the body's balance, and when the body seems unbalanced you feel dizzy. The other options are not related to balance.

10. (Analysis) **(2) motion sickness** Motion sickness can result from continuous rhythmic motion (as in a car), which moves the fluid in the semicircular canals. The other conditions are not related to the semicircular canals.

LESSON 6 Analysis Skill:
Fact or Opinion

A fact is something that is true. It can be proved. An opinion is someone's belief. An opinion may or may not be true, and it cannot be proved.

You deal with facts and opinions all the time. For example, the AIDS virus is transmitted in body fluids such as blood and semen. People at high risk of contracting AIDS are homosexuals, drug users who share needles, and people who have had infected blood transfusions. These are all facts. People who are not in these high-risk groups often believe they have no chance of getting AIDS. This point of view about the risk of contracting AIDS is an opinion. It is probably not true. Even low-risk groups have some chance of getting AIDS.

Much of what you read in science is fact. However, you will also read about scientists' opinions. Scientists observe things and then form an opinion, or **hypothesis**, to explain how or why things are the way they are. Then they experiment to learn if the hypothesis is correct.

Read this brief passage and distinguish the facts from the opinions.

> In Lake Saima in Finland, visibility is only about six feet, even on a sunny day. How do the seals there manage to catch the six pounds of fish they need to eat each day? Sight is probably useless. One possibility is echolocation — the animal version of radar. Dolphins give off pressure waves and monitor the echoes that bounce back. But seals do not produce such waves. They do make barely audible clicks, but their ears are not good at hearing the echoes. According to one biologist, seals may be listening underwater through another type of antenna, the membranes in their whiskers.

Fact	Opinion
♦ Visibility—6 feet or less	♦ Sight probably useless
♦ Seals catch 6 lb fish a day	♦ Echolocation may be used
♦ Dolphins use echolocation	♦ Seals may use the membranes in their whiskers to listen
♦ Seals make barely audible clicks	
♦ Seals' ears not good at hearing under water	

Were you able to distinguish the facts from the opinions? Certain words or phrases signal an opinion. These include <u>according to</u>, <u>it is possible</u>, <u>believe</u>, <u>think</u>, <u>feel</u>, <u>may</u>, <u>might</u>, <u>seem</u>, <u>agree</u>, and <u>disagree</u>.

Practicing Analysis

Read the following passage.

> In laboratory experiments, scientists have found that animals live much longer than expected if they are fed a diet with all the necessary ingredients but with only 60 to 65 percent of the usual calories. These animals continue to appear healthy and strong after the well-fed animals in the experiment have died. For example, mice fed the special diet lived as long as 4 1/2 years. Mice who ate as much as they wanted lived an average of 3 years. The same increased life span has been shown with fish, spiders, worms, and protozoa.
>
> A low-calorie diet may work to increase life span only in normally short-lived creatures like mice and insects. These animals may have a mechanism that helps them survive a few years of famine and still be able to reproduce. Longer-lived animals, such as monkeys and humans, are fertile for many years. Thus they may not have a mechanism to prolong life through lean years. Some scientists, however, think that restricting calories can extend the life span of all forms of life, including humans.

Questions 1 and 2 refer to the passage. Circle the best answer for each question.

1. Which of the following statements is a fact?

   (1) Laboratory mice lived as long as 4 1/2 years when fed a low-calorie, nutritious diet.

   (2) Short-lived animals have a mechanism that helps them survive famine years and then reproduce.

   (3) Longer-lived animals do not have a mechanism that helps them prolong life during famine years and then reproduce.

   (4) Humans can lengthen their maximum life span by eating a reduced-calorie diet.

   (5) None of the above is a fact.

2. What do scientists believe about the effect of a reduced-calorie diet?

   A. The diet can extend the maximum life span of short-lived animals.

   B. The diet cannot extend the maximum life span of long-lived animals.

   C. The diet can extend the maximum life span of all animals.

   (1) A only
   (2) B only
   (3) C only
   (4) A and B only
   (5) A, B, and C

To check your answers, turn to page 98.

Topic 6: Disease

Read the following passage.

Diseases that can be spread from one person to another are called **infectious diseases**. Today, we know that infectious diseases are caused by microorganisms called **pathogens**. The most common pathogens are **bacteria** and **viruses**.

Long ago, people thought that disease was caused by evil spirits that entered the body. The disease could be cured if the spirits were driven out by prayer or foul-tasting medicines. The invention of the microscope allowed scientists to see microorganisms, but over two hundred years passed before scientists started to connect microorganisms and disease. Louis Pasteur, a French scientist working in the nineteenth century, proved that yeasts cause fermentation of beet juice. He also thought that the rod-shaped organisms he found in sour juice were responsible for its souring. If such organisms could sour juice, Pasteur thought, perhaps they could also cause disease in humans.

About the same time, a German scientist named Robert Koch discovered that a specific kind of bacterium caused anthrax in sheep, cattle, and humans. While examining organs of animals that had died of anthrax, Koch found many rod-shaped bacteria in the blood vessels. He transferred some of these into the cut skin of a healthy mouse, which then developed anthrax and died. Koch found many of the same bacteria in the blood of the dead mouse. However, Koch was not satisfied until he could watch the bacteria multiply. He set up an experiment to grow the bacteria and infect laboratory animals. His experiments provided significant proof for Pasteur's idea that microorganisms can cause disease.

Anthrax bacillus

Koch's procedure for studying disease-causing organisms is still used today. The steps in his method, known as Koch's postulates, are listed below.

1. Isolate the organism believed to cause the disease.
2. Grow the organism outside the animal in a sterile food medium called a **culture**.
3. Produce the same disease by injecting a healthy animal with organisms from the culture.
4. Examine the sick animal and recover the organisms that caused the disease.

Exercises

Questions 1 to 7 refer to the previous passage. Circle the best answer for each question.

1. Before Louis Pasteur, what opinion was held about the nature of disease?

   (1) Diseases are caused by evil spirits.
   (2) Diseases are caused by microorganisms called pathogens.
   (3) Anthrax is caused by rod-shaped bacteria.
   (4) Diseases are caused by yeast.
   (5) Diseases are caused by viruses.

2. At the time he thought of it, Pasteur's idea that microorganisms cause disease was an opinion. What contributed to this idea being accepted as fact today?

   (1) the development of foul-tasting medicines
   (2) the invention of the microscope
   (3) Pasteur's experiments with yeast and fermentation
   (4) Koch's experiments with anthrax bacteria
   (5) the discovery of microorganisms

3. Based on Koch's discovery of the microorganism that causes anthrax, what can you conclude?

   (1) All diseases are caused by microorganisms.
   (2) Improved types of yeasts would improve fermentation.
   (3) Yeast can be used to ferment juice.
   (4) Some diseases are not contagious.
   (5) The spread of anthrax can be stopped by stopping the spread of anthrax bacteria.

4. Which of the following was Pasteur's main contribution to the study of disease?

   (1) the invention of the microscope
   (2) the discovery that yeast causes fermentation
   (3) the discovery of microorganisms
   (4) the idea that microorganisms cause disease
   (5) a cure for anthrax

5. If the fourth step of Koch's postulates were not carried out during an experiment, what would be the result?

   (1) The injected animal would not get sick.
   (2) The injected animal would not die.
   (3) The cause of the disease would not be proved.
   (4) The culture would become contaminated.
   (5) The injected animal would die.

6. Why did Koch think that a specific kind of bacteria was responsible for anthrax?

   (1) Humans also get anthrax.
   (2) He found many of the bacteria in animals that had died of anthrax.
   (3) He knew that yeast causes fermentation of juice.
   (4) He knew that anthrax was highly contagious.
   (5) He was able to grow anthrax bacteria in the laboratory.

7. What is a culture?

   (1) a type of pathogen
   (2) a type of bacteria
   (3) a sterile food medium
   (4) a laboratory with animal subjects
   (5) the result of fermentation

To check your answers, turn to page 98.

Reviewing Lesson 6

Read the following passage and look at the diagram.

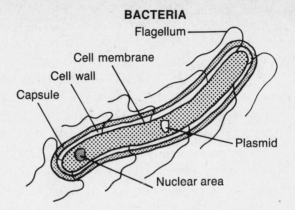

BACTERIA

Bacteria are small one-celled organisms that live almost everywhere. The largest bacteria are only about $\frac{1}{65,000}$ centimeter across, and many are smaller. One of the ways bacteria are identified is by their shape. Bacteria shaped like spheres or circles are called **cocci**. Rod-shaped bacteria are called **bacilli**. Corkscrew-shaped or coiled bacteria are called **spirilla**. Some bacteria have **flagella**, whiplike structures that move the cell through water or other fluids.

Bacteria are surrounded by a **slime layer**. Some bacteria have a thick slime layer called a **capsule**. Bacteria that can cause serious infection usually have a capsule. Scientists think that the capsule may protect the bacteria from the body's defenses against infection.

Beneath the capsule is a stiff cell wall. Within the cell wall is a cell membrane that allows substances to pass into and out of the cell. Bacteria have a nuclear area that contains genetic material, but there is no nuclear membrane around it. Genetic material may also be found in small structures called **plasmids**.

To be active, bacteria need a suitable temperature, moisture, darkness, and food. Those that cause infections in humans grow best at 98.6° F, which is normal body temperature. Bacteria can remain alive but become inactive when conditions for growth are lacking. When the environment becomes favorable again, the bacteria may grow and reproduce rapidly.

Questions 1 and 2 refer to the passage and diagram. Circle the best answer for each.

1. What is the shape of the bacteria <u>Diplococcus pneumoniae</u>, which causes pneumonia?

    (1) rod
    (2) circular
    (3) coiled
    (4) corkscrew
    (5) whiplike

2. What is likely to happen to bacteria if the temperature of the environment drops?

    (1) They will grow rapidly.
    (2) They will reproduce rapidly.
    (3) They will become inactive.
    (4) They will cause fermentation.
    (5) They will get a disease.

Read the following passage.

A virus is made up of a single molecule of genetic material surrounded by a coat of protein called a **capsid**. The capsid protects the genetic material of the virus from substances that could damage it.

A virus shows no signs of life as long as it is outside a living cell. It does not grow, reproduce, or perform any chemical activity necessary for life. In order to become active, a virus must invade a living cell called a **host cell**. Once inside the host cell, the virus transfers its own genetic material into the cell. The virus takes over all the cell's activities. The host cell is then used to duplicate the virus's own genetic material and viral protein. These are combined into new virus particles called **virions**. Each virion is exactly like the parent virus. It can leave the host cell to infect other cells and begin the cycle over again.

Most viruses can multiply only in certain kinds of cells. Some infect specific plants like potato or tobacco plants. Viruses that attack animals often attack specific tissues. Cold viruses, for example, attack the tissues that line the nose and throat. Polio viruses attack certain nerve cells in the brain and spinal cord.

Questions 3 to 6 refer to the passage. Circle the best answer for each question.

3. If a virus's capsid were destroyed, what would be the most likely outcome?

   (1) The virus would reproduce.
   (2) The virus would invade a host cell.
   (3) The viral genetic material would be damaged.
   (4) The virus would become active.
   (5) Virions would invade a host cell.

4. If a host cell contains virions, and the cell dies, the virions would most likely

   (1) die
   (2) be released to invade other cells
   (3) use the host cell for food
   (4) take the place of the host cell
   (5) develop capsids

5. A virus takes over the activities of its host cell by

   (1) transferring its own genetic material into the host cell
   (2) reproducing itself in the host cell
   (3) releasing virions into the host cell
   (4) releasing its capsid into the host cell
   (5) breaking down the cell's membrane

6. Scientists suspected that viruses existed before seeing them. This was possible because scientists had

   (1) identified the viral nucleus
   (2) seen viral capsids
   (3) seen virions
   (4) identified host cells
   (5) recognized diseases viruses caused

To check your answers, turn to pages 98–99.

GED Mini-Test

Directions: Choose the best answer to each item.

Items 1 to 4 refer to the following passage.

A cure for the common cold may be forthcoming as scientists discover more about viruses. Rhinoviruses, the viruses that cause most colds, look like miniature planets covered with steep mountains and deep valleys. The body's immune system makes antibodies, substances that fight disease. The antibodies attach to the mountaintops of the viruses. But the antibodies are easily fooled by the viruses' ability to change shape from one generation to the next. In addition, there are about 100 types of rhinoviruses, and each has mountaintops with a different shape.

Some scientists are studying the valleys rather than the mountains, because these change less. The valleys attach to receptors on cells in the nose and throat. Several substances are being tested that will block the virus from attaching to a cell. Other scientists are trying to block the receptors on the cells' surface to keep the virus from latching on. In yet another approach, millions of synthetic copies of the human receptors are placed in the nose to act as decoys for the virus.

Most of these approaches do not try to kill the virus; they only block it to keep an infection from starting. However, some scientists think that it is not a good idea to prescribe a drug that might be taken by millions of healthy people to prevent a minor illness like a cold.

---

1. According to the passage, which of the following statements reflects the opinion some scientists have about current research on the common cold?

    (1) A drug to prevent colds may be too risky for so many people to take for such a minor illness.
    (2) A cure for the cold will never be found.
    (3) Rhinoviruses change their shape from one generation to another.
    (4) There are about 100 varieties of rhinoviruses.
    (5) Rhinoviruses attach to cells in the nasal passages.

2. Which of the following is an unstated assumption of the passage?

    (1) In the future it may be possible to cure the common cold.
    (2) Rhinoviruses are simple pathogens.
    (3) Synthetic receptors can act as decoys for viruses.
    (4) Viruses can be destroyed with existing drugs.
    (5) About 100 types of rhinoviruses cause colds.

3. What do rhinoviruses look like?

(1) decoys
(2) comets with tails
(3) planets with mountains and valleys
(4) corkscrews
(5) rods

4. What are antibodies?

(1) viruses
(2) receptors on human cells
(3) substances in the nasal passages
(4) disease-fighting substances
(5) drugs

Items 5 to 8 refer to the following passage.

The human body is constantly being invaded by bacteria and viruses. The immune system fights off these invaders and resists infection and disease. Certain vitamins, especially vitamin C, may help the immune system fight disease.

Linus Pauling created a stir when he proposed that large doses of vitamin C could prevent or reduce the severity of colds and flu. Other researchers have claimed that vitamin C can speed recovery from mononucleosis, viral pneumonia, and viral hepatitis. They believe that every infectious disease involves a severe loss of vitamin C throughout the body. Other experts doubt that vitamin C can be a useful treatment. They point to studies showing that vitamin C has no effect on colds after they start.

5. Which of the following is presented by the author as a fact?

(1) Vitamin C can prevent the common cold.
(2) Vitamin C improves the immune response.
(3) The immune system resists invading bacteria and viruses.
(4) The immune system cannot resist mononucleosis.
(5) Recovery from viral pneumonia can be speeded by vitamin C.

6. If vitamin C does have a significant role in helping the immune system, what should people do?

(1) Eat a diet low in vitamin C.
(2) Eat foods high in vitamin C.
(3) Take huge doses of vitamin C after a disease is established.
(4) Eat more meats and dairy products.
(5) Do nothing special until they get sick.

7. What is the role of vitamin C in preventing disease?

(1) It may help the immune system.
(2) It can cure viral pneumonia.
(3) It can cure the flu.
(4) It can cure a cold after it starts.
(5) It can cure mononucleosis.

8. What did Linus Pauling propose?

(1) that the immune system fights disease
(2) that the common cold can be prevented or eased with large doses of vitamin C
(3) that mononucleosis and viral hepatitis can be cured with large doses of vitamin C
(4) that all infectious diseases cause losses of vitamin C throughout the body
(5) that the common cold can be cured with large doses of vitamin C

To check your answers, turn to page 99.

Answers and Explanations

1. (Analysis) **(1) Laboratory mice lived as long as 4 1/2 years when fed a low-calorie, nutritious diet.** Option (1) is the only fact; it describes the result of an experiment. Options (2), (3), and (4) are all opinions, or hypotheses, that scientists have formed about the results of experiments and their application to other forms of life. Option (5) is not true.

2. (Analysis) **(5) A, B, and C** These options, even though they may be opposites of one another, are all held by some scientists interested in this research. The second paragraph of the passage describes the different ideas the scientists have suggested to explain the results and importance of the experiments.

Exercises (page 93)

1. (Analysis) **(1) Diseases are caused by evil spirits.** The relationship between microorganisms and disease had not yet been proved, so all views about disease were opinions. Options (2) and (3) are incorrect because they were discovered during Pasteur's time. Options (4) and (5) are incorrect because they were not known until after Pasteur's time.

2. (Analysis) **(4) Koch's experiments with anthrax bacteria** Koch's experiments proved that a specific kind of bacteria was the cause of anthrax. This fact supported Pasteur's idea. Option (1) has nothing to do with the cause of disease. Options (2) and (5) happened before Pasteur's time. Option

(3), the yeast experiments, did not help prove the cause of disease.

3. (Comprehension) **(5) The spread of anthrax can be stopped by stopping the spread of anthrax bacteria.** Since a specific organism causes the disease in animals, if you can prevent the organism from getting into another animal, you can stop the spread of the disease. Option (1) is not true; some diseases such as cancer are not caused by microorganisms. Options (2) and (3) are incorrect because fermentation is not a disease. Option (3) is not true; yeast causes fermentation. Option (4) is true but does not follow from Koch's discovery of the anthrax bacteria, since anthrax is contagious.

4. (Comprehension) **(4) the idea that microorganisms cause disease** This idea forms the basis of all later discoveries about infectious diseases. Options (1), (3), and (5) are not contributions of Pasteur. Option (2) is one of Pasteur's contributions, but it is not directly related to the study of disease.

5. (Analysis) **(3) The cause of the disease would not be proved.** Since the fourth step of Koch's postulates involves confirming that the pathogens are indeed present in the diseased animal, leaving out this step means you cannot prove that the disease was caused by the pathogens. Option (1) is not true. Options (2) and (5) would be possible even without step 4 of the postulates. Option (4) is not true.

6. (Comprehension) **(2) He found many of the bacteria in animals that had died of anthrax.** That is what led him to think that the bacteria might be the cause of anthrax. Options (1), (4), and (5) are true but are not reasons for thinking anthrax is caused by a specific microorganism. Option (3) is incorrect because fermentation is not related to anthrax.

7. (Comprehension) **(3) a sterile food medium** According to the passage, a culture is used to grow bacteria in the laboratory. Options (1), (2), (4), and (5) are therefore incorrect.

Reviewing Lesson 6 (pages 94–95)

1. (Comprehension) **(2) circular** The clue to the answer is in the word Diplococcus, which indicates that the bacteria is a type of coccus, or circular-shaped bacteria. The other options refer to bacilli, option (1); spirilla, options (3) and (4); and the shape of the flagella, option (5).

2. (Analysis) **(3) They will become inactive.** A suitable temperature is one of the conditions bacteria need in order to grow and reproduce. So when the temperature drops, they are likely to stop growing and reproducing, and they become inactive. Options (1) and (2) are incorrect because they describe what bacteria do under good conditions. Options (4) and (5) are not related to drops in temperature.

3. (Analysis) **(3) The viral genetic material would be damaged.** Since the capsid protects the genetic material, if the capsid were destroyed, the genetic material could be damaged. Options (1) and (4) are incorrect because neither can happen until a virus invades a host cell. Option (2) would be impossible once the virus had lost its capsid. Option (5) is incorrect because the question is asking about a virus, not about virions.

4. (Analysis) **(2) be released to invade other cells** The most likely answer to the question is that the virions would be released to invade other cells. There is no reason to suspect they would die, option (1). Virions, being viruses, do not take in food, option (3). Virions could not accomplish option (4) since they cannot function by themselves. Virions already have capsids, option (5).

5. (Comprehension) **(1) transferring its own genetic material into the host cell** The passage states that this is how a virus takes over the activities of a host cell. Options (2), (3), and (4) cannot happen until the virus has already taken over the cell's activities. Option (5) is incorrect because breaking down the cell's membrane would kill the cell.

6. (Analysis) **(5) recognized diseases viruses caused** The only possible way scientists could have known about viruses before seeing them was by observing their effect on living things—studying the diseases viruses caused. Options (1), (2), and (3) could not occur unless scientists could already see viruses. Without seeing viruses, option (4) would be unknown.

GED Mini-Test (pages 96–97)

1. (Analysis) **(1) A drug to prevent colds may be too risky for so many people to take for such a minor illness.** Because the research focuses on drugs that would prevent colds, some scientists think the research is going in the wrong direction and that focusing on a cure would be better. Option (2) is unlikely. Options (3), (4), and (5) are facts rather than opinions.

2. (Analysis) **(1) In the future it may be possible to cure the common cold.** This assumption underlies all research in this area. Option (2) is not true. Options (3) and (5) are true but are not unstated assumptions. Option (4) is not true.

3. (Comprehension) **(3) planets with mountains and valleys** This is stated in the first paragraph of the passage; therefore the other options are incorrect.

4. (Comprehension) **(4) disease-fighting substances** Antibodies are defined in the first paragraph of the passage. The other options are not definitions of antibodies.

5. (Analysis) **(3) The immune system resists invading bacteria and viruses.** The other options are all opinions of various scientists. These opinions have not yet been proved.

6. (Analysis) **(2) Eat foods high in vitamin C.** If vitamin C helps the immune system, then it would make sense to keep the body's supply of vitamin C high. Option (1) states the opposite. Options (3) and (5) do not make sense because the immune system would need more vitamin C before a disease started, in order to fight off the disease. Option (4) is incorrect because these foods are not high in vitamin C.

7. (Comprehension) **(1) It may help the immune system.** The other options are incorrect because they refer to curing disease, not preventing disease.

8. (Comprehension) **(2) that the common cold can be prevented or eased with large doses of vitamin C** Option (1) is incorrect because that fact had already been established. Options (3) and (4) are the claims of other researchers. Option (5) is not true; Pauling claimed that colds could be prevented, not cured.

LESSON 7 Evaluation Skill: Evaluating Information

How do you decide whether something is true? You apply what you know about recognizing implications, figuring out unstated assumptions, and distinguishing facts from opinions. Then you can determine how accurate a given statement is.

Suppose, for example, that you want to do your part to make a better environment. Should you buy plastic garbage bags that are advertised as degradable? Something that is degradable breaks down and is reabsorbed into the environment. So, by implication, the plastic garbage bags sound all right. But what is the manufacturer's unstated assumption about the degradability of plastic? The manufacturer is assuming that the right conditions for degrading will exist. In the news, they tell us that this type of plastic needs sunlight to degrade. But most trash ends up in a dark pile in a landfill. So the manufacturer's assumption is not likely to be true. The so-called degradable garbage bags probably will not degrade in the landfill. You will have to figure out some other way to help the environment.

When you try to evaluate the accuracy of information, ask yourself the following questions:

♦ Is the information based on fact or on opinion?

♦ Does the information follow logically from the facts presented?

♦ Is the information based on an unstated assumption? If so, what is the assumption and is the assumption true?

The answers to these questions will help you decide whether the information is accurate or not. Remember, when answering multiple-choice test items, eliminating inaccurate choices will help you identify the correct answer.

Practicing Evaluation

Read the following passage.

> In medicine, infertility is defined as the inability of a couple to conceive or to carry a pregnancy to a live birth. The trend toward postponing parenthood into the thirties and forties age bracket has doubled the amount of infertility in the United States over the past 20 years. Today almost 20 percent of the couples in the United States cannot have a child.
>
> Increased age contributes to problems of infertility. Difficulty producing eggs increases with a woman's age. The chance of miscarriage increases dramatically after age 35. Contrary to the myth that infertility is always the woman's problem, infertility affects both sexes equally.

Questions 1 and 2 refer to the passage. Circle the best answer for each question.

1. Which of the following statements is true?

(1) Infertility is always the woman's problem.
(2) Most couples in their thirties and forties cannot have children.
(3) The best time to have a child is when you are in your early twenties.
(4) Most couples in the United States are fertile.
(5) The chance of miscarriage decreases with increased age.

2. Given the information in the passage, which of the following statements is the most accurate?

(1) When deciding when to have children, couples should take into account that the risk of infertility increases with age.
(2) Age is not a factor in infertility.
(3) People should have children when they are as young as possible.
(4) There is no risk of infertility for couples who already have had one child.
(5) More men than women have infertility problems.

To check your answers, turn to page 108.

Topic 7: Mendel and Genetics

Read the following passage and look at the diagram.

How often have you heard someone say of a child, "She has her mother's eyes and her father's chin"? Statements like this refer to **traits**, or characteristics, that we inherit from our parents. The study of traits and how they are inherited is called **genetics**.

Gregor Mendel, an Austrian monk, has been called the father of genetics. He began his experiments in 1857. He bred pea plants because they have many traits that are easy to identify.

First, Mendel developed **purebred** pea plants. These plants always produced the same traits. Next, he crossed, or bred, plants with opposite traits. For example, he crossed purebred tall plants with purebred dwarf plants. Much to Mendel's surprise, all the plants that resulted were tall! Mendel called this F_1 generation of plants **hybrids** because they contained a mixture of the traits for tallness and dwarfness.

Then Mendel crossed tall hybrid plants from the F_1 generation. He found that some plants in the resulting F_2 generation were tall and some were dwarf. Each time he repeated the experiment, he found that there were three tall plants for one dwarf plant.

How did Mendel explain his results? He reasoned that the offspring inherited traits from both parents. However, some traits were more powerful than others. He called these traits **dominant**. The traits that did not show up in the F_1 generation he called **recessive**. In the F_2 generation, the plants that show the recessive trait must be plants that did not inherit any dominant trait for tallness.

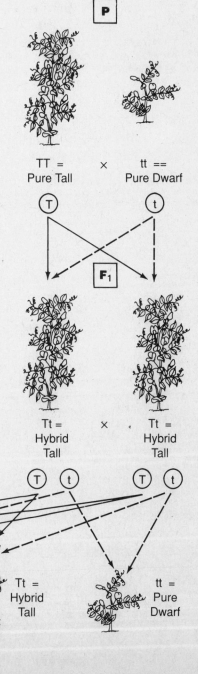

Exercises

Questions 1 to 6 refer to the previous passage and diagram. Circle the best answer for each question.

1. How did Mendel make sure that the results of his breeding experiments would be accurate?

   (1) He started with purebred plants.
   (2) He used hybrid plants having opposite traits.
   (3) He selected plants with several opposite traits.
   (4) He used only dwarf plants.
   (5) He used only tall plants.

2. If you cross purebred pea plants having inflated pods with purebred pea plants having wrinkled pods, all the F_1 generation will have inflated pods. Which statement about the F_1 generation is true?

   (1) The plants are all hybrids.
   (2) The plants are all purebred.
   (3) Inflated pods is a recessive trait.
   (4) Wrinkled pods is a hybrid trait.
   (5) None of the above statements is true.

3. What is a dominant trait?

   (1) a trait that will appear only in purebred offspring
   (2) a trait that will appear only in hybrid offspring
   (3) a trait that will appear in some purebred and all hybrid offspring
   (4) a trait that will appear in every offspring
   (5) a trait that will disappear when bred

4. What evidence supports Mendel's conclusion that the recessive trait was carried by the F_1 generation even though none of them showed the recessive trait?

   (1) All of the F_1 generation plants showed the dominant characteristic.
   (2) Some of the F_2 generation plants showed the recessive trait.
   (3) Some of the F_2 generation plants showed the dominant trait.
   (4) All of the F_2 generation plants showed the recessive trait.
   (5) None of these statements supports Mendel's conclusion.

5. Why did Mendel use pea plants in his experiments?

   (1) They are the only type of plant that shows how genetics works.
   (2) They have characteristics that are easy to identify .
   (3) The rules of genetics apply only to plants.
   (4) They were the only plants available in Austria at that time.
   (5) They were available in dwarf varieties.

6. When two hybrid plants are bred, what is the chance that one offspring will show the recessive trait?

   (1) no chance
   (2) 1 out of 4
   (3) 2 out of 4
   (4) 3 out of 4
   (5) 4 out of 4

To check your answers, turn to page 108.

Reviewing Lesson 7

Read the following passage and look at the diagram.

A **gene** is a portion of the genetic molecule that determines a particular trait. A plant or animal inherits half of its genes from each parent. The set of genes that a plant or animal inherits is called its **genotype**.

To predict which traits will be inherited, scientists use a **Punnett square**. The Punnett square below shows the crossing of pea plants with two sets of contrasting traits — tall and dwarf, and red and white flowers. The genotype of both parents is written as TtRr. Letters representing the genes from the male parent are placed across the top of the square. Letters representing the genes from the female parent are placed along the left.

Male Genes

		TR	tR	Tr	tr	
Female Genes	TR	TTRR	TtRR	TTRr	TtRr	T = tall, dominant
	tR	TtRR	ttRR	TtRr	ttRr	R = red, dominant
	Tr	TTRr	TtRr	TTrr	Ttrr	t = dwarf, recessive
	tr	TtRr	ttRr	Ttrr	ttrr	r = white, recessive

Each square represents one offspring. Each offspring inherits the genes at the top of the column and the genes at the left of each row.

Questions 1 and 2 refer to the passage and diagram. Circle the best answer for each question.

1. What is the offspring of crossing a female (YYTT) having dominant genes for yellow peas (Y) and dominant genes for tallness (T) with a male (yytt) having recessive genes for green peas (y) and recessive genes for dwarfness (t)?

   (1) YyTt
   (2) YYTt
   (3) Yytt
   (4) YyTT
   (5) yyTT

2. Which of the following statements is true of the offspring of a male of genotype ttrr and a female of genotype ttrr?

   (1) All offspring will inherit both dominant traits.
   (2) All offspring will inherit both recessive traits.
   (3) All offspring will inherit one dominant and one recessive trait.
   (4) Some offspring will inherit one dominant and one recessive trait.
   (5) Some offspring will inherit both dominant traits.

To check your answers, turn to pages 108–109.

Read the following passage and study the diagram.

The Punnett square shows the genotypes of offspring, but what does each offspring look like? To figure out the **phenotype**, or appearance, you must look at the dominant traits of each individual. For example, the phenotype of an individual with a genotype of YyIi is yellow with inflated pods. The recessive genes for green (y) and wrinkled pods (i) do not show in the individual's appearance. If there are no dominant genes for a particular trait, then the recessive trait is the phenotype. For example, an individual with genotype Yyii has a phenotype of yellow with wrinkled pods.

Wrinkled Pod Genes

Green Genes		YI	yI	Yi	yi
	YI	YYII	YyII	YYIi	YyIi
	yI	YyII	yyII	YyIi	yyIi
	Yi	YYIi	YyIi	YYii	Yyii
	yi	YyIi	yyIi	Yyii	yyii

Y = yellow, dominant

y = green, recessive

I = inflated pods, dominant

i = wrinkled pods, recessive

Questions 3 to 6 refer to the passage and diagram. Circle the best answer for each question.

3. If someone gave you a pea plant that resulted from a genetic experiment, what could you tell with greatest accuracy?

(1) the plant's phenotype
(2) the plant's genotype
(3) which recessive genes the plant has
(4) which generation the plant is from
(5) none of the above

4. According to the diagram, how many phenotypes will be green with wrinkled pods?

(1) 8
(2) 4
(3) 3
(4) 2
(5) 1

5. How many different phenotypes are shown in the diagram?

(1) 16
(2) 5
(3) 4
(4) 3
(5) 2

6. If one of the male parent's recessive genes were changed, and the changed gene were passed on to all his offspring, this would

(1) always result in a change in the genotype of the offspring
(2) always result in a change in the phenotype of the offspring
(3) never result in a change in the genotype of the offspring
(4) never result in a change in the phenotype of the offspring
(5) have no effect on the offspring

To check your answers, turn to page 109.

GED Mini-Test

Directions: Choose the best answer to each item.

Items 1 to 4 refer to the following passage and diagram.

DNA is a molecule that contains hereditary information and controls the activities of each cell. The DNA in every cell of the body bears the genetic code, which determines the traits of an organism.

DNA MOLECULE

Key

A = adenine
C = cytosine
G = guanine
P = phosphate
S = sugar
T = thymine

The structure of DNA was not known until 1953, when an American biologist, James D. Watson, and a British biophysicist, F. H. C. Crick, worked it out. They described DNA as a double helix, or spiral, made up of two strands wound around each other and connected by crosspieces. The molecule looks like a twisted ladder.

The side pieces of DNA are made of alternating units of sugar and phosphate. The rungs of the DNA ladder are composed of pairs of nitrogen bases. Adenine is always paired with thymine, and guanine is always paired with cytosine. Their positions (right and left) may vary, and their sequence along the ladder may vary. These variations in position and sequence account for the many genetic traits an organism may have.

1. Which of the following statements is always true?

(1) Adenine is paired with cytosine.
(2) Adenine is on the right side of the pair.
(3) Adenine follows guanine.
(4) Adenine is paired with thymine.
(5) The sequence of nitrogen bases repeats itself precisely.

2. What would be the effect of straightening out the DNA molecule?

(1) It would become shorter.
(2) It would become longer.
(3) It would become circular.
(4) It would become spherical.
(5) It would remain a helix.

3. How do offspring inherit traits from both parents?

(1) They receive DNA from the female parent.
(2) They receive DNA from the male parent.
(3) The offspring create entirely new DNA.
(4) They get some DNA from each parent.
(5) DNA controls the activities of each cell.

4. What substances make up the side pieces of DNA?

(1) adenine and thymine
(2) thymine and guanine
(3) guanine and sugar
(4) sugar and phosphate
(5) phosphate and adenine

Items 5 to 8 refer to the following passage.

Scientists know more about inheritance in lower forms of life than they do in humans. One reason for this is that people have a longer life cycle. A researcher can study several generations of insects in a few months or many generations of bacteria in a week. At most, a researcher can see five or six human generations in a lifetime. The number of humans presents another problem. A single mating in a plant or animal may produce hundreds of offspring. Each human family is small in comparison, and it represents a very tiny sampling of genetic possibilities. A third problem in studying human genetics is separating the effect of heredity from the effect of environment. Since people are complex, thinking, responsive creatures, their environment influences their characteristics.

To solve these problems, scientists who study human genetics have a different approach. Rather than tracing genetic traits through families over time, they study the frequency of genetic traits across a large sample of the population. When they know the frequency of a trait in the population, they can predict the probability of any given trait appearing in offspring.

5. Which of the following is a matter of opinion rather than fact?

(1) the number of generations of bacteria in a given amount of time
(2) the relatively small number of offspring produced by two human mates
(3) the extent to which the environment influences characteristics of humans
(4) the frequency of a particular trait in a given population
(5) the life span of humans compared to that of insects and bacteria

6. Why is it difficult to study human genetics from one generation to the next?

(1) There is too much time between generations.
(2) There is too much genetic variation between generations.
(3) Environments influence people.
(4) Humans have identifiable traits.
(5) Researchers study gene frequency in large populations.

7. In order to ensure that the predictions about gene frequency in the general population are accurate, what must researchers do?

(1) use a sample consisting of one family
(2) use a sample consisting of a large and varied group of people
(3) use a sample consisting only of children
(4) use a sample consisting of several generations of one family
(5) use a sample consisting of only one ethnic group

8. Some people can taste a chemical called PTC, and others cannot. About 30 percent of the people in the United States cannot taste PTC. What is the probability that the next American child born will not be able to taste PTC?

(1) 0
(2) 30 percent
(3) 50 percent
(4) 70 percent
(5) 100 percent

To check your answers, turn to page 109.

Answers and Explanations

1. (Evaluation) **(4) Most couples in the United States are fertile.** Since the passage states that about 20 percent of couples are infertile, it follows that about 80 percent are fertile. Options (1), (2), and (5) are contradicted by the passage. Option (3) is not necessarily true, although peak fertility is an important factor in the timing of having children. However, the trend toward postponing parenthood implies that other factors such as career development, finances, and marital status can enter into such a decision.

2. (Evaluation) **(1) When deciding when to have children, couples should take into account that the risk of infertility increases with age.** Options (2) and (5) are contradicted by the passage. Option (3) is an opinion and is not necessarily true, since many factors affect the timing of having children. Option (4) is not true, since the risk of miscarriage increases with age, regardless of whether the couple has had children.

Exercises (page 103)

1. (Evaluation) **(1) He started with purebred plants.** By starting the experiment with plants that always produced the same characteristics, Mendel made sure that the results of cross-breeding would be reliable. If he had started with hybrid plants, option (2), he would not have been sure why he was getting a mix of traits. If he had selected plants with several contrasting traits, option (3), there would have been so many combinations of characteristics in the offspring that it would have been difficult to keep track of what was going on. Using only one type of plant, options (4) and (5), would have meant that the experiment would not show the results of breeding plants with contrasting traits.

2. (Evaluation) **(1) The plants are all hybrids.** Since the offspring receive a different trait from each parent, they will all be hybrids. Option (2) is not correct because contrasting traits are being bred. Option (3) is not correct because all the offspring had inflated pods, making that the dominant trait. Option (4) is not true; wrinkled pods is a recessive trait. Option (5) is not correct because option (1) is correct.

3. (Analysis) **(3) a trait that will appear in some pure-bred and all hybrid offspring** Since a dominant trait is more powerful than a recessive trait, it will appear in all hybrids. It will also appear when two dominant traits are combined in a purebred offspring. Option (1) is incorrect because it leaves out hybrid offspring. Option (2) is incorrect because it leaves out some purebred offspring. Option (4) is incorrect because some offspring will not inherit the dominant trait. Option (5) is incorrect because the dominant trait will always be more powerful when combined with a recessive trait.

4. (Evaluation) **(2) Some of the F_2 generation plants showed the recessive trait.** Although the recessive trait was hidden in their F_1 parents, some of the F_2 plants received two recessive traits from the parents and showed the recessive characteristic. This proved that the recessive trait was carried by the hybrid F_1 parents even though it did not show. Option (1) does not provide evidence that the recessive trait is present. Option (3) is incorrect because offspring with a dominant trait do not show that the recessive trait was present in the parents. Option (4) is not true; only one offspring showed the recessive trait. Option (5) is not true because option (2) is true.

5. (Comprehension) **(2) They have characteristics that are easy to identify.** This is stated in the second paragraph of the passage. Options (1), (3), and (4) are not true. Option (5) is true but was not the reason he selected pea plants.

6. (Comprehension) **(2) 1 out of 4** As the diagram shows, when two hybrids are bred, one of the four offspring will show the recessive trait. The other three have inherited the dominant trait and will show the dominant trait.

Reviewing Lesson 7 (pages 104–105)

1. (Analysis) **(1) YyTt** The offspring gets YT genes from the female parent and yt genes from the male parent. The resulting genotype is a combination of the two: YyTt.

The other options are incorrect because those genotypes cannot be produced by the parents in the question.

2. (Evaluation) **(2) All offspring will inherit both recessive traits.** Since the dominant traits are not present in either of the parents, they cannot show up in any of the off-spring, thus eliminating options (1), (3), (4), and (5).

3. (Evaluation) **(1) the plant's phenotype** You could tell only the phenotype because the phenotype is the plant's appearance. You would need to have records from the experiment in order to tell the plant's genotype, because the plant might have some recessive genes that do not show up in the phenotype. You could not tell what genes the plant has with great accuracy just from its appearance, options (2) and (3). You also could not tell what generation it is from just by looking at it, option (4). Option (5) is incorrect because it is possible to tell the phenotype, option (1).

4. (Comprehension) **(5) 1** Only one phenotype is green with wrinkled pods (yyii). The others all have at least one dominant gene, which would change either the color or the pod shape.

5. (Comprehension) **(3) 4** There are four phenotypes, or appearances, in the diagram. They are yellow and inflated; green and inflated; yellow and wrinkled; green and wrinkled. Each of these four phenotypes may have different genotypes.

6. (Analysis) **(1) always result in a change in the genotype of the offspring** Since the question states that the gene

is always passed on to the offspring, it follows that the gene will affect the genotypes of all offspring. Option (3) is therefore eliminated. The new gene may or may not show up in the phenotype, depending on the gene inherited from the female parent, which is unknown. Therefore options (2) and (4) are incorrect. Option (5) is incorrect because the genotype of the offspring would be affected.

GED Mini-Test (pages 106–107)

1. (Evaluation) **(4) Adenine is paired with thymine.** This is the only statement that is always true. Options (2) and (3) are sometimes true. Options (1) and (5) are not true.

2. (Analysis) **(2) It would become longer.** If the spiral were straightened out, the molecule would become longer, not shorter, option (1). Options (3), (4), and (5) are incorrect because straightening the molecule would make it flat. The other shapes would not result from straightening the helix.

3. (Analysis) **(4) They get some DNA from each parent.** This is an unstated assumption of the passage. DNA, which transmits hereditary information, must come from both parents. Options (1), (2), and (3) are therefore incorrect. Option (5) is true but does not explain how traits are inherited.

4. (Comprehension) **(4) sugar and phosphate** According to the passage and diagram, sugar and phosphate units alternate on the side pieces of DNA. All other options include at least one nitrogen base, which is not part of the side pieces of DNA.

5. (Analysis) **(3) the extent to which the environment influences characteristics of humans** Since the effect of the environment is not really known, scientists have different opinions on the extent of its influence. The other options are observable, measurable facts and therefore are not opinions.

6. (Comprehension) **(1) There is too much time between generations.** The human life cycle is fairly long—usually more than 20 years between generations—so to see the results of experiments would take too long. Option (2) is not true. Options (3) and (4) are true, but they have nothing to do with the difficulty of studying generations of humans. Option (5) is a result, not a cause, of the difficulty of studying human generations over time.

7. (Evaluation) **(2) use a sample consisting of a large and varied group of people** To ensure the accuracy of research results, the sample used must be representative of the population as a whole. Therefore, a large and varied group is necessary. Option (1) is too small a sample. Option (3) is not a varied sample. Option (4) is not practical in terms of time and would also not give a large enough sample. Option (5) is not a varied sample.

8. (Comprehension) **(2) 30 percent** Since 30 percent of the people in the general population cannot taste PTC, the likelihood of the next child born not being able to taste PTC is also 30 percent.

LESSON 8 Analysis Skill: Conclusions and Supporting Statements

Understanding what you read often involves telling the difference between conclusions and supporting statements. A conclusion is a logical result or generalization. Supporting statements offer details and facts that prove the conclusion.

In science, laws, theories, and hypotheses are often conclusions. Observations, events, conditions, measurements, and other details are often supporting statements. Frequently you will be asked to explain why certain things have happened or to predict what will happen under certain circumstances. In both instances, you must be able to tell the difference between a conclusion and the information that proves the conclusion is true.

Telling the difference between conclusions and supporting statements draws on skills you have already learned. Sometimes you must distinguish a main idea (conclusion) from supporting details. Sometimes you must decide which facts support an opinion. At other times you must figure out cause and effect.

Read this brief passage and decide what is the conclusion and what are the supporting statements.

> The most common water pollutants are organic materials (substances that were once living). Most organic material is attacked by bacteria and broken down into simpler substances. The greater the supply of organic matter, the larger the population of bacteria in the water, and the faster the oxygen is used up. Since all animals in a stream need oxygen, the oxygen level indicates which forms of life a polluted stream can support. Fish have the highest oxygen need. Invertebrates need less oxygen, and bacteria need still less.

What is the conclusion of this passage? From the information provided, we can conclude that polluted water cannot support certain animal life. To support this conclusion, we can point to several facts. Organic pollutants are food for bacteria, which reproduce rapidly and use up a lot of oxygen. When the oxygen level drops, fish cannot survive because they have the highest oxygen need. The invertebrates die next, and the oxygen-using bacteria die last.

When you look for conclusions as you read, pay attention to key words such as <u>for that reason</u>, <u>therefore</u>, <u>since</u>, <u>so</u>, and <u>thus</u>. These words often signal a conclusion.

Practicing Analysis

Read the following paragraph.

> Extinction, or the disappearance of a species, is a natural occurrence. All organisms are replaced sooner or later by better-adapted or newly evolved forms. However, some wild species have abruptly disappeared because humans eat or use them. The passenger pigeon and great auk were hunted to extinction during the 1800s. The buffalo was hunted almost to extinction. Today certain types of whales are likely to become extinct unless whaling nations can agree to reduce their catch.

Passenger pigeon

Great auk

Questions 1 and 2 refer to the paragraph. Circle the best answer for each question.

1. Which of the following statements supports the conclusion that wild animals that can be used by humans are more likely to face sudden extinction than other animals?

   (1) The passenger pigeon and great auk were hunted to extinction during the 1800s.
   (2) Extinction is a natural occurrence.
   (3) Species adapt to their environments.
   (4) Whales are no longer hunted.
   (5) All species eventually become extinct.

2. "All organisms are replaced sooner or later by better-adapted or newly evolved forms." Which conclusion does this statement support?

   (1) Whales are a protected species.
   (2) The extinction of the passenger pigeon could have been prevented.
   (3) People have no influence on which species become extinct.
   (4) Humans will eventually be replaced by another form of life.
   (5) No species lasts longer than 50 million years.

To check your answers, turn to page 118.

Topic 8: Ecosystems

Read the following passage and look at the illustration.

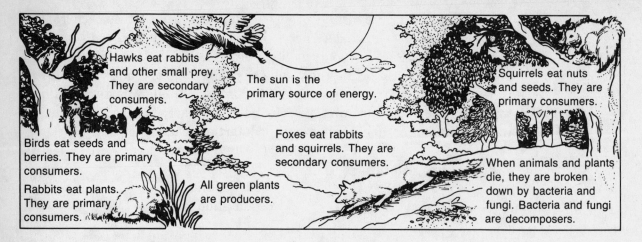

Hawks eat rabbits and other small prey. They are secondary consumers.

The sun is the primary source of energy.

Squirrels eat nuts and seeds. They are primary consumers.

Birds eat seeds and berries. They are primary consumers.

Foxes eat rabbits and squirrels. They are secondary consumers.

When animals and plants die, they are broken down by bacteria and fungi. Bacteria and fungi are decomposers.

Rabbits eat plants. They are primary consumers.

All green plants are producers.

An **ecosystem** is a natural community in which the living and nonliving things interact. Lakes, rivers, meadows, ponds, and swamps are all ecosystems. The most important relationships in an ecosystem involve the flow of food and energy.

Food and the Ecosystem. The movement of food through an ecosystem is called a **food chain**. A food chain begins with producers, which in most ecosystems are green plants. Animals that eat plants are called **primary consumers**. Animals that eat primary consumers are called **secondary consumers**. Some ecosystems have **tertiary consumers**, which are animals that eat secondary consumers.

When plants and animals die, bacteria and fungi in the soil break them down into substances that can be used by plants. Organisms that break down plants and animals are called **decomposers**. Decomposers recycle the matter in an ecosystem.

Energy and the Ecosystem. The Sun is the main source of energy for an ecosystem. The Sun's energy enters the ecosystem through green plants, which store some of the energy. The stored energy is passed along the food chain as consumers eat producers and other consumers. However, since living things use energy to carry out life processes, less and less energy is available to each level of consumer. For example, only 10 out of 1,000 units of energy are transferred from plants to primary consumers. Only 1 out of every 10 units of energy is transferred from primary consumers to secondary consumers.

Exercises

Questions 1 to 7 refer to the previous passage and illustration. Circle the best answer for each question.

1. Of the sentences below, four are supporting statements and one is a conclusion. Which of the following is a conclusion?

   (1) Plants produce food.
   (2) Matter in an ecosystem is being constantly recycled.
   (3) Decomposers break down dead plants and animals into substances that can be used by plants.
   (4) Secondary consumers eat primary consumers.
   (5) Primary consumers eat plants.

2. What would be the immediate result if all the primary consumers were removed from the ecosystem?

   (1) Plants would stop producing food.
   (2) Decomposers would increase their activity.
   (3) Secondary consumers would have nothing to eat.
   (4) Tertiary consumers would have more to eat.
   (5) A new source of energy would be needed.

3. Some birds eat insects as well as seeds and berries. Which of the following most accurately describes their role in the ecosystem?

   (1) producers
   (2) primary consumers
   (3) primary consumers and secondary consumers
   (4) secondary consumers
   (5) producers and primary consumers

4. An insect that feeds on other insects that are crop pests is a

   (1) producer
   (2) primary consumer
   (3) secondary consumer
   (4) tertiary consumer
   (5) decomposer

5. Which of the following is an unstated assumption of the passage?

   (1) All living things are part of an ecosystem.
   (2) Only humans are not part of an ecosystem.
   (3) The Sun is the main source of energy in an ecosystem.
   (4) All matter is being constantly recycled.
   (5) Squirrels are secondary consumers.

6. What is the role of decomposers in an ecosystem?

   (1) to produce food for primary consumers
   (2) to break down dead plants and animals into substances plants can use
   (3) to provide food for secondary consumers
   (4) to provide food for tertiary consumers
   (5) to provide energy

7. Which organisms are the highest-level consumers in a food chain?

   (1) hawks
   (2) seed-eating birds
   (3) rabbits
   (4) squirrels
   (5) primary consumers

To check your answers, turn to page 118.

Reviewing Lesson 8

Read the following passage and look at the graph.

In the year 10,000 B.C., there were about 10 million people in the world. By the year A.D. 1, that number had grown to about 300 million. The population remained fairly constant for the next 1600 years. By the year 1650, the population had risen to only about 510 million people. During the next 200 years, however, the population more than doubled. Between 1850 and 1950, the population doubled again. The next doubling of the population took only 40 years.

Today the world's birth rate is three times higher than the death rate. The result is that the population continues to rise. The rapid increase in population is a serious problem. Overpopulated areas suffer from disease, poverty, and crime. Natural resources such as water and food are limited.

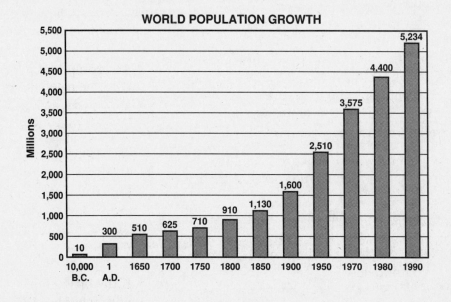

WORLD POPULATION GROWTH

Questions 1 and 2 refer to the passage and graph. Circle the best answer for each question.

1. Which of the following does not support the conclusion that world population is growing rapidly?

   (1) The birth rate is three times the death rate.
   (2) The last doubling of world population took only 40 years.
   (3) World population remained fairly stable until 1650.
   (4) Between 1900 and 1990, world population more than tripled.
   (5) Between 1950 and 1990, world population more than doubled.

2. Which factor is probably most responsible for the increased world population?

   (1) decrease in birth rate
   (2) increase in death rate
   (3) increase in disease
   (4) better weather conditions
   (5) increase in birth rate

Study the following chart and read the passage.

The number of people living in a specific area is called **population density**. Population density is a ratio. It is calculated by dividing the number of people living in an area by the amount of usable land available in that area. If an area's population increases, so does its population density.

WORLD POPULATION DENSITY PER SQUARE MILE

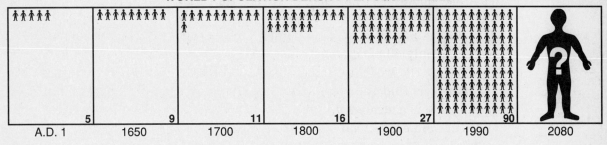

Questions 3 to 6 refer to the chart and passage. Circle the best answer for each question.

3. About how many times greater was the world population density in 1990 than in 1650?

   (1) 80
   (2) 35
   (3) 10
   (4) 5
   (5) 3

4. Which of the following supports the statement, "If an area's population increases, so does its population density"?

   (1) The population density in the year A.D. 1 was about 5 people per square mile.
   (2) Population density is calculated by dividing the number of people living in an area by the amount of land available in the area.
   (3) The death rate has no effect on population density.
   (4) Population figures are estimates.
   (5) The population density in the year 1650 was 11 people per square mile.

5. According to the information in the chart, what is likely to happen to the population density by the year 2080 if all conditions remain the same?

   (1) It is likely to triple.
   (2) It is likely to decrease by one-half.
   (3) It is likely to double.
   (4) It is likely to remain the same.
   (5) It is likely to go back to the 1900 density.

6. If the birth rate and death rate remain at present levels, how can the human population density be decreased?

   (1) Increase the food supply.
   (2) Decrease the supply of fossil fuels.
   (3) Increase the amount of usable land.
   (4) Decrease the population density of selected animals.
   (5) Decrease the amount of usable land.

To check your answers, turn to pages 118–119.

GED Mini-Test

Directions: Choose the best answer to each item.

Items 1 to 4 refer to the following passage.

In stable ecosystems, for each plant and animal there is usually at least one consumer that serves to check its population growth. That is why there are usually constant numbers of plants and animals from year to year. When you take a species of plant or animal from its natural ecosystem and introduce it into another ecosystem, it might die out quickly. But if it survives, the species may reproduce at an incredible rate.

How can a new species be so successful so quickly? The food relationships in an ecosystem have evolved slowly over time and they change very slowly. Just as many people are reluctant to eat an unfamiliar food, animals are unlikely to eat something they have never seen before. If there is no consumer willing to eat the new species, it will reproduce rapidly. For example, in 1859, 24 rabbits were imported from Europe to Australia, which had no rabbit problem. Today, Australia has well over a billion rabbits despite many efforts to control the population.

The best solution to the overpopulation of new species is prevention. However, if new organisms do overrun an area, sometimes population growth can be controlled by importing another species that will prey on the first.

1. Which of the following statements supports the conclusion that overpopulation of a new species can sometimes be controlled by the introduction of an animal that will prey on it?

   (1) The introduction of new organisms cannot always be prevented.
   (2) Rabbits were successfully wiped out in Australia.
   (3) In stable ecosystems, there is usually at least one consumer that checks an organism's population growth.
   (4) In time, the population will decrease.
   (5) The overpopulated species cannot find anything to eat.

2. What is the best way to solve the problems associated with introducing new organisms into an ecosystem?

   (1) Introduce an animal to prey on the new organism.
   (2) Ensure that there will be something the new organism will eat.
   (3) Prevent the introduction.
   (4) Ensure that the new ecosystem has nothing the new organism will eat.
   (5) There is no way to solve these problems.

3. What can you assume the rabbits found in Australia?

(1) an ecosystem with no food and no predators
(2) an ecosystem with food and no predators
(3) an ecosystem with no food and predators
(4) an ecosystem with many meat-eating animals
(5) an ecosystem similar to Europe's

4. Which is the best title for this passage?

(1) Stable Ecosystems
(2) The Introduction of Rabbits to Australia
(3) Preventing the Introduction of New Species
(4) The Introduction of New Species into Ecosystems
(5) The Food Chain in a Stable Eco-system

Items 5 and 6 refer to the following passage.

When fossil fuels such as coal, oil, and gas are burned, they release gases into the air. The three most common gases include:

1. Sulfur dioxide. Sulfur dioxide results from burning coal and oil that contain sulfur. When sulfur dioxide is inhaled, it usually causes choking, and in large doses it can be fatal. When it combines with water, it produces sulfuric acid, which can damage the eyes and lungs.
2. Nitric oxide. A byproduct of burning gasoline, nitric oxide combines with oxygen in the air to form nitrogen dioxide. Nitrogen dioxide combines with water in the eyes and lungs and forms nitric acid, which can cause permanent damage.
3. Carbon monoxide. This gas is produced when fossil fuels do not burn completely. The major source of carbon monoxide is the exhaust from automobile engines. Carbon monoxide combines with blood much faster than oxygen does. This is poisonous and can be fatal.

Sulfur dioxide and nitrogen dioxide combine with water vapor in the atmosphere to cause acid rain. Acid rain is very harmful to plant and animal life. In industrial areas, the gases also combine with fog to produce smog, a gray or rust-colored haze. Smog can be harmful to the lungs and eyes of people and animals.

5. To prevent acid rain, the production of which of the following gases would have to be controlled?

(1) nitric oxide and sulfur dioxide
(2) nitric oxide and carbon monoxide
(3) carbon monoxide and sulfur dioxide
(4) water vapor and carbon monoxide
(5) smog and sulfur dioxide

6. The amount of nitrogen dioxide in the atmosphere could be reduced by lowering which emissions from cars?

(1) nitrogen dioxide
(2) carbon monoxide
(3) sulfur dioxide
(4) nitric oxide
(5) nitric acid

To check your answers, turn to page 119.

Answers and Explanations

1. (Analysis) **(1) The passenger pigeon and great auk were hunted to extinction during the 1800s.** This statement provides supporting evidence in the form of examples of species that over a short period of time became extinct. Options (2), (3), and (5) are true but they do not support the conclusion. Option (4) is not true; whales are still hunted.

2. (Analysis) **(4) Humans will eventually be replaced by another form of life.** Since all species are subject to extinction, it follows that humans are, too. Options (1) and (2) are true but have nothing to do with the general fate of all species. Options (3) and (5) are not true.

Exercises (page 113)

1. (Analysis) **(2) Matter in an ecosystem is being constantly recycled.** This is the generalization that is supported by the other statements, each of which describes one step in the food chain.

2. (Analysis) **(3) Secondary consumers would have nothing to eat.** This is the immediate effect of the loss of primary consumers, because they are the food for secondary consumers. Option (1) is incorrect; plants would continue to function. Options (2) and (4) are incorrect because the activities of decomposers and tertiary consumers would not be immediately affected. Option (5) is not true; the Sun would continue to provide energy.

3. (Comprehension) **(3) primary consumers and secondary consumers** In their role as berry- and seed-eaters, birds are primary consumers (eating plants). In their role as insect-eaters, the birds are secondary consumers (eating animals). Option (1) refers to plants. Options (2) and (4) are only partly correct. Option (5) refers to plants and animals.

4. (Comprehension) **(3) secondary consumer** The crop pests, which feed on plants, are primary consumers. Anything that eats a primary consumer (the crop pest) is a secondary consumer.

5. (Analysis) **(1) All living things are part of an ecosystem.** Although the passage describes what an ecosystem is, it never explicitly states that all living things are part of an ecosystem. Option (2) is incorrect because humans are part of many ecosystems. Options (3), (4), and (5) are not unstated assumptions because they are clearly stated in the passage or illustration.

6. (Comprehension) **(2) to break down dead plants and animals into substances plants can use** This is stated in the third paragraph of the passage. Option (1) is the role of producers. Option (3) is the role of primary consumers. Option (4) is the role of secondary consumers. Option (5) is the role of the Sun.

7. (Comprehension) **(1) hawks** Of the listed organisms, all are primary consumers except the hawk, which is a secondary consumer. Since secondary consumers are farther up the food chain, they are higher-level consumers than the others.

Reviewing Lesson 8 (pages 114–115)

1. (Analysis) **(3) World population remained fairly stable until 1650.** This is the only statement not related to the present rapid growth of world population. The other options all support the conclusion that the population is growing quickly.

2. (Analysis) **(5) increase in birth rate** Option (1) is incorrect because a decrease in birth rate would slow population growth. Option (2) is incorrect because an increase in the death rate would also slow population growth. Options (3) and (4) are not true.

3. (Comprehension) **(3) 10** Since the density in 1990 was 90 and the density in 1650 was 9, 90 ÷ 9 = 10, or a tenfold increase.

4. (Analysis) **(2) Population density is calculated by dividing the number of people living in an area by the amount of land available in the area.** Application of this general rule leads to the conclusion that when there are more people living in the same amount of space, the density increases. Option (1) has nothing to do with population increases. Option (3) is not true; when the death rate is higher than the birth rate, density decreases. Option (4) is true but has nothing to do with increased population density. Option (5) is not true.

5. (Comprehension) **(1) It is likely to triple.** In the last 90 years, the population density tripled; if conditions remain the same, it follows that the density will triple in the next 90 years.

6. (Analysis) **(3) Increase the amount of usable land.** Since the rate of population growth will remain the same, the only way to decrease population density is to spread the population into previously unused areas. Options (1), (2), and (4) have nothing to do with changing the population density of humans. Option (5) would have the effect of increasing population density since there would be less available land.

GED Mini-Test (pages 116–117)

1. (Analysis) **(3) In stable ecosystems, there is usually at least one consumer that checks an organism's population growth.** Application of this principle leads to the conclusion that introducing a predator may control the organism that is overrunning its new environment. Option (1) is true but has nothing to do with possible solutions to the problem. Option (2) is not true. Option (4) may or may not be true, but the process would take a very long time. Option (5) is a contradiction; if the organism could not find anything to eat, it would not be able to reproduce at a fast rate.

2. (Comprehension) **(3) Prevent the introduction.** According to the passage, prevention is the best solution. Option (1) is a solution, but not the best one. Options (2) and (4) are generally beyond human control. Option (5) is incorrect because there is a way to solve

these problems, which is stated in option (3).

3. (Analysis) **(2) an ecosystem with food and no predators** Since rabbits thrived in Australia, it can be assumed that they found plenty to eat and no animals to fear. Options (1) and (3) are incorrect because without food the rabbits would have starved. Option (4) is incorrect because if there were many meat-eating animals, it is likely that the rabbits would have been eaten. Option (5) cannot be true, since if the ecosystem were similar to Europe's, the rabbit population would not have gotten out of control.

4. (Analysis) **(4) The Introduction of New Species into Ecosystems** The passage deals generally with what happens when a new species is introduced into an ecosystem. It does not describe stable ecosystems at length, option (1). The example of rabbits in Australia is a supporting detail, option (2). The passage does not say how to prevent the introduction of new species, option (4). It is not primarily about the food chain in stable ecosystems, option (5).

5. (Analysis) **(1) nitric oxide and sulfur dioxide** Nitric oxide reacts with oxygen to form nitrogen dioxide. Nitrogen dioxide and sulfur dioxide react with rainwater to form acid rain. The other options contain gases that do not form acid rain.

6. (Analysis) **(4) nitric oxide** Car exhaust is one of the main sources of nitric oxide in the atmosphere. When nitric oxide combines with oxygen, it produces nitrogen dioxide. When the amount of nitric

oxide from cars is reduced, it follows that the nitrogen dioxide content of the atmosphere would also be reduced. Option (1) is incorrect because nitrogen dioxide is not part of automobile exhaust. Options (2), (3), and (5) are incorrect because they are substances that do not form nitrogen dioxide.

Review: Life Science

In this section you have read about the structure of living organisms and about some of the processes living organisms carry out. You have also studied how living things interact with one another and with their environment. In addition, you have learned a little about genetics and evolution.

You have also learned some skills for understanding and thinking about what you read. The following exercises make use of the comprehension, analysis, and evaluation skills you have been practicing. These exercises also expand on the life science subjects you have read about so far.

Directions: Choose the best answer for each item.

Items 1 and 2 refer to the following passage.

When disease-causing microorganisms enter the body, white blood cells called phagocytes find them. Phagocytes surround, or engulf, microorganisms, much as an amoeba eats its food. One kind of phagocyte can enlarge to become a macrophage. Macrophages are so large that they can engulf a hundred or more bacteria at one time.

Other kinds of white blood cells, called lymphocytes, produce antibodies. Antibodies are protein substances that react with specific foreign organisms in the body and make them ineffective. Lymphocytes produce many different antibodies to attack different disease organisms.

---

1. How do phagocytes destroy disease-causing organisms?

    (1) by producing antibodies
    (2) by preventing them from entering the body
    (3) by engulfing them
    (4) by splitting them
    (5) by piercing their cell walls

2. Why does the number of phagocytes and lymphocytes in the body increase during an infection?

    (1) More phagocytes and lymphocytes are needed to fight the infection.
    (2) Disease-causing microorganisms produce phagocytes and lymphocytes.
    (3) Antibodies produce more phagocytes and lymphocytes.
    (4) Red blood cells are decreasing.
    (5) The body has become immune to the infection.

Items 3 to 6 refer to the following diagram and paragraph.

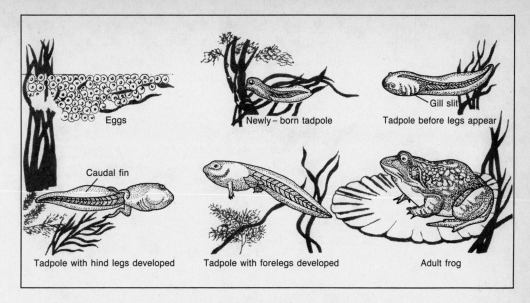

Eggs

Newly – born tadpole

Gill slit

Tadpole before legs appear

Caudal fin

Tadpole with hind legs developed

Tadpole with forelegs developed

Adult frog

The life cycle of the frog is an example of the process of metamorphosis. The word <u>metamorphosis</u> means change. In this process, the immature form, called a tadpole, gradually changes into an adult frog. The tadpole lives in water and breathes through gills. As it matures, the tadpole loses the gills and develops lungs. The adult frog can survive out of the water because it can breathe through its lungs.

3. Which of the following is the best title for the paragraph and diagram?

(1) Metamorphosis in the Frog
(2) The Structure of the Tadpole
(3) The Life Cycle of Water-Dwelling Animals
(4) Metamorphosis in Animals
(5) Reproduction in Frogs

4. Which of the following statements does the author take for granted?

(1) All animals go through metamorphosis.
(2) All plants go through metamorphosis.
(3) Different animal structures are suited to different types of environments.
(4) Almost all eggs develop into adults.
(5) Metamorphosis occurs in humans.

5. What is metamorphosis?

(1) the process of reproduction
(2) the process by which an immature form changes into a different adult form
(3) the growth of any young organism into an adult
(4) changes in an adult organism caused by aging
(5) the process by which tadpoles absorb oxygen from water

6. Which of the following must develop before a tadpole becomes a land-dwelling adult?

(1) legs
(2) lungs
(3) gills
(4) tail
(5) mouth

To check your answers, turn to page 128.

Items 7 and 8 refer to the following map and definitions.

Ecosystems that cover large areas of the world are called <u>biomes</u>. Biomes have different types of climates, plants, and animals. Six biomes are described and located on the map.

MAJOR BIOMES OF THE WORLD

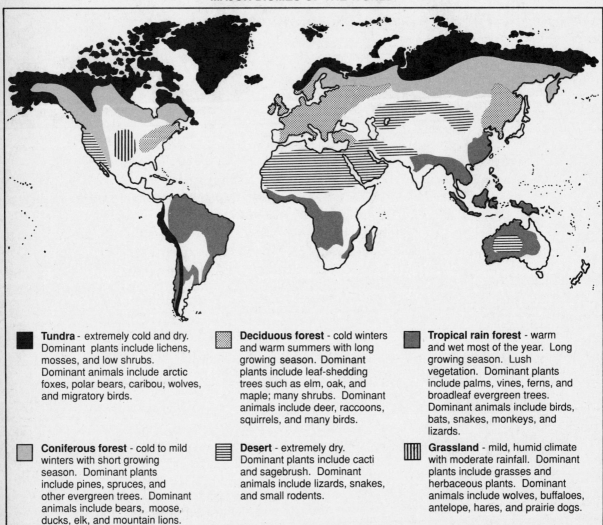

Tundra - extremely cold and dry. Dominant plants include lichens, mosses, and low shrubs. Dominant animals include arctic foxes, polar bears, caribou, wolves, and migratory birds.

Deciduous forest - cold winters and warm summers with long growing season. Dominant plants include leaf-shedding trees such as elm, oak, and maple; many shrubs. Dominant animals include deer, raccoons, squirrels, and many birds.

Tropical rain forest - warm and wet most of the year. Long growing season. Lush vegetation. Dominant plants include palms, vines, ferns, and broadleaf evergreen trees. Dominant animals include birds, bats, snakes, monkeys, and lizards.

Coniferous forest - cold to mild winters with short growing season. Dominant plants include pines, spruces, and other evergreen trees. Dominant animals include bears, moose, ducks, elk, and mountain lions.

Desert - extremely dry. Dominant plants include cacti and sagebrush. Dominant animals include lizards, snakes, and small rodents.

Grassland - mild, humid climate with moderate rainfall. Dominant plants include grasses and herbaceous plants. Dominant animals include wolves, buffaloes, antelope, hares, and prairie dogs.

7. According to the map, which kind of biome occurs farthest north?

(1) grassland
(2) deciduous forest
(3) tundra
(4) coniferous forest
(5) desert

8. Which of the following kinds of plants would you <u>not</u> expect to find in a tropical rain forest?

(1) palm trees
(2) vines
(3) cacti
(4) ferns
(5) evergreen trees

Items 9 to 12 refer to the following diagram and paragraph.

THE WATER CYCLE

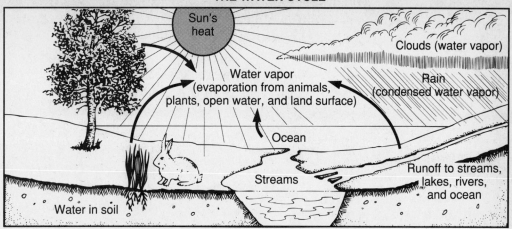

All plants and animals must have water in order to live. Cells are made mostly of water, and the chemical reactions in living things must take place in water. However, the amount of the Earth's water is limited. It must be used again and again. The constant circulation of the Earth's water is called the water cycle.

9. Which of the following statements is the most accurate?

(1) Animals return water to the air only by eliminating liquid wastes.
(2) The only water vapor in the air is found in clouds.
(3) Plants take in water mostly by breathing.
(4) On the Earth, water is found mainly in the soil.
(5) Water in animals' breath is returned to the air.

10. Which of the following statements supports the conclusion that water is used again and again?

(1) Water is recirculated.
(2) The water cycle circulates the Earth's water supply.
(3) Evaporation from the ocean returns water to the atmosphere.
(4) Cells are composed mainly of water.
(5) Water is needed to carry on life processes.

11. What causes water to evaporate in the water cycle?

(1) clouds
(2) Sun
(3) runoff
(4) oceans
(5) soil

12. Which of the following statements is an opinion?

(1) Since the supply of water is limited, it must be used again and again.
(2) Water conservation is not useful since the Earth will never run out of water.
(3) All living organisms must have water in order to live.
(4) The circulation of the Earth's water is called the water cycle.
(5) The chemical reactions in living things require water.

To check your answers, turn to pages 128–129.

Items 13 to 16 refer to the following passage and diagram.

The human brain consists of three parts: the cerebrum, the cerebellum, and the brain stem.

The cerebrum is the largest area of the brain. It consists of two halves, or hemispheres. The hemispheres are connected by fibers and nerves. Because these fibers and nerves cross, the left half of the cerebrum controls the right side of the body, and the right half of the cerebrum controls the left side of the body. Each hemisphere also seems to specialize in certain functions. For example, the left hemisphere is largely responsible for language and logical thinking. The right hemisphere is largely responsible for artistic and perceptual expression.

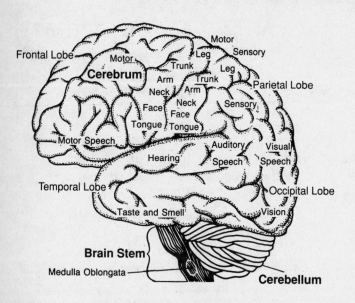

The cerebellum helps the cerebrum control movement. It also coordinates information from the eyes, inner ears, and muscles to maintain balance.

The brain stem contains the medulla. This part of the brain controls the organs of the body.

13. What would be the best title for this passage and diagram?

(1) The Human Nervous System
(2) The Human Brain
(3) The Cerebrum
(4) The Brain
(5) The Nervous System

14. Which of the following functions is most likely to be affected by a sharp blow to the back of the head?

(1) taste
(2) vision
(3) heart rate
(4) hearing
(5) leg movement

15. A patient has an injury to the cerebellum. What symptoms is the patient likely to show?

(1) loss of leg movement
(2) blurred vision
(3) loss of hearing
(4) speech difficulties
(5) loss of balance

16. Which of the following statements is the most accurate?

(1) As long as the cerebellum is functioning, motor activity will be normal.
(2) The brain stem controls leg muscles.
(3) Injury to the left hemisphere results in a total loss of speech.
(4) Motor activity is controlled by the cerebrum and cerebellum.
(5) The two hemispheres of the brain are not connected.

Items 17 to 20 refer to the following illustration and paragraph.

VESTIGIAL STRUCTURES

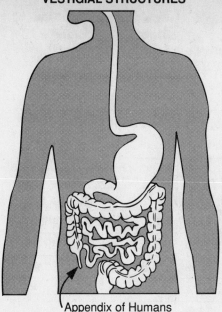

Many animals have well-developed structures that perform important functions. In other animals, a similar structure may be poorly developed. These poorly developed structures are called vestigial structures. Scientists believe that vestigial structures are the remains of organs that were well-developed in ancestors of present-day organisms. Vestigial structures offer evidence to support the theory that many animals evolved from common ancestors.

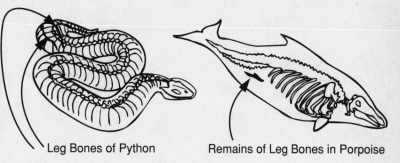

Appendix of Humans   Leg Bones of Python   Remains of Leg Bones in Porpoise

17. Which statement best summarizes the paragraph?

   (1) Present-day snakes have vestigial leg bones.
   (2) All animals have vestigial structures.
   (3) Vestigial structures offer support for the theory of evolution.
   (4) Many nonwalking animals once had functioning legs.
   (5) Vestigial structures are well-developed in today's animals.

18. Which of the following statements is scientific opinion rather than fact?

   (1) Vestigial structures are the remains of organs that were well-developed in common ancestors.
   (2) A vestigial structure may be poorly developed.
   (3) Porpoises and pythons have vestigial leg bones.
   (4) A vestigial structure may have no apparent function.
   (5) Humans have vestigial appendixes.

To check your answers, turn to page 129.

19. Which of the following statements is the most accurate?

   (1) The appendix of humans is a poorly developed organ.
   (2) Pythons once walked.
   (3) The vestigial leg bones of porpoises serve an important function.
   (4) Pythons and porpoises have similar skeletons.
   (5) The existence of vestigial structures proves without a doubt that the theory of evolution is a fact.

20. Which statement supports the conclusion that present-day organisms evolved from common ancestors?

   (1) The appendix has no apparent function in humans.
   (2) Some animals have a well-developed structure, and others have a corresponding vestigial structure.
   (3) Some animals have no vestigial structures.
   (4) Porpoises once had legs.
   (5) The leg bones of a python have no apparent function.

Items 21 to 24 refer to the following passage and charts.

There are three different genes—A, B, and O—that determine a person's blood type. Six different genotypes are possible, and they result in four different blood types, as shown in the chart at the right. For example, a person with blood type A may have genotype AA or AO. The type O gene is recessive, while type A and B genes are dominant. A person who inherits two dominant genes, A and B, has type AB blood.

Since a person's blood type depends on the two genes inherited from his or her parents, different populations show different frequencies of blood types. The chart on the bottom shows how the blood types are distributed in different population groups.

Blood Type	Genotype	Can Get Blood From	Can Give Blood To
A	AA, AO	O, A	A, AB
B	BB, BO	O, B	B, AB
AB	AB	A, B, AB, O	AB
O	OO	O	A, B, AB, O

Blood Type	A	B	AB	O
U.S.A.—White	41.0%	10.0%	4.0%	45.0%
U.S.A.—Black	26.0%	21.0%	3.7%	49.3%
Swedish	46.7%	10.3%	5.1%	37.9%
Japanese	38.4%	21.8%	8.6%	31.2%
Polynesian	60.8%	2.2%	0.5%	36.5%
Chinese	25.0%	35.0%	10.0%	30.0%
North American Indian	7.7%	1.0%	0.0%	91.3%

21. People with type O blood are sometimes called universal donors. A universal donor is someone who

(1) can receive blood from anyone
(2) cannot give blood to anyone
(3) has a genotype AO
(4) can give blood to anyone
(5) can receive only type A blood

22. In Japan, the blood banks ran out of type AB blood. What effect would this have on blood transfusions for people there with AB blood?

(1) They would not be able to have a blood transfusion.
(2) They would have to get type AB blood from another country.
(3) They could have a transfusion of type A, B, or O blood.
(4) They could have a transfusion of type A blood only.
(5) They could have a transfusion of type B blood only.

23. Which of the following would make the best title for the second chart?

(1) Blood Type Frequency in Selected Population Groups
(2) Human Blood Types
(3) Receiving and Giving Blood
(4) Genotypes for Each Blood Type
(5) Blood Type Frequency in the U.S.

24. Which of the following statements is most accurate?

(1) There are an equal number of blood types and genotypes.
(2) The genes A, B, O, and AB determine which blood type you have.
(3) Each individual inherits one gene for blood type from each of his or her parents.
(4) Blood types are evenly distributed in different populations.
(5) A person with type O blood has the genotype AO or BO.

Items 25 to 28 are based on the following passage.

Herbicides are chemicals that kill plants. When herbicides are used on crops, it is important that the herbicide kill only the weeds and not the crops. Scientists have been developing crops that are not affected by specific herbicides, enabling farmers to use herbicides to control weeds. For example, scientists have developed a strain of cotton that is resistant to the herbicide bromoxynil. When this herbicide is used, it kills weeds but not the resistant cotton plants. Other crops that have varieties resistant to certain herbicides are soybeans, tobacco, tomatoes, and sugar beets.

Some environmental groups are opposed to the development of herbicide-resistant crops. They say that these crops encourage farmers to continue to use chemicals that pollute the environment and that may be unsafe. These groups favor practices such as improved cultivation techniques and creative planting plans to make chemicals unnecessary. They prefer scientists to concentrate on developing strains of crops that are naturally resistant to disease and pests.

25. What is a herbicide-resistant crop?

(1) a variety of crop that cannot have a herbicide applied to it
(2) a variety of crop that is resistant to disease
(3) a variety of crop that is resistant to insect pests
(4) a variety of crop that is resistant to weeds
(5) a variety of crop that is not affected by particular chemical plant killers

26. What is an unstated assumption of the passage?

(1) Herbicides are safe to use.
(2) Crops cannot be grown without herbicides.
(3) Weeds are a problem in large farming areas.
(4) All cotton plants are resistant to herbicides.
(5) Most crops are naturally resistant to pests.

27. Which of the following is an opinion rather than a fact?

(1) Herbicides are used to kill weeds.
(2) Tobacco and soybean varieties are resistant to certain herbicides.
(3) Bromoxynil is a chemical herbicide.
(4) Improved cultivation techniques are preferable to herbicides in controlling weeds.
(5) Scientists have developed strains of herbicide-resistant crops.

28. Applying chemical herbicides to herbicide-resistant crops is likely to

(1) kill the crops
(2) damage the crops
(3) result in herbicides remaining on the crops after harvest
(4) kill more crop plants than weeds
(5) kill insects that are crop pests

To check your answers, turn to pages 129–130.

Answers and Explanations

1. (Comprehension) **(3) by engulfing them** The passage states that phagocytes work by surrounding or engulfing bacteria. Option (1) is true of lymphocytes, not phagocytes. Option (2) is incorrect because the phagocytes cannot work until the disease-causing organism is in the body. Options (4) and (5) do not relate to the way phagocytes work.

2. (Analysis) **(1) More phagocytes and lymphocytes are needed to fight the infection.** When infectious organisms enter the body, the body produces more phagocytes and lymphocytes in order to destroy the infection. Option (2) is incorrect because the body produces phagocytes and lymphocytes. Option (3) is incorrect because lymphocytes produce antibodies. Option (4) has nothing to do with fighting infection. Option (5) makes no sense; if the body were immune, there would be no infection.

3. (Comprehension) **(1) Metamorphosis in the Frog** The passage and diagram describe the process of metamorphosis, using the frog as an example. Option (2) is too specific. Option (3) is too general, and includes animals, such as fish, that do not undergo metamorphosis. Option (4) is also too general; the passage describes metamorphosis only in terms of the frog. Option (5) is not the subject of the paragraph.

4. (Analysis) **(3) Different animal structures are suited to different types of environments.** For example, in showing the change from tadpole to frog, the author shows gill slits for breathing underwater and legs for moving about on land. The author takes for granted the usefulness of these structures in their environment. Options (1) and (2) are incorrect, since many plants and animals do not undergo metamorphosis. Option (4) may or may not be true, but it does not have anything to do with the process of metamorphosis. Option (5) is not true.

5. (Comprehension) **(2) the process by which an immature form changes into a different adult form** Option (1) is incorrect because reproduction is a different function from growth. Option (3) is incorrect because it is too general. Option (4) is incorrect because it describes what happens to the mature adult form after metamorphosis. Option (5) refers to respiration, not to metamorphosis.

6. (Comprehension) **(2) lungs** Lungs are structures that are essential to supporting life on land, because without them the animal could not breathe. The other structures are less vital for land-dwelling animals.

7. (Comprehension) **(3) tundra** The black areas farthest north are shown by the map key to be tundra. Options (1), (2), (4), and (5) are not as far north as the tundra, option (3).

8. (Analysis) **(3) cacti** Of the plants listed, cacti are the only ones that can live in a desert biome, which suggests they would not do well in a tropical rain forest, where it is very wet.

9. (Evaluation) **(5) Water in animals' breath is returned to the air.** When animals exhale, they release water vapor into the atmosphere. Option (1) is not true because animals also release water as they breathe. Option (2) is not true; invisible water vapor is found throughout the atmosphere. Option (3) is not true; plants take in water from their roots. Option (4) is not true; the oceans contain most of the water on Earth.

10. (Analysis) **(3) Evaporation from the ocean returns water to the atmosphere.** This statement provides evidence by giving an example of water changing form during the water cycle. Options (1) and (2) just restate the conclusion. Options (4) and (5) are true but do not offer the evidence that water is used again and again.

11. (Analysis) **(2) Sun** The diagram shows that the Sun's heat causes water to evaporate from various places. Options (1), (3), and (4) are incorrect because they are forms of water. Option (5) is also incorrect; the soil serves to store water.

12. (Analysis) **(2) Water conservation is not useful since the Earth will never run out of water.** Although it is true that water is recycled on

Earth, water conservation can be useful because water may run low in particular places. The other options are all facts stated in the passage.

13. (Comprehension) **(2) The Human Brain** The passage focuses on the structure of the human brain. It does not describe the entire nervous system, options (1) and (5). It describes more than just the cerebrum, option (3). It describes the human brain, not brains in general, option (4).

14. (Analysis) **(2) vision** Since the perception of vision is located in the occipital lobe, a blow to the back of the head is likely to affect it. Options (1) and (4) are located at the side. Option (3) is controlled by the medulla at the base of the brain. Option (5) is controlled in the upper part of the cerebrum.

15. (Analysis) **(5) loss of balance** Since the cerebellum coordinates information from the eyes, inner ears, and muscles to maintain balance, a loss of balance is likely to indicate damage to the cerebellum. Option (1), loss of leg movement, is a specific movement partially controlled by the cerebrum. Options (2), (3), and (4) are likely to be caused by damage to the cerebrum.

16. (Evaluation) **(4) Motor activity is controlled by the cerebrum and cerebellum.** Option (1) is partially true, but the cerebrum also needs to be functioning for motor activity to be normal. Option (2) is not true. Option (3) is not likely to be true because the right hemisphere would be able to control at least some speech. Option (5) is contradicted by the passage.

17. (Comprehension) **(3) Vestigial structures offer support for the theory of evolution.** Options (1) and (4) are too specific to summarize the passage. Option (2) is not true. Option (5) is incorrect because, by definition, vestigial structures are poorly developed.

18. (Analysis) **(1) Vestigial structures are the remains of organs that were well-developed in common ancestors.** This is a theory that attempts to explain the existence of vestigial structures. It has not been proved. (Note that the passage says, "Scientists believe . . . .") Options (2), (3), (4), and (5) are all facts as presented by the passage and diagram.

19. (Evaluation) **(1) The appendix of humans is a poorly developed organ.** This is a statement of fact. Option (2) is not likely to be true; it is possible, however, that the ancestors of pythons walked. Option (3) is not true since vestigial organs do not serve an important purpose. Option (4) is not true since the two skeletons are different. Option (5) is not true. Vestigial structures are evidence to support the theory of evolution, but they do not prove that evolution is fact rather than theory.

20. (Analysis) **(2) Some animals have a well-developed structure, and others have a corresponding vestigial structure.** The presence of similar structures in these animals lends support to the idea that they had a common ancestor. Options (1), (3), and (5) are true, but they do not indicate descent from a common ancestor, merely that these organisms have changed over time. Option (4) is not true; the ancestors of porpoises probably had legs.

21. (Comprehension) **(4) can give blood to anyone** A donor is someone who donates, or gives; thus a universal donor is someone who can give blood to anyone. If you look at the first chart, you can see that type O blood can be given to all other types.

22. (Analysis) **(3) They could have a transfusion of type A, B, or O blood.** According to the chart, people with type AB blood can receive any of the blood types; thus a shortage of type AB blood would not affect them. They could still receive types A, B, or O.

23. (Comprehension) **(1) Blood Type Frequency in Selected Population Groups** The chart provides information on the percent of the population that has a particular blood type in various population groups. Option (2) is too general. Options (3) and (4) have nothing to do with the information in the second chart. Option (5) is too specific, since other groups are represented.

24. (Evaluation) **(3) Each individual inherits one gene for blood type from each of his or her parents.** The passage states that a person's blood type depends on two genes inherited from his or her parents. It follows from this that one gene must come from each parent. Option (1) is incorrect because there are more genotypes than blood types. Option (2) is incorrect because there is no AB gene. Options (4) and (5) are not true.

25. (Comprehension) **(5) a variety of crop that is not affected by particular chemical plant killers** This is explained in the first paragraph of the passage. Option (1) is incorrect because <u>herbicide-resistant</u> means that it is not affected by herbicides; therefore herbicides can be applied. Options (2), (3), and (4) describe crops that are resistant to other types of agricultural problems.

26. (Analysis) **(3) Weeds are a problem in large farming areas.** The passage gives both pro-herbicide and anti-herbicide points of view. Both sides agree that weeds must be controlled; they differ as to how. Option (1) is incorrect because the author presents the environmentalists' view that herbicides are not safe. Option (2) is not correct because the passage indicates some practices that do not require the use of herbicides to grow crops. Options (4) and (5) are not true.

27. (Analysis) **(4) Improved cultivation techniques are preferable to herbicides in controlling weeds.** This statement reflects the opinions of some environmental groups. Not everyone shares this point of view. Options (1), (2), (3), and (5) are statements of fact presented in the passage.

28. (Analysis) **(3) result in herbicides remaining on the crops after harvest** Herbicides are chemicals that remain in the environment until they break down; therefore they also remain on the plants. Options (1), (2), and (4) are incorrect because the crops are herbicide-resistant and would not be affected by the chemical. Option (5) is incorrect because herbicides kill plants, not insects.

Performance Analysis
Life Science Review

Use the chart below to identify your strengths and weaknesses in each thinking skill area in the Life Science unit.

Circle the number of each item that you answered correctly on the Life Science Review.

Thinking Skill Area	Life Science	Lessons for Review
Comprehension	1, **3**, 5, **6**, **7**, **13**, 17, **21**, **23**, 25	1, 2, 4
Analysis	2, 4, **8**, 10, **11**, 12, **14**, **15**, 18, 20, **22**, 26, 27, 28	3, 5, 6, 8
Evaluation	**9**, 16, **19**, **24**	7

Boldfaced numbers indicate items based on charts, graphs, illustrations, and diagrams.

If you answered 25 or more of the 28 items correctly, congratulations! You are ready to go on to the next section.

If you answered 24 or fewer items correctly, determine which skill areas are most difficult for you. Then go back and review the Life Science lessons for those areas.

It took the Colorado River 6 million years to create the Grand Canyon.

◆ **era**
*the largest division
of geologic time*

◆ **atmosphere**
*the layers of gases
that surround the
Earth*

Earth science is the branch of science that studies the Earth and
the space around the Earth. Knowledge of Earth science can be
useful in many ways. It helps us understand and use the environment in which we live.

In Lesson 9 you will learn about the structure of the Earth. You will
also learn how the Earth has changed over time. Major changes
have signaled the beginning or end of an **era**, which is the largest
division of geologic time.

In Lesson 10 you will learn about the air that surrounds the Earth.
The gases that surround the Earth make up the **atmosphere**. The
atmosphere, which consists of four main layers, extends from the
surface of the Earth into outer space.

☞ *See Also: GED Exercise Book Science, pages 15–21*

♦ water cycle
continuous movement of water from the Earth's surface to the air, then back to the surface again

♦ air mass
large body of air that has the same temperature and moisture throughout

♦ nonrenewable resource
a resource that cannot be replaced once it is used

♦ erosion
the gradual wearing away and moving of rock materials

♦ mass wasting
the downhill movement of rocks and soil caused by gravity

In this lesson you will also learn about the water that covers approximately 70 percent of the Earth's surface. The Earth's water supply is constantly being renewed through a series of steps called the **water cycle**. You will come to understand how moisture from the Earth forms clouds and how this moisture eventually falls back to Earth in the form of rain, snow, sleet, or hail.

Air and water create changing weather patterns. You will discover how our weather is caused by the movement of **air masses**, which are large areas of air that have the same temperature and humidity throughout.

In Lesson 11 you will learn about the Earth's resources. You will discover that one of the world's most important resources is petroleum. Petroleum provides fuel for energy. It also provides the raw materials needed for many useful chemicals. The fact that consumers use so much petroleum, however, has presented a problem. Petroleum is a **nonrenewable resource**—that is, it cannot be replaced once it is used. Many scientists fear that the world may run out of petroleum before other resources can be found to take its place.

In Lesson 12 you will learn how the Earth is constantly changing. Some of these changes, such as earthquakes and volcanoes, are dramatic and dangerous. Others, such as **erosion** and **mass wasting**, can occur so slowly that many years pass before we see their effects.

LESSON 9 Comprehension Skill: Identifying Implications

When you identify an implication, you figure out something that is likely to be true based on what you have read.

An **implication** is a fact that is not stated directly by the author. Rather, it is suggested by the author's words. For example, if you read, "The oceans cover about 70 percent of the Earth's surface," you would understand what the author said about oceans. But you would also be able to **infer**, or figure out, another fact: since the Earth's surface is either water or land, about 30 percent of the Earth's surface must be land.

You can increase your ability to identify implications by following three suggestions as you read.

1. **Think about consequences.** A consequence is an effect or result. If something happens, what is likely to happen next? If certain conditions exist, what effect does this have? If a tornado occurs, for example, damage is the likely result.
2. **Use common sense.** By using your common sense, you can make reasonable assumptions based on the information given. For example, if the sky is clear, you can assume that it will not rain.
3. **Look for generalizations.** If a statement is true in general, it must be true for specific items as well. For example, if ocean water is salty (general statement), then water from the Pacific Ocean must be salty, too (specific instance).

Practicing Comprehension

Read the following paragraph.

A galaxy is a group of millions or billions of stars. The galaxy in which our Sun is located is called the Milky Way. The Milky Way is a disk-shaped galaxy with two or more spiral arms that extend from the center. Our Sun and the solar system are located in one of the spiral arms. It takes the Sun about 225 million years to make one turn around the center of the galaxy. Measurements of the movements of different stars in the galaxy show that they move with respect to one another while turning around the center of the galaxy. Thousands of years ago, people saw the stars in slightly different patterns than how we see stars today.

Questions 1 and 2 refer to the previous paragraph. Circle the best answer for each question.

1. A star located farther from the center of the galaxy than the Sun is likely to

 (1) take longer than 225 million years to make one turn around the center of the galaxy
 (2) take 225 million years to make one turn around the center of the galaxy
 (3) take fewer than 225 million years to make one turn around the galaxy
 (4) turn around the center of the galaxy first in one direction and then in the other
 (5) remain in the same place

2. In 50,000 B.C., the group of stars called the Big Dipper looked like an arrow. Today, the group looks like a cup with a long handle. What can be said about the appearance of the Big Dipper in another 50,000 years?

 (1) It will look like an arrow.
 (2) It will look like a cup with a long handle.
 (3) Its appearance will not change.
 (4) Its appearance will have changed slightly.
 (5) It will have disappeared.

Questions 3 and 4 refer to the following chart. Circle the best answer for each question.

Planet	Distance from Sun in Astronomical Units*
Mercury	0.39
Venus	0.72
Earth	1.0
Mars	1.5
Jupiter	5.2
Saturn	9.2
Uranus	19.2
Neptune	30.0
Pluto	39.4

*One astronomical unit is the distance from the Earth to the Sun.

3. A planet that is more than one astronomical unit from the Sun is

 (1) closer to the Sun than the Earth is
 (2) the same distance from the Sun as the Earth is
 (3) farther from the Sun than the Earth is
 (4) likely to receive more solar energy than the Earth does
 (5) a planet with no moon

4. It takes the Sun's light 8 minutes to reach the Earth. How long does it take the Sun's light to reach Neptune?

 (1) 15 seconds
 (2) 8 minutes
 (3) 24 minutes
 (4) 240 minutes
 (5) 30 hours

To check your answers, turn to page 142.

Topic 9: The Planet Earth

Read the following passage and look at the diagram.

For many years, scientists have been gathering information about the Earth's interior. Intense heat and high pressure make human exploration of this region impossible. Thus, most of what is known about the interior structure of the Earth has been learned by studying the movement of seismic waves, or vibrations, produced by earthquakes. From this indirect evidence, scientists have concluded that the Earth is made up of four different layers: the crust, the mantle, the outer core, and the inner core.

1. The **crust** is the part of the Earth familiar to us, for it includes the surface of the Earth. This layer is made up of many kinds of rock. The crust is about 8 kilometers thick under the oceans and about 32 kilometers thick under the continents.
2. The **mantle** lies beneath the crust. The mantle is composed of rock that contains mainly oxygen, iron, and silicon. The temperatures in this region range from 870°C to 2,200°C and cause some of the solid rock to flow like a liquid.
3. The **outer core** is below the mantle. The outer core is composed of molten iron and nickel. Temperatures in the outer core range from 2,200°C to 5,000°C.
4. The **inner core** is at the center of the Earth. This region, which has a temperature of about 5,000°C, is solid iron and nickel. Although iron and nickel usually melt at this temperature, great pressure in the inner core pushes the particles together so tightly that they remain solid.

EARTH'S LAYERS

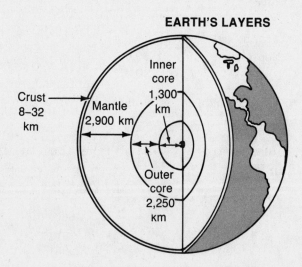

Exercises

Questions 1 to 6 refer to the previous passage and diagram. Circle the best answer for each question.

1. What is the most likely thickness of the Earth's crust under Africa?

   (1) 8 kilometers
   (2) 32 kilometers
   (3) 1,300 kilometers
   (4) 2,250 kilometers
   (5) 2,900 kilometers

2. What led scientists to conclude that the Earth has four different layers?

   (1) direct evidence from human observation
   (2) rock samples from all four layers
   (3) unchanging seismic wave patterns
   (4) changing behavior of seismic waves as they reach each layer of the Earth's interior
   (5) temperature readings from the Earth's interior

3. Which layer of the Earth's surface has been explored directly by humans?

   (1) crust
   (2) mantle
   (3) outer core
   (4) inner core
   (5) all four layers

4. Like a magnet, the Earth has a magnetic field. What is the most likely cause of the Earth's magnetism?

   (1) high temperatures in the core
   (2) oxygen in the Earth's crust
   (3) iron in the inner and outer cores
   (4) high pressure on the core
   (5) continents

5. What are seismic waves?

   (1) vibrations produced by earthquakes
   (2) light waves from the Sun that penetrate the Earth's surface
   (3) movements of liquid rock in the mantle
   (4) deep ocean waves
   (5) volcanic eruptions

6. Which of the following does the author of the passage take for granted?

   (1) It is reasonable to draw scientific conclusions from indirect evidence.
   (2) Indirect evidence is superior to human observation.
   (3) Someday humans will explore the inner core.
   (4) The Earth is made of five different layers.
   (5) Each of the Earth's layers is thicker than the one above it.

To check your answers, turn to page 142.

Reviewing Lesson 9

Read the following passage.

A **fossil** is the evidence or remains of a living thing. Most fossils form when plants or animals die and are buried in layers of crumbled rock particles called **sediments**, which later harden.

The chances of an organism leaving a fossil are actually small. The soft parts of a dead organism usually decay or are eaten before a fossil can form. Fossils that do form are often incomplete. The plants and animals most likely to be preserved as fossils are ones that lived in or near water. There, sand and mud provide quick burial for these organisms.

Some fossils show only the mark or evidence of a living thing. Called **trace fossils**, these include footprints, tracks, and burrows. Much of what is known about dinosaurs has come from footprints found in rock.

Fossils can show changes in the Earth's surface and climate. For example, fossils of coral found in Antarctica show that the climate of this region was once much warmer.

Fossils can also help date layers of rock. If a particular type of organism lived on the Earth for only a brief period of time, the rock containing its fossil must be from approximately the same time period.

Questions 1 and 2 refer to the passage. Circle the best answer for each question.

1. In a mountainous region of Canada, fossils of fish are found. Which is the most likely explanation?

   (1) The region was once much colder.
   (2) The region was once much warmer.
   (3) The region was once under water.
   (4) Fish that lived in mountain streams formed fossils.
   (5) Dead fish were taken up the mountain.

2. Which of the following would be most similar to a trace fossil?

   (1) human footprints made in concrete before it hardened
   (2) a fly trapped in concrete before it hardened
   (3) tire tracks in the sand
   (4) footprints of a deer in the snow
   (5) tire tracks made in concrete before it hardened

To check your answers, turn to pages 142–143.

Read the following information.

Scientists have developed a geologic time line to record the history of the Earth. Geologic time is often described in terms of four eras.

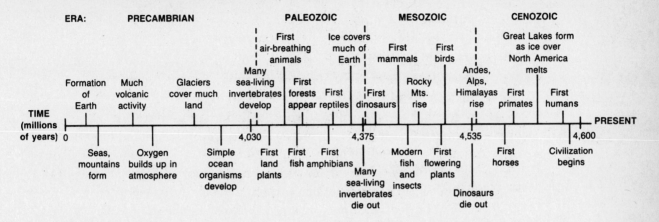

Questions 3 to 6 refer to the information. Circle the best answer for each question.

3. According to the time line, where did the first living things originate?

(1) in the Precambrian era
(2) in the atmosphere
(3) in the ocean
(4) in the Paleozoic era
(5) in a swamp

4. What might have been eaten by the first fish?

(1) flowering plants
(2) sea-living invertebrates
(3) insects
(4) amphibians
(5) land plants

5. Which of the following statements is most accurate?

(1) Human civilization has been in existence for over half the history of the Earth.
(2) Dinosaurs were the chief form of life during the Paleozoic era.
(3) The Great Lakes are younger than the Rocky Mountains.
(4) Dinosaurs died out 5 million years ago.
(5) The first life forms appeared on land.

6. For approximately how many years did dinosaurs inhabit the Earth?

(1) 160 years
(2) 4,375 years
(3) 160 million years
(4) 160 billion years
(5) 4,535 million years

To check your answers, turn to page 143.

GED Mini-Test

Directions: Choose the <u>best answer</u> to each item.

<u>Items 1 to 4</u> refer to the following passage.

The Earth's crust contains three main types of rock: sedimentary, igneous, and metamorphic. Sedimentary rocks are formed when particles of sand, mud, pieces of rock, or remains of dead organisms harden. Igneous rocks are formed when hot liquid rock, called magma, hardens into crystals. Metamorphic rocks are formed when rocks are subjected to extreme heat and pressure. Sedimentary rocks are formed near the surface of the Earth, usually in water. Igneous and metamorphic rocks are formed in the lower part of the Earth's crust or in the upper part of the mantle.

Magma can cool and form crystals above or below the surface of the Earth. Large mineral crystals form when magma cools slowly. Heat is lost very slowly beneath the Earth's surface. Magma can break through the Earth's crust in a volcanic eruption. It is then called lava. As lava comes in contact with the air, cooling can be very rapid. This sudden cooling leads to the formation of barely visible crystals.

1. According to the passage, which of the following could explain why igneous and metamorphic rocks are formed deep beneath the Earth's surface?

   (1) Temperatures in the Earth's interior are high.
   (2) Pressure in the Earth's interior is low.
   (3) The rocks are protected from wind and weather.
   (4) Chemical reactions can occur beneath the Earth's crust.
   (5) Magma hardens more quickly beneath the Earth's crust.

2. Which best explains the presence of large mineral crystals?

   (1) the force of a volcano
   (2) the slow cooling of molten rock in the Earth's crust
   (3) the effect of air on the magma
   (4) the rapid cooling of lava
   (5) the pressure on magma

3. An igneous rock reaches the Earth's surface and is broken down by weathering and erosion. The pieces of rock would then be likely to

   (1) melt and become magma
   (2) become sediments and harden into sedimentary rock
   (3) be pushed together to form a metamorphic rock
   (4) form another igneous rock
   (5) be carried into the ocean and form metamorphic rock

4. Which is most accurate?

   (1) Igneous rock is always formed from metamorphic rock.
   (2) Sedimentary rock is always formed from organic material.
   (3) Sedimentary rock is always formed from particles of rock.
   (4) Metamorphic rock is always formed from igneous, sedimentary, or other metamorphic rock.
   (5) Molten rock is always found under the surface of the Earth.

Items 5 to 8 refer to the following passage.

In order to be called a mineral, a substance must have five basic properties, or characteristics.

1. It must be found naturally in the Earth.
2. It must be a solid.
3. It must never have been alive.
4. It must be made of particular elements.
5. Its particles must be arranged in a definite pattern called a crystal.

Minerals are often found mixed in with other rocks. Rock deposits that contain valuable minerals are called ores. Removing the mineral from the ore involves mining the ore and then smelting it. During smelting, the ore is heated in such a way that the mineral is separated from the other substances.

---

5. Which of the following could <u>not</u> be a mineral?

  (1) a crystalline salt
  (2) a substance formed in the mantle
  (3) a substance consisting of 18% oxygen and 82% silicon
  (4) a solid that is shiny
  (5) a rock formed from plant remains

6. Which of the following would be involved in removing iron from its ore?

  (1) adding impurities
  (2) polishing
  (3) freezing
  (4) high temperatures
  (5) strong odors

7. What is smelting?

  (1) mining the ore
  (2) identifying minerals
  (3) crystalizing noncrystalline substances
  (4) heating ore to separate a mineral
  (5) changing the properties of a mineral

8. Someone looking at a piece of granite noticed that it had been formed from tiny particles of three different colors. Another sample of granite was formed from particles of four different colors. What could lead someone to conclude that granite is not a mineral?
  A. Granite is manmade.
  B. Granite is not made of particular elements.
  C. Granite is a mineral.

  (1) A only
  (2) B only
  (3) C only
  (4) A and B
  (5) B and C

To check your answers, turn to page 143.

Answers and Explanations

1. (Comprehension) **(1) take
   longer than 225 million
   years to make one turn
   around the center of the
   galaxy** Since our Sun takes
   225 million years to make one
   turn around the galaxy, it
   follows that a star farther from
   the center would take longer,
   since it has a greater distance
   to travel. Therefore options (2)
   and (3) are incorrect. Option
   (4) is incorrect because
   heavenly bodies do not change
   direction while they are
   revolving around another
   body. Option (5) is incorrect
   because all stars are moving.

2. (Comprehension) **(4) Its
   appearance will have
   changed slightly.** Since the
   paragraph indicates that the
   patterns of stars have changed
   slightly over thousands of
   years, it is reasonable to
   assume that the patterns will
   continue to change slightly.
   Option (1) is what the Big
   Dipper used to look like.
   Options (2) and (3) are its
   present appearance, which is
   not likely to remain unchanged. Option (5) is more
   than a slight change, and
   therefore is not likely to occur.

3. (Comprehension) **(3) farther
   from the Sun than the
   Earth is** Since the distance
   from the Earth to the Sun is 1
   astronomical unit, it follows
   that a planet more than one
   astronomical unit from the
   Sun will be farther from the
   Sun than the Earth is. Options
   (1) and (2) are therefore
   incorrect. Option (4) is incorrect because a planet farther
   from the Sun than the Earth is
   likely to receive less, not more,

solar energy than the Earth.
Option (5) has nothing to do
with distance from the Sun.

4. (Analysis) **(4) 240 minutes** It
   takes the Sun's light 8 minutes
   to travel 1 astronomical unit, so
   it would take 30 x 8, or 240
   minutes, to travel 30 astronomical units — the distance
   from the Sun to Neptune.

Exercises (page 137)

1. (Comprehension) **(2) 32 kilometers** The passage states that
   average thickness of the crust is
   32 kilometers under the
   continents. Africa is a continent, so its crust is about 32
   kilometers thick.

2. (Comprehension) **(4) changing
   behavior of seismic waves as
   they reach each layer of the
   Earth's interior** The passage
   indicates that the study of
   seismic waves led to the
   conclusion that the Earth has
   four layers. If the seismic waves
   did not change, option (3),
   scientists would have concluded
   that the Earth's interior was
   basically the same all the way
   to the core. Options (1), (2), and
   (5) are incorrect; they are
   evidence that requires direct
   observation, which is impossible in Earth's interior.

3. (Comprehension) **(1) crust**
   Since the crust includes the
   Earth's surface, it follows that
   it has been explored by people.
   The other parts of the Earth's
   interior are out of reach
   because of high temperature
   and pressure. Option (5), all
   four layers, is incorrect because
   only option (1), the crust, has
   been explored by humans.

4. (Analysis) **(3) iron in the
   inner and outer cores** Since
   magnets are often made of iron,

it follows that the iron in the
Earth's core would exert
magnetic force. The other
options have nothing to do
with magnetic substances.

5. (Comprehension) **(1) vibrations produced by earthquakes** The other options do
   not define or describe seismic
   waves.

6. (Analysis) **(1) It is reasonable to draw scientific
   conclusions from indirect
   evidence.** The structure of the
   Earth's interior was figured
   out by studying the indirect
   evidence of seismic waves. The
   author does not indicate that
   conclusions about the Earth's
   interior are wrong but rather
   accepts them as reasonable.
   Option (2) is incorrect; direct
   evidence is better than indirect
   evidence. Option (3) is not
   likely. Option (4) is not true.
   Option (5) is not supported by
   the diagram.

Reviewing Lesson 9 (pages 138–139)

1. (Comprehension) **(3) The
   region was once under
   water.** Since fossils are most
   likely to be formed by organisms in or near the water, the
   best option is (3). Options (1)
   and (2) have nothing to do
   with the location of fish.
   Option (4) is remotely possible,
   but since mountain streams
   are usually shallow and rocky,
   it is unlikely that enough soft
   sediment would be available to
   bury the organism quickly.
   Option (5) is unlikely.

2. (Comprehension) **(1) human
   footprints made in concrete
   before it hardened** The best
   option is (1) because it describes evidence of living

activity (human footprints) in a substance that hardens and becomes permanent (concrete). Option (2) is incorrect; a fly is a whole organism, not a trace. Options (3) and (4) are incorrect; sand and snow do not hold a permanent print. Option (5) is incorrect; a tire is not a living thing.

3. (Comprehension) **(3) in the ocean** The first reference to living things mentions the ocean. Options (1) and (4) are incorrect because they refer to time periods, not places. Options (2) and (5) are not indicated by the time line.

4. (Analysis) **(2) sea-living invertebrates** Options (1), (3), and (4) are incorrect because the time line shows that these organisms appeared after the first fish. Option (5) is incorrect; land plants existed before the first fish, but fish would not have been able to leave the water to eat land plants.

5. (Evaluation) **(3) The Great Lakes are younger than the Rocky Mountains.** The time line shows that the Rocky Mountains appeared before the Great Lakes. Option (1) is incorrect; according to the time line, human civilization arose very recently. Option (2) is incorrect; dinosaurs lived during the Mesozoic era. Option (4) is incorrect; the time line shows that dinosaurs died out about 100 million years ago. Option (5) is incorrect; the first life forms appeared in the ocean.

6. (Comprehension) **(3) 160 million years** The time line is measured in millions of years. Dinosaurs appeared slightly more than 4,375 million years after the Earth formed. Subtracting 4,375 from 4,535 gives 160, which on the time line represents 160 million

years. Options (1) and (4) are misreadings of the scale of the time line. Options (2) and (5) are incorrect because they represent time passed since the beginning of the time line.

GED Mini-Test (pages 140–141)

1. (Comprehension) **(1) Temperatures in the Earth's interior are high.** According to the passage, igneous rock is formed from hot liquid rock, and metamorphic rock is formed in the presence of high temperature and pressure. Since both types of rock require high temperatures, which are found in the Earth's interior, the correct option is (1). Option (2) contradicts the information presented about metamorphic rock and is also an untrue statement about the Earth's interior. Options (3) and (4) do not explain the formation of igneous and metamorphic rocks. Option (5) is related only to igneous rock.

2. (Comprehension) **(2) the slow cooling of molten rock in the Earth's crust** Options (1), (3), and (4) do not contribute to the slow cooling of molten rock, or magma, which is necessary for the formation of large crystals. Option (5) is contradicted by the passage.

3. (Analysis) **(2) become sediments and harden into sedimentary rock** Since only sedimentary rock is formed on or near the Earth's surface, only option (2) is correct. Options (1), (3), and (4) could be correct if the rock fragments were buried deep in the Earth. Option (5) is incorrect because it is sedimentary, not metamorphic, rock that is formed in the ocean.

4. (Evaluation) **(4) Metamorphic rock is always formed from igneous, sedimentary, or other metamorphic rock.**

Option (1) is incorrect because igneous rock is formed from magma. Options (2) and (3) are partially correct; sedimentary rock can be formed from both organic and inorganic particles. Option (5) is not true; sometimes molten rock appears on the surface of the Earth as lava.

5. (Comprehension) **(5) a rock formed from plant remains** Since plant remains were once alive, a rock formed from plant remains could not be classified as inorganic. (See statement 3.) Options (1), (2), (3), and (4) could all be minerals since they include the basic properties.

6. (Comprehension) **(4) high temperatures** Since smelting is a process that involves heating, one can assume that high temperatures are involved. Option (1) is incorrect; smelting removes impurities. Option (2) might be involved eventually in preparing iron for use, but not in separating the iron from its ore. Option (3) is incorrect; the process requires heating, not cooling. Option (5) is not necessarily a result of the smelting process, but it may appear correct to some who think that smelting has to do with the sense of smell.

7. (Comprehension) **(4) heating ore to separate a mineral** None of the other options has anything to do with smelting as defined in the passage.

8. (Analysis) **(2) B only** Since granite can be made of three or four types of materials, it follows that it is not made of particular elements. Statement A is not true; granite is a naturally occurring substance. Statement C is also not true; since granite is not made of particular elements, it cannot be called a mineral.

LESSON 10 Application Skill: Other Contexts

When you put your knowledge to use in new situations, you are applying ideas to another context.

Some people have jobs that involve a lot of application skills. For example, the person who tells you the weather report on television is probably a meteorologist. A meteorologist's job is to use general information about climate and weather patterns to predict the weather at a specific place and time. Another job that involves application skill is that of an engineer. An engineer applies the laws of physics or chemistry to build bridges or develop new plastics, for example. What other jobs that involve the skill of application can you think of?

When you study science, you are learning a subject that has many practical applications. The principles discovered by scientists can be applied to most areas of life. When you read about a science topic, think about how this knowledge might be used in other contexts, or situations. You can increase your ability to apply science knowledge to new situations by asking yourself:

♦ What is being described or explained?

♦ What situations might this information relate to?

♦ How would this information be used in those situations?

Read this brief paragraph and see how the information can be applied to other situations.

> The average weather over a long period of time is called climate. One of the most important factors in climate is temperature. In general, temperature is determined by how much sunlight a location receives. The area near the equator receives the greatest amount of sunlight, while areas near the North and South poles receive the least. Average temperature usually decreases as you move from the equator to the poles.

This information about temperature is very general. It states that the average temperature decreases as you move north and south from the equator. However, we can apply this information to specific situations. For example, we can compare two cities and decide which is likely to have the higher average temperature. Mexico City, Mexico is likely to be warmer than Toronto, Canada. New Orleans, Louisiana is warmer on average than Portland, Maine.

Practicing Application

Read the following paragraph.

Rocks and other materials on the Earth's surface can be broken down into small bits. This process is called weathering. Mechanical weathering occurs when water freezes in the cracks of rocks. The freezing water expands, causing the cracks to open wider. Chemical weathering occurs when substances in the atmosphere combine with materials on Earth and slowly change them into something else. Biological weathering occurs when plants or other organisms live on rocks and wear them down.

Questions 1 and 2 refer to the paragraph. Circle the best answer for each question.

1. Which of the following is an example of chemical weathering?

    (1) acid rain wearing down limestone
    (2) potholes in streets after a cold winter
    (3) lichens on a rock
    (4) grass growing in pavement cracks
    (5) pipes freezing and bursting

2. A situation similar to mechanical weathering exists when

    (1) weeds grow between bricks on a patio
    (2) detergent and water remove dirt from clothes
    (3) acid rain falls on forested areas
    (4) a glass jar full of soup placed in the freezer bursts
    (5) moss grows in shady areas under trees

To check your answers, turn to page 152.

Topic 10: Air and Water

Read the following passage and look at the diagram.

The air that surrounds the Earth is called the **atmosphere**. The Earth's atmosphere provides us with a safe environment. It gives us moisture and oxygen, a comfortable temperature, and protection from the Sun's ultraviolet rays.

The atmosphere contains gases necessary for the survival of all living things. Gases in the atmosphere include nitrogen, oxygen, carbon dioxide, water vapor, and argon. There are also small amounts of neon, helium, krypton, and xenon.

The atmosphere is divided into four main layers. The lowest layer is called the **troposphere.** This is the layer in which we live. The troposphere extends to a height of about 10 miles. As you go higher, the air becomes colder and less dense, or "thinner." For example, at an altitude of 3.5 miles, there is only half as much oxygen as there is at the Earth's surface.

Above the troposphere is the **stratosphere**. The stratosphere extends to a height of about 31 miles. A form of oxygen called **ozone** is found in the stratosphere and the mesosphere. Ozone shields the Earth from the Sun's harmful ultraviolet rays.

Above the stratosphere is the **mesosphere**. The mesosphere extends to a height of about 53 miles. The temperature in the mesosphere drops to about -100°C.

The uppermost region of the atmosphere is the **thermosphere**. The thermosphere does not have a well-defined upper limit. It is the hottest layer of the atmosphere, with temperatures as high as 2,000°C.

LAYERS OF EARTH'S ATMOSPHERE

Thermosphere = 50 mi. & up
Mesosphere = 31 to 50 mi.
Stratosphere = 10 to 31 mi.
Troposphere = 0 to 10 mi.
31 mi.
10 mi.
Ozone Layer
Earth's Surface

Exercises

Questions 1 to 7 refer to the previous information. Circle the best answer for each question.

1. Which of the following people would need the most extra protection from the Sun's ultraviolet rays?

   (1) a teacher
   (2) a mountain climber
   (3) an astronaut
   (4) a commercial airline pilot
   (5) a meteorologist

2. Someone at the top of Mt. Everest, which is about 29,000 feet high, would be in the

   (1) troposphere
   (2) stratosphere
   (3) ozone layer
   (4) mesosphere
   (5) thermosphere

3. Which of the following people is likely to experience dropping temperatures and thinner air during the course of a day?

   (1) a lifeguard
   (2) a mountain climber
   (3) an airline attendant
   (4) a landscape gardener
   (5) a farmer

4. What would be the effect of destroying the ozone layer?

   (1) People would have trouble breathing.
   (2) The Sun's ultraviolet rays would be reflected back into space.
   (3) More ultraviolet rays would reach the Earth's surface.
   (4) The atmosphere would extend higher than it does now.
   (5) There would be no effect.

5. Which of the following gases is present in the atmosphere in very small amounts?

   (1) nitrogen
   (2) oxygen
   (3) carbon dioxide
   (4) water vapor
   (5) krypton

6. Which of the following is an unstated assumption of the passage?

   (1) Gravity holds the atmosphere in place around the Earth.
   (2) The higher up you go, the denser the atmosphere becomes.
   (3) Helium does not occur naturally in the atmosphere.
   (4) The uppermost region of the atmosphere is the thermosphere.
   (5) No human being has gone higher than the mesosphere.

7. A runner from Boston, Massachusetts, at sea level, traveled to Denver, Colorado, in the Rocky Mountains. While jogging in Denver, the runner had trouble breathing. Which of the following statements is supported by this experience?

   (1) Physical activity is more difficult in tropical weather.
   (2) The mesosphere extends to about 53 miles above the Earth's surface.
   (3) The higher you go in the troposphere, the less oxygen is available.
   (4) The ozone layer protects the Earth from the Sun's ultraviolet rays.
   (5) The air becomes colder as you go higher in the troposphere.

To check your answers, turn to page 152.

Reviewing Lesson 10

Read the following passage.

Changes in weather are caused by movements of **air masses**. An air mass has the same temperature and humidity throughout. An air mass may cover thousands of square miles.

Air masses are named according to where they form. There are four major types of air masses that affect the United States: maritime tropical, maritime polar, continental tropical, and continental polar. They are called maritime if they come from the sea, and continental if they form over land.

A **maritime tropical** air mass forms over the ocean near the equator. Its air is warm and moist. In the summer it brings hot, humid weather to the United States, but in the winter it may come in contact with a cold air mass and cause rain or snow.

A **maritime polar** air mass forms over the Pacific Ocean in both winter and summer and over the North Atlantic Ocean in summer. It contains cool, moist air. During the summer this air mass brings fog to the western coastal states and cool weather to the eastern states. In the winter this air mass produces heavy snow and very cold temperatures.

A **continental tropical** air mass forms over Mexico during the summer. It brings hot, dry air to the southwestern United States.

A **continental polar** air mass forms over land in northern Canada. It contains cold, dry air. It brings very cold weather to the United States in winter.

Questions 1 to 4 refer to the passage. Circle the best answer for each question.

1. Which of the following statements describes air masses that form over land?

    (1) They contain cold, moist air.
    (2) They contain warm, moist air.
    (3) They contain dry air.
    (4) They contain moist air.
    (5) They contain moist air in winter and dry air in summer.

2. A blizzard in the Northern Pacific states could be caused by which of the following air masses?

    (1) maritime polar
    (2) continental tropical
    (3) continental polar
    (4) maritime polar or continental polar
    (5) continental polar or continental tropical

3. Which of the following is a false statement concerning air masses?

(1) An air mass is called maritime if it originates over water.
(2) An air mass is called continental if it originates over land.
(3) Air masses called tropical originate over tropical seas.
(4) An air mass can cause a change in humidity.
(5) Air masses cover a very large area.

4. A maritime tropical air mass meets a continental polar air mass over the East Coast in January. What is the weather likely to be?

(1) sunny
(2) snowy
(3) hurricane
(4) dry and cold
(5) dry and warm

Questions 5 to 7 refer to the diagram. Circle the best answer for each question.

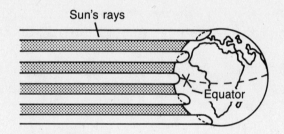

Sun's rays

Equator

5. What change should a person traveling from the middle of the Northern Hemisphere to the equator be prepared for?

(1) shorter days
(2) shorter nights
(3) cooler temperatures
(4) a greater possibility of rainy weather
(5) a greater possibility of sunburn

6. According to the diagram, the Sun's rays

(1) strike all parts of the Earth evenly
(2) are most direct at the North Pole
(3) are most direct at the South Pole
(4) are most direct in the Northern Hemisphere
(5) are most direct at the equator

7. Climates near the equator are warmer than climates near the poles. Based on the diagram, which of the following statements could explain this observation?

(1) Days are longer near the equator than near the poles.
(2) The Sun's rays are more concentrated near the equator than near the poles.
(3) Weather patterns produce more sunny days near the equator than near the poles.
(4) There are fewer oceans near the equator than near the poles.
(5) There is less wind near the equator than near the poles.

To check your answers, turn to pages 152–153.

GED Mini-Test

Directions: Choose the best answer to each item.

Items 1 to 4 refer to the following passage.

The Earth's supply of fresh water is always being renewed by the water cycle. The water cycle consists of three steps.

The first step is evaporation. Heat from the Sun causes large amounts of water on the surface of the Earth to change into water vapor. Much of this water comes from the oceans; the rest comes from soil, plants, and freshwater sources. The water vapor is carried by winds over land and oceans.

The next step in the cycle is condensation. Condensation is the process by which vapor turns back into liquids. This happens when warm air near the surface of the Earth rises and is cooled. Once the air cools, it can no longer hold as much water vapor. The extra water vapor condenses into droplets that form clouds.

The final step is precipitation. When the droplets of water become too heavy to float in air, water falls in the form of rain or snow. Much of this water reaches oceans, rivers, lakes, and soil. Now this water is warmed by the Sun. As some of it evaporates, the water cycle begins all over again.

---

1. Which of the following statements best explains why clouds form?

(1) Water vapor in the air condenses.
(2) Droplets of water in the air become heavy.
(3) Droplets of water in the air evaporate.
(4) Heat from the Sun reacts with gases in the atmosphere.
(5) Warm air above the Earth reacts with cool air on the surface of the Earth.

2. What is the best explanation of why the water cycle is always being renewed?

(1) Clouds release precipitation.
(2) Oceans and rivers carry water.
(3) Evaporation and condensation cause precipitation.
(4) Precipitation causes condensation.
(5) Evaporation causes condensation.

3. Which of the following series of events causes rain or snow?

(1) Water vapor condenses and then evaporates.
(2) Air becomes cold and then becomes warm.
(3) Water evaporates and then condenses.
(4) Water condenses and then is carried by wind.
(5) Air rises and then sinks.

4. Which of the following is most similar to the water cycle?

(1) snow falling and then melting
(2) water running down the bathroom walls after someone takes a hot shower
(3) plants taking in carbon dioxide and releasing oxygen
(4) automobile engines using high-octane gasoline
(5) making ice cubes

Approximately 70 percent of the Earth's surface is covered by oceans. Ocean water makes up 97 percent of all the water on Earth. Ocean water is a mixture of gases and solids dissolved in pure water. The main solid dissolved in ocean water is sodium chloride. Sodium chloride is common table salt.

Scientists use the term salinity to describe the saltiness of ocean water. Salinity is a measure of the amount of salt for each 1,000 kilograms of water. Most ocean water has a salinity of between 33 and 37, meaning that between 33 and 37 grams of salt will be found in 1,000 kilograms of water. The salinity tends to be highest in places where there is a great deal of evaporation or little rainfall.

The Sun is the major source of heat for the ocean. As a result, temperatures are highest at the surface of the ocean. Waves and currents transfer the heat downward to a depth of about 300 meters. Below 300 meters, the temperature of ocean water drops rapidly. At a depth of 1,500 meters, the temperature of the water is just a few degrees above freezing.

5. Many living things cannot use water that contains salt. According to the passage, which of the following statements must be true?

   (1) About 70 percent of the water on Earth can be used to meet the needs of living things.
   (2) A very small amount of the water on Earth can be used to meet the needs of living things.
   (3) Pollution has severely reduced the amount of water that can be used by living things.
   (4) The amount of water that can be used by living things is steadily increasing.
   (5) Lack of rainfall has decreased the amount of water available to living things.

6. About how much water on the Earth is drinkable?

   (1) 3 percent
   (2) 33 percent
   (3) 37 percent
   (4) 70 percent
   (5) 97 percent

7. Which of the following situations is most like the heating of ocean water?

   (1) A frozen dinner placed in a microwave oven is heated evenly throughout.
   (2) People sitting near a fireplace feel warm, while people sitting far from the fireplace feel chilly.
   (3) Water heated on a stove begins to boil when it reaches a certain temperature.
   (4) An electrical wire becomes hot when too much current flows through it.
   (5) A thermos bottle keeps liquids hot for many hours on a cold day.

8. What is the average salinity of ocean water?

   (1) 33–37
   (2) 70
   (3) 97
   (4) 100
   (5) 1,000

To check your answers, turn to page 153.

Answers and Explanations

1. (Application) **(1) acid rain wearing down limestone** When limestone is exposed to acids, a chemical reaction results. Options (2) and (5) are examples of mechanical weathering because they are the result of the freezing and thawing of water. Options (3) and (4) are examples of biological weathering.

2. (Application) **(4) a glass jar full of soup placed in the freezer bursts** Since mechanical weathering is caused when water expands as it freezes, the bursting jar is most similar to the process of mechanical weathering. Option (1) is an example of biological weathering. Options (2), (3), and (5) are not related to weathering.

Exercises (page 147)

1. (Application) **(3) an astronaut** An astronaut travels above the ozone layer and has no natural protection against the Sun's ultraviolet rays. The other options involve people whose occupations do not take them above the ozone layer.

2. (Comprehension) **(1) troposphere** Since the troposphere extends about ten miles up (52,800 feet), the top of Mt. Everest is still well below its upper limit. The other layers of the atmosphere are above the troposphere and are therefore incorrect.

3. (Application) **(2) a mountain climber** As a mountain climber goes higher, the air becomes colder and thinner. Options (1), (4), and (5) are occupations that do not require

changes in altitude. Option (3) is incorrect because airline attendants work in a man-made environment, the airplane, in which temperature and air pressure are controlled.

4. (Analysis) **(3) More ultraviolet rays would reach the Earth's surface.** Since the ozone layer prevents most ultraviolet rays from reaching the surface of the Earth, its destruction would remove a natural barrier. Destruction of the ozone layer would not affect the breathing of oxygen in the troposphere, option (1). Option (2) describes the function of the ozone layer. Option (4) is incorrect because the ozone layer is part of the stratosphere and mesosphere. Option (5) is incorrect because there would be an effect.

5. (Comprehension) **(5) krypton** The first four options are gases that make up a large portion of the atmosphere. The passage indicates that krypton is one of the gases present in small amounts.

6. (Analysis) **(1) Gravity holds the atmosphere in place around the Earth.** The author does not state but assumes that the atmosphere remains around the Earth because of the force of gravity. Options (2) and (3) are not true. Option (4) is true, but it is not an unstated assumption; it is stated in the passage and shown in the diagram. Option (5) is also not true; human beings have been to the Moon, which is beyond the mesosphere.

7. (Analysis) **(3) The higher you go in the troposphere, the**

less oxygen is available. The runner had difficulty breathing in Denver because it is much higher than Boston, and there is less oxygen. Option (1) makes no sense because neither place has tropical weather. Options (2) and (4) are true but not related to the change in altitude between Boston and Denver. Option (5) is true, but the change in temperature was not the cause of the runner's breathing problem.

**Reviewing Lesson 10
(pages 148–149)**

1. (Comprehension) **(3) They contain dry air.** Options (1), (2), (4), and (5) are incorrect because continental air masses contain dry air only.

2. (Comprehension) **(1) maritime polar** Options (2), (3), (4), and (5) are incorrect because continental air masses do not cause precipitation. The only other type of air mass that might have caused the blizzard would be a maritime tropical in winter.

3. (Evaluation) **(3) Air masses called tropical originate over tropical seas.** Since a continental tropical air mass originates over land, all tropical air masses do not originate over tropical seas. Options (1) and (2) explain why the terms maritime and continental are used. Options (4) and (5) are accurate descriptions of air masses.

4. (Application) **(2) snowy** When two such air masses meet in winter, precipitation is the likely result. Snow is the only form of precipitation among the options.

5. (Application) **(5) a greater possibility of sunburn** Options (1) and (2) are incorrect because length of day and night depend on the season. Option (3) is incorrect because direct sunlight at the equator produces warmer temperatures. Option (4) cannot be predicted based on the diagram.

6. (Evaluation) **(5) are most direct at the equator** If you look at the diagram, you can see that, except at the equator, the Sun's rays are spread over a wider area because of the curve in the Earth's surface. Therefore, options (1), (2), (3), and (4) are incorrect.

7. (Analysis) **(2) The Sun's rays are more concentrated near the equator than near the poles.** Options (1), (4), and (5) are incorrect because they are not shown in the diagram. Option (3) might be true, but there is no information about weather patterns in the diagram.

GED Mini-Test (pages 150–151)

1. (Comprehension) **(1) Water vapor in the air condenses.** Option (2) is incorrect because heavy droplets in clouds result in rain. Option (3) is incorrect because water on the surface of the Earth evaporates to form water vapor in the air. Option (4) is not related to cloud formation. Option (5) is incorrect because warm air, not cool air, is near the surface of the Earth.

2. (Evaluation) **(3) Evaporation and condensation cause precipitation.** This is the only option that includes all three steps of the water cycle. Options (1), (2), and (5) are only parts of this cycle. Option (4) has the steps in the wrong order.

3. (Analysis) **(3) Water evaporates and then condenses.** Options (1) and (2) are incorrect because the steps are in the wrong order. Option (4) is incorrect because it is water vapor that is carried by the wind. Option (5) is incorrect because once warm air rises, the water vapor condenses.

4. (Application) **(2) water running down the bathroom walls after someone takes a hot shower** Steam condenses on the cooler surfaces of the walls, and drops of water start running down the walls. Option (1) is a small part of the water cycle. Options (3) and (4) have nothing to do with water. Option (5) is turning water from a liquid to a solid, which is not part of the water cycle.

5. (Evaluation) **(2) A very small amount of the water on Earth can be used to meet the needs of living things.** Option (1) is incorrect because 70 percent is a statistic taken out of context—something to beware of on multiple-choice tests. Options (3), (4), and (5) could be true, but they are not related to the information provided.

6. (Comprehension) **(1) 3 percent** Since undrinkable ocean water accounts for 97 percent of the water on Earth, it follows that only about 3 percent of the Earth's water is drinkable.

7. (Application) **(2) People sitting near a fireplace feel warm, while people sitting far from the fireplace feel chilly.** This is correct because heat from the fire is similar to heat from the Sun. The most warmth will be provided to things closest to the source of heat. Option (1) is incorrect because ocean water is not

heated evenly throughout. Option (3) is incorrect because ocean water is not heated from the bottom up. Option (4) is incorrect because it deals with electrical energy, not water. Option (5) is incorrect because the passage does not suggest that ocean water is like an insulator.

8. (Comprehension) **(1) 33–37** The passage indicates that the average salinity is between 33 and 37 grams per thousand kilograms.

LESSON 11 Analysis Skill: Fact or Opinion

When you recognize that some statements are beliefs and cannot be proved, you are distinguishing opinions from facts.

In Lesson 6, you learned how scientists find out if their opinions, or hypotheses, are facts. They do this by performing experiments whose results may support their ideas. Many of the facts we take for granted were once scientists' opinions.

When you read about science, you may come across opinions. You may read that scientists have different points of view about the meaning of certain facts. Sometimes the opinion is that of a scientist or another expert. Sometimes it is held by the author.

Read this brief passage and look for opinions.

> In 1979 an accident occurred at the Three Mile Island nuclear power plant. The accident caused heat to build up in the reactor vessel — the steel structure that contains the nuclear reaction and prevents the release of radiation. Fuel rods and metal pipes in the reactor vessel melted and then hardened. The steel reactor vessel showed little damage. This is important since the vessel has no backup system. If the reactor vessel fails, a nuclear reaction can melt its way through the vessel into the Earth. The fact that the heavy metal reactor vessel remained intact suggests that the chance of a catastrophic nuclear power plant accident might be smaller than once thought. However, some in the nuclear power industry believe that another accident similar to that at Three Mile Island would be the last. They say that pressure from the general public would force the closing of remaining nuclear power plants.

In this paragraph, the description of the accident at Three Mile Island is factual. For example, the author states that the fuel rods melted and hardened. This is a fact that can be proved. There are also several opinions in this passage.

◆ The author's opinion that the chance of a catastrophic nuclear power plant accident might be smaller than once thought

◆ The opinion of people in the nuclear power industry that another accident would force the closing of all nuclear plants

◆ The opinion of the general public that nuclear power plants are dangerous

The opinions in the passage may or may not be true, but the facts can be proved to be true.

When you read, watch for words like <u>believe</u>, <u>seem</u>, <u>feel</u>, <u>think</u>, <u>agree</u>, <u>may</u>, and <u>could</u>. These words often signal an opinion rather than a fact.

Practicing Analysis

Read the following paragraph.

> Changes in the Sun's activity, such as sunspots, may affect the weather. For example, years of low sunspot activity seem to correlate with periods of drought in North America, as shown by rainfall records and tree rings. It is doubtful, however, that there is a direct cause and effect relationship between sunspots and droughts. Perhaps sunspots cause some change in the atmosphere that affects the weather.

Questions 1 and 2 refer to the paragraph. Circle the best answer for each question.

1. Which of the following is a true, provable statement?

    (1) Sunspots cause changes in the weather.
    (2) There are variations in sunspot activity.
    (3) High sunspot activity is associated with periods of drought.
    (4) Low sunspot activity is the direct cause of drought in North America.
    (5) Sunspots affect only North America.

2. Which of the following statements is an opinion?

    (1) The Sun is the main source of energy for the Earth.
    (2) The Sun's activity varies over time.
    (3) Periods of high sunspot activity are followed by periods of low sunspot activity.
    (4) Sunspots affect the weather in North America.
    (5) The relationship between sunspots and weather is not completely understood.

To check your answers, turn to page 162.

Topic 11: The Earth's Resources

Read the following passage.

The things we need to live, such as water, food, and energy, are called **resources**. Sometimes resources are in short supply. In the 1970s, shipments of **petroleum**, or oil, to our country had been cut. All petroleum products—including gasoline—were in short supply. According to energy planners and Earth scientists, an oil shortage could occur again, and if so, it could be permanent.

Oil is a **nonrenewable resource**, one that cannot be replaced once it is used up. Earth scientists disagree about how much oil is left in the Earth. Some say that, given our present rate of consumption, the United States will be out of oil by the year 2060. Others feel that there is enough oil to last another 300 years. All agree, however, that at some point the supply of oil will be gone.

Energy planners are trying to find ways to prevent a major oil shortage. Most believe that the best solution would be to look for a **renewable resource** that could be used in place of oil. A renewable resource is one that can be replaced. The most abundant renewable resource on Earth is energy from the Sun. Scientists have developed solar cells that can convert sunlight into electricity. They have also developed ways to heat homes with solar energy. At present, solar energy cannot be produced at a low enough cost or on a large enough scale to meet much of our energy needs.

Another way to postpone an oil shortage is to use less oil. During the oil shortage of the 1970s, the United States passed laws to reduce the speed limit on highways and to lower the thermostats in public buildings. Citizens were shown ways to conserve energy, but lifelong habits are hard to break. Many people continue their old, energy-wasting habits.

Exercises

Questions 1 to 8 refer to the passage. Circle the best answer for each question.

1. Which of the following statements is an opinion of some scientists?

   (1) The Earth's supply of oil is limited.
   (2) The reduction of speed limits conserves energy.
   (3) There is enough oil to last 300 years.
   (4) Oil became more plentiful in the 1980s than it had been in the 1970s.
   (5) Solar energy cannot currently meet our demands for energy.

2. Which of the following statements is an opinion of the author?

   (1) Americans are wasteful of energy.
   (2) An oil shortage could easily happen again.
   (3) By the year 2060, the U.S. will be out of oil.
   (4) Solar energy is a renewable resource.
   (5) Information about energy conservation is available in the U.S.

3. Which of the following actions would help conserve a <u>nonrenewable</u> resource?

    (1) driving to work instead of taking public transportation
    (2) turning off the water while brushing your teeth
    (3) driving a small, fuel-efficient car
    (4) raising shades or blinds to let in sunlight in warm weather
    (5) converting a heating system from oil to natural gas

4. Which of the following is implied by the passage?

    (1) Vehicles will eventually be powered by something other than a petroleum product.
    (2) Oil will be the best choice for home heating in 500 years.
    (3) Oil supplies will last at least 1,000 years.
    (4) If people would conserve oil, it would last forever.
    (5) Solar energy will never be inexpensive enough to use widely.

5. What would be the effect on the supply of oil if people cut down their use of it?

    (1) The supply would be used up sooner.
    (2) The supply would be used more slowly and would eventually run out.
    (3) The supply would be used more slowly and would last forever.
    (4) The supply of oil would be used up the same as it would if people did not cut down their use.
    (5) New sources of oil would be found.

6. Which of the following is a renewable resource?

    (1) oil
    (2) natural gas
    (3) coal
    (4) gasoline
    (5) solar energy

7. Which of the following statements is the most accurate?

    (1) Solar energy is an inexpensive alternative to petroleum.
    (2) A gasoline shortage is not likely to occur again.
    (3) The supply of oil will last until the year 2060.
    (4) Conserving oil will postpone the time when supplies run out.
    (5) New sources of petroleum will be found to meet our energy needs.

8. A geologist studying the layers of the Earth would be most interested in which of the following resources?

    (1) sunlight
    (2) ocean water
    (3) food
    (4) nuclear energy
    (5) petroleum

To check your answers, turn to page 162.

Reviewing Lesson 11

Read the following article.

The air around the Earth is always moving. Moving air is called **wind**. Throughout history, people have used energy from the wind to move ships, turn mill wheels, and pump water.

Windmills began to appear on American farms around 1860. The energy from these windmills was used to pump water out of the ground for crops and farm animals. In 1890 a windmill was invented that could generate, or make, electricity. Wind generators became very popular with American farmers.

One problem with wind generators was that they did not always work. On calm days they could not work. On stormy days they were often knocked down or blown apart. Because of this, most wind generators were set aside in the 1940s, when electricity became available from electric power plants.

The need to find energy sources other than fossil fuels such as coal, oil, and natural gas has sparked new interest in wind energy. In recent years, new materials and designs have been used to make several tough, efficient wind generators. These machines can adjust to changing wind conditions and withstand storms.

Energy planners do not expect wind energy ever to meet all our needs. But they do think that the use of wind energy can help to conserve fossil fuels and reduce air pollution.

Questions 1 to 4 refer to the article. Circle the best answer for each question.

1. According to energy planners, what is the importance of wind energy?

    (1) Wind energy is a nonrenewable resource.
    (2) Wind energy can replace fossil fuels as the leading energy resource.
    (3) Wind energy was abandoned in the 1940s.
    (4) Wind energy can reduce the need for fossil fuels.
    (5) Wind energy can be used on farms.

2. Which of the following aspects of wind energy is not discussed in the article?

    (1) how wind generators were used in the 1800s
    (2) how modern wind generators are used
    (3) the reliability of wind generators
    (4) the ability of wind energy to meet energy needs
    (5) the use of wind energy on American farms

3. According to the article, modern wind generators differ from earlier wind generators in that they are

(1) larger
(2) less expensive
(3) able to generate electricity
(4) more reliable
(5) used mainly on farms

4. According to the article, what caused a new interest in wind energy?

(1) a desire to return to a simpler lifestyle
(2) an interest in American history
(3) a need to find more energy sources
(4) a desire to better understand wind
(5) an increased need for electricity on American farms

Read the following passage.

One source of air pollution is the burning of coal and oil by factories and power plants. Coal and oil contain sulfur. When they are burned, sulfur is released into the atmosphere. The sulfur reacts with oxygen to form sulfur oxides. Some of these sulfur oxides combine with water in the air to form acids. Eventually these acids fall to the Earth as acid rain.

Acid rain is nearly as acidic as pure lemon juice. When acid rain falls into a lake, much of the plant and animal life dies. Today many lakes look clear and blue because the water is nearly empty of wildlife. Animals living near such a lake may die of starvation.

What can be done about acid rain? Factories and power plants must stop burning sulfur-containing fuels. But fuels with a low sulfur content are often expensive and hard to find.

Questions 5 to 8 refer to the passage. Circle the best answer for each question.

5. The main cause of acid rain is

(1) a lack of oxygen in the air
(2) a lack of moisture in the air
(3) high temperatures
(4) the burning of certain fuels
(5) oil released into the atmosphere

6. Which of the following is a fact?

(1) Factories will stop burning coal and oil.
(2) Research could lead to the discovery of a sulfur-free fuel.
(3) Acid rain destroys all living things on contact.
(4) Acid rain is formed by sulfur oxides combining with water in the air.
(5) Dangers of acid rain are exaggerated.

To check your answers, turn to pages 162–163.

7. According to the passage, some lakes look clear and blue because they have

(1) too much oxygen
(2) many plants and animals
(3) had above-average rainfall
(4) been polluted by oil
(5) been polluted by acid rain

8. Which of the following statements can be inferred from the passage?

(1) Sulfur is poisonous to living things.
(2) Plants and animals live well in water that is clear and blue.
(3) High levels of acidity can be harmful to living things.
(4) Plants and animals die when acid is removed from rainwater.
(5) Plants and animals live best when sulfur is present in the air.

GED Mini-Test

Directions: Choose the best answer to each item.

Items 1 to 4 refer to the following passage.

Most of the energy we use every day comes from fossil fuels. Fossil fuels were formed in the Earth millions of years ago when the remains of dead plants and animals were buried beneath layers of mud. The chief fossil fuels are coal, petroleum, and natural gas.

Coal is solid fossil fuel. It was the first fossil fuel to be used by industry. In the United States today, coal is burned mainly to produce electric power.

Petroleum, or oil, is liquid fossil fuel. Petroleum is presently the leading fuel in the United States and other industrialized nations. Raw petroleum taken from the Earth is called crude oil. The refining of crude oil produces gasoline, fuel oil for home heating, kerosene, plastics, synthetic fibers, and cosmetics.

Natural gas is a fossil fuel in a gaseous state. It is less dense than liquid petroleum, so it is usually found on top of oil deposits. Natural gas is a "clean-burning" fuel compared to coal and oil, because it produces less air pollution. Some homes use natural gas for cooking and heating.

1. A shortage of petroleum would probably not affect the availability of which of the following products?

    (1) 100% cotton shirt
    (2) lipstick
    (3) dress made of synthetic linen
    (4) plastic kitchen utensils
    (5) high-octane gasoline

2. Based on the passage, which of the following best describes the relationship between fossil fuels and industry?

    (1) Fossil fuels have always been of little importance to industry.
    (2) Fossil fuels were once important but are now of little importance.
    (3) Fossil fuels were once of little importance but are now important.
    (4) Fossil fuels have been and are still of great importance to industry.
    (5) Fossil fuels are of little importance now but will be of great importance in the future.

3. According to the passage, air quality would probably improve if

    (1) oil were to replace coal in the production of electric power
    (2) natural gas were to replace oil as the leading fuel
    (3) coal were to replace oil in home heating
    (4) coal were to replace natural gas in home heating
    (5) coal were to replace oil in industry

4. Which of the following phrases describes the origin of fossil fuel?

    (1) produced from chemicals in laboratories
    (2) produced from fossilized materials in laboratories
    (3) formed in the Earth from the remains of plants and animals
    (4) formed in the Earth from pieces of rock
    (5) produced from crude oil in oil refineries

During an ice age, weather becomes colder and the ice caps and glaciers spread south and north from the poles. The Earth has gone through many ice ages, the last of which ended about 10,000 years ago.

No one is certain what causes ice ages, but there are many opinions on the subject. Scientists have pointed to changes in energy output from the Sun, the amount of volcanic dust in the atmosphere, the Earth's magnetic field, the amount of carbon dioxide in the atmosphere, and circulation of deep ocean currents as possible causes of ice ages. A recent hypothesis links ice ages with slight but regular changes in the Earth's orbit around the Sun.

5. Which of the following is <u>not</u> an opinion?

(1) Changes in the Earth's magnetic field result in ice ages.
(2) The circulation of deep ocean currents causes ice ages.
(3) The last ice age ended about 10,000 years ago.
(4) During an ice age, the amount of volcanic dust in the atmosphere is greater than usual.
(5) Solar energy output varies, leading to ice ages.

6. What would happen to the polar ice cap in Antarctica during an ice age?

(1) It would become thinner and smaller.
(2) It would spread northward.
(3) It would spread southward.
(4) It would be warmed by increased energy from the Sun.
(5) It would stay the same.

7. If, during an ice age, glaciers hundreds of feet thick reached the middle of the United States, what would be the likely result?

(1) People would move to the Southern Hemisphere.
(2) People would move toward the equator.
(3) People would remain in the northern part of the United States.
(4) The human race would die out.
(5) The days would become shorter.

8. Which of the following would be a good title for this passage?

(1) Ice Ages Past and Present
(2) The Spread of the Polar Ice Caps
(3) What Causes an Ice Age?
(4) The Effect of Ice Ages on Human Civilization
(5) A New Ice Age Is on the Way

To check your answers, turn to page 163.

Answers and Explanations

1. (Analysis) **(2) There are variations in sunspot activity.** This is the only statement that has so far been proved. Option (1) is an opinion that has not been proved. Option (3) is contradicted by the passage. Option (4) is also an opinion. Option (5) does not make sense; sunspots, if they affect North America, would affect the rest of the world also. In any case, option (5) is not a provable fact.

2. (Analysis) **(4) Sunspots affect the weather in North America.** This is an opinion, not a provable statement. Options (1), (2), (3), and (5) are all facts.

Exercises (pages 156–157)

1. (Analysis) **(3) There is enough oil to last 300 years.** Option (3) is one of two estimates given in the passage. Neither opinion can be proved. Option (1) is a fact because oil is known to be a nonrenewable resource. Options (2) and (4) are incorrect because they are facts. Both energy use and the availability of oil can be measured. Option (5) is also a fact.

2. (Analysis) **(1) Americans are wasteful of energy.** Options (2) and (3) are incorrect because they reflect the opinions of scientists and energy planners, not the author. Options (4) and (5) are incorrect because they are facts.

3. (Application) **(3) driving a small, fuel-efficient car** This would save gasoline, a nonrenewable resource. Option (1) would use more fuel, not less.

Option (2) involves conserving water, a renewable resource. Option (4) does not make sense; in warm weather no fuel would be used to heat a home, and added sunlight would just make the building warmer. If air conditioners were in use, this would waste energy. Option (5) involves switching from one nonrenewable resource to another and thus would not help conserve a nonrenewable resource.

4. (Comprehension) **(1) Vehicles will eventually be powered by something other than a petroleum product.** The passage suggests that renewable resources will come to replace nonrenewable resources such as petroleum. Options (2) and (3) are contradicted by the passage. Option (4) makes no sense. Since the supply of oil is limited, it will eventually run out, even if people conserve it. Option (5) is not implied by the passage, which states that at present solar energy is too expensive for widespread use.

5. (Analysis) **(2) The supply would be used more slowly and would eventually run out.** Cutting down on the use of oil will stretch the supply for a longer time, but since the supply is limited, it will eventually run out. Option (1) is the opposite of what happens when a resource is conserved. Option (3) is only partially correct. The supply would be used more slowly, but it would not last forever since it is limited. Option (4) is incorrect because the supply would last longer. Option (5) may be true, but it does not relate to the effect of conservation on supply.

6. (Comprehension) **(5) solar energy** Of the forms of energy mentioned, only solar energy is renewable because there is a constant supply of it from the Sun. The other resources are all found on Earth in limited supply.

7. (Evaluation) **(4) Conserving oil will postpone the time when supplies run out.** Option (1) is not true at the present time. Option (2) is incorrect because another shortage could happen, since the supply of gasoline is limited. Option (3) is an opinion about the size of the oil supply. Option (5) may be true, but whatever new supplies are found, they will not be enough to meet long-term energy needs, since the new supplies will be limited also.

8. (Application) **(5) petroleum** Since petroleum is found in deposits on Earth, it would be of most interest to a geologist. Geologists are not primarily concerned with the study of sunlight, ocean water, food, or nuclear energy.

Reviewing Lesson 11 (pages 158–159)

1. (Analysis) **(4) Wind energy can reduce the need for fossil fuels.** This is the most important aspect of wind energy. Option (1) is not true; wind energy is a renewable resource. Option (2) is contradicted by the article. Option (3) is true, but it is not important. Option (5) may or may not be true, but it is not the most important aspect of wind energy.

2. (Comprehension) **(2) how modern wind generators are used** Modern wind

generators are described, but their uses are not specified. The other options are incorrect because these topics are all covered in the article.

3. (Comprehension) **(4) more reliable** The article states that modern wind generators are tough and can withstand storms. Options (1) and (2) are incorrect because the article makes no mention of size or cost. Options (3) and (5) are incorrect because both are characteristics of earlier wind generators.

4. (Comprehension) **(3) a need to find more energy sources** This option is clearly stated in the article. The other options may be related to an interest in wind energy, but they are not mentioned in the article.

5. (Comprehension) **(4) the burning of certain fuels** The passage states that sulfur is released when coal and oil are burned. Options (1) and (2) are incorrect because oxygen and moisture help form acid rain. Options (3) and (5) are incorrect because they are not related to the formation of acid rain.

6. (Analysis) **(4) Acid rain is formed by sulfur oxides combining with water in the air.** This statement can be proved. Options (1), (2), and (5) are incorrect because they are opinions. Option (3) is not a fact.

7. (Comprehension) **(5) been polluted by acid rain** This is stated in the second paragraph. Option (2) is incorrect because it is a lack of plants and animals that makes the lake look blue. Options (1), (3), and (4) are incorrect because they have nothing to do with information given in the passage.

8. (Comprehension) **(3) High levels of acidity can be harmful to living things.** This is correct because both acidic rain and acidic soil damage or kill plants and animals. Options (1) and (5) are incorrect because no mention is made of the effect of pure sulfur on living things. Option (2) is incorrect because the only mention of clear, blue water in this passage relates to a lake polluted by acid rain. Option (4) is incorrect because it is the presence of acid in rainwater that can kill plants and animals.

GED Mini-Test (pages 160–161)

1. (Application) **(1) 100% cotton shirt** This is correct because cotton is a natural fiber. Option (2) is incorrect because cosmetics are produced from petroleum. Option (3) is incorrect because synthetic fibers are produced from petroleum. Options (4) and (5) are incorrect because plastic and gasoline are produced from petroleum.

2. (Evaluation) **(4) Fossil fuels have been and are still of great importance to industry.** The other options are incorrect because the passage clearly refers to the use of fossil fuels in industry both in the past and the present.

3. (Comprehension) **(2) natural gas were to replace oil as the leading fuel** The last paragraph of the passage describes natural gas as a "clean-burning" fuel compared to coal and oil. All other options are incorrect because coal and oil produce more air pollution than natural gas.

4. (Comprehension) **(3) formed in the Earth from the remains of plants and animals** Options (1), (2), and (5) are incorrect because fossil fuels were formed naturally in the Earth. Option (4) is incorrect because fossil fuels did not form from pieces of rock.

5. (Analysis) **(3) The last ice age ended about 10,000 years ago.** This statement is one of several facts about ice ages presented in the first paragraph. The other options are all opinions of scientists about the possible causes of ice ages.

6. (Analysis) **(2) It would spread northward.** Since Antarctica is located over the South Pole, the spread of its ice cap would be northward into the Southern Hemisphere. Options (1) and (4) are the opposite of what happens during an ice age. Option (3) is incorrect because Antarctica is located over the South Pole. Option (5) is incorrect because option (2) is correct.

7. (Application) **(2) People would move toward the equator.** The area around the equator would remain the warmest area on the Earth. Option (1) is incorrect because the glaciers would be spreading from the South Pole northward into the Southern Hemisphere. Option (3) does not make sense since the glaciers would cover that area. Option (4) is not likely, since people can move toward the equator and many would survive. Option (5) is not related to ice ages, but to seasons.

8. (Comprehension) **(3) What Causes an Ice Age?** This is a good title because the passage discusses various ideas that have been proposed to explain ice ages, although none of them has been proved. Option (1) is incorrect because we are not in an ice age at present. Option (2) refers to a detail from the passage. Options (4) and (5) are not discussed in the passage.

LESSON 12 Analysis Skill: Cause and Effect

You can identify a cause by asking yourself, "Why did this happen?" You can identify an effect by asking yourself, "What happened?"

As you learned in Lesson 3, a cause is something that makes something else happen. An effect is what happens as a result of a cause. When there is a long-term shortage of rain, for example, grass and other vegetation turn brown and dry. The lack of rain is a cause, and the drying out of plants is an effect.

Often an effect goes on to become the cause of another effect. For example, the brown and dry vegetation — the effect of a drought — can become the cause of starvation for animals that depend on vegetation for food. Or an earthquake — the result of movement in the Earth's interior — goes on to become the cause of further effects such as collapsed buildings and roads.

As you read, it is important to be aware of cause and effect relationships. Watch for words and phrases such as because, since, thus, effect, affect, result, occurs when, was caused by, led to, and due to. These often signal cause-and-effect relationships.

Practicing Analysis

Read the following passage.

Beaches are always changing. Sand is moved around by wind, waves, and currents. Beaches can get smaller during the winter when high storm waves carry sand from the beach out to sea. In the summer, calmer waves and currents move the sand back to the beach, making it larger.

When offshore currents slow down, they deposit sand. In this way, certain shore features are formed. Sand bars are long piles of sand that are mostly under water. If they are connected to the shore, the sand bars are called spits. Very long sand bars that are above the water line are called barrier bars or barrier islands. Barrier islands run parallel to the coast and are separated from it by a narrow body of water. During storms, barrier islands protect the coast on the mainland from the full force of wind and waves. Atlantic City, New Jersey and Miami Beach, Florida are cities built on barrier islands.

Questions 1 and 2 refer to the previous passage. Circle the best answer for each question.

1. What would be the effect on a beach if there were several mild winters without severe storms?

(1) The beach would get smaller.
(2) The beach would get smaller and then larger.
(3) The beach would get larger.
(4) The beach would get larger and then smaller.
(5) The beach would remain unchanged.

2. Why are people who live on a barrier island evacuated when a hurricane is approaching?

(1) Roads are likely to collapse.
(2) Flooding and wind damage are likely to occur.
(3) Telephone service may be temporarily interrupted.
(4) Food will be in short supply.
(5) Drinking water will be scarce.

Questions 3 and 4 refer to the diagrams. Circle the best answer for each question.

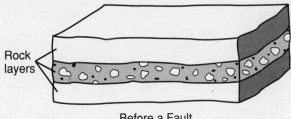

Before a Fault

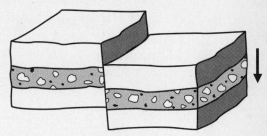

Normal Fault

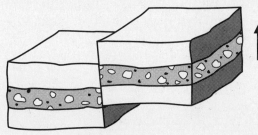

Reverse Fault

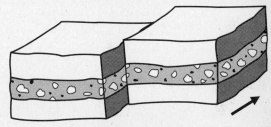

Horizontal Fault

3. What is the result of both normal and reverse faults?

(1) Rocks move sideways.
(2) A steep face of rock is exposed.
(3) A river valley is formed.
(4) Layers of rock are lined up.
(5) Rock layers curve.

4. What is a likely result of normal and reverse faulting over a wide area and long time period?

(1) formation of lakes
(2) formation of rivers
(3) formation of mountains
(4) formation of level plains
(5) formation of deserts

To check your answers, turn to page 172.

Topic 12: The Changing Earth

Read the following passage.

Changes in the Earth's surface can be caused when the force of gravity pulls rocks and soil down mountain slopes. This is called **mass wasting**. As sediments come to rest at the bottom of a slope, they form what is called a **talus slope**.

Mass wasting can occur rapidly or slowly. One type of rapid mass wasting is a landslide. During a **landslide**, huge quantities of soil, small stones, and large rocks tumble down a mountain. A landslide can be caused by an earthquake, volcanic eruption, or heavy rain—any natural event that weakens the supporting rock.

Another type of rapid mass wasting is a mudflow. A **mudflow** is usually caused by a heavy rain. As rain mixes with soil to form mud, gravity begins to pull the mud downhill. As the mud moves, it picks up more soil and becomes thicker. It is difficult to imagine the power of a mudflow—it can move just about anything in its path, including a whole house!

Slow mass wasting can occur in an earthflow. Usually caused by a heavy rain, an **earthflow** consists of the slow, downhill movement of soil and plant life.

Soil creep is the slowest form of mass wasting. Soil particles that have been disturbed by heavy rain, alternate periods of freezing and thawing, or animal activity are pulled downhill by gravity. Soil creep is so slow that its effects often go unnoticed for quite some time. Evidence of a long period of soil creep includes tilted trees and telephone poles along the side of a steep slope.

Exercises

Questions 1 to 8 refer to the passage. Circle the best answer for each question.

1. Which is <u>not</u> a cause of mass wasting?

   (1) heavy rain
   (2) animal activity
   (3) earthquakes
   (4) wind
   (5) volcanic eruptions

2. A row of tilted trees along a steep slope could be a result of

   (1) a talus slope
   (2) soil creep
   (3) a mudflow
   (4) a landslide
   (5) rapid mass wasting

3. What is the basic cause of each case of mass wasting?

   (1) heavy rain
   (2) volcanic eruption
   (3) animal activity
   (4) earthflows
   (5) gravity

4. Which area is likely to experience the least rapid mass wasting?

   (1) The Great Plains
   (2) The Rocky Mountains
   (3) The Allegheny Mountains
   (4) The Grand Canyon
   (5) Mt. St. Helens

5. Which of the following conditions is necessary in order for a mudflow to occur?

   (1) an earthquake
   (2) a volcanic eruption
   (3) heavy rain
   (4) animal activity
   (5) all of the above

6. Which of the following is likely to cause the most severe property damage?

   (1) earthflow
   (2) soil creep
   (3) slow mass wasting
   (4) landslide
   (5) talus slope

7. Which of the following statements is always true?

   (1) Mass wasting is the result of earthquakes and volcanic eruptions.
   (2) Rapid mass wasting occurs only in coastal areas.
   (3) The general movement in mass wasting is from high ground to low ground.
   (4) Heavy rains are present in all cases of mass wasting.
   (5) Mountain ranges can be formed by mass wasting.

8. What is a talus slope?

   (1) a mountainside without topsoil
   (2) a mountain valley
   (3) sediments piled at the bottom of a slope
   (4) a rocky outcrop on a slope
   (5) tilted landscape features

To check your answers, turn to page 172.

Reviewing Lesson 12

Read the following article.

An **earthquake** is the shaking and trembling that results from the sudden movement of rock in the Earth's crust. When a strong earthquake hits a populated area, there can be tremendous destruction and hundreds of deaths.

The most common cause of earthquakes is faulting. A **fault** is a break in the Earth's crust. During faulting, rocks along the fault begin to move. They break and slide past each other. Parts of the Earth's crust may be pushed together or pulled apart. During this process, energy is released.

The point beneath the Earth's surface where the rocks break and move is called the **focus** of the earthquake. Directly above the focus, on the Earth's surface, is the **epicenter**. The most violent shaking occurs at the epicenter.

When rocks in the Earth's crust break, vibrations travel out in all directions from the focus. These vibrations are known as **seismic waves**. There are three main types of seismic waves.

The seismic waves that travel the fastest are called **primary waves**. These waves can travel through solids, liquids, and gases. Primary waves are push-pull waves. They cause pieces of rock to move back and forth in the same direction as the wave is moving.

The seismic waves that travel the next fastest are **secondary waves**. Secondary waves can travel through solids but not through liquids or gases. Rock pieces disturbed by secondary waves move from side to side at right angles to the direction the wave is traveling.

The slowest seismic waves are **surface waves**. Surface waves travel from the focus directly up to the epicenter. Surface waves cause the Earth to bend and twist, sometimes causing whole buildings to be swallowed up by the ground.

The more energy an earthquake releases, the stronger and more destructive it is. The strength of an earthquake is measured on a scale called the Richter scale. The **Richter scale** measures how much energy an earthquake releases by assigning the earthquake a number from one to ten. Any number above six on the Richter scale indicates a very strong earthquake.

Questions 1 and 2 refer to the previous article. Circle the best answer for each question.

1. According to the article, the strength of an earthquake is directly related to which of the following?

    (1) length of the fault
    (2) amount of energy released
    (3) speed of the seismic waves
    (4) distance from the focus to the epicenter
    (5) amount of rock broken

2. Which of the following statements about seismic waves can be inferred from the article?

    (1) Surface waves are the most destructive.
    (2) Primary waves are the most destructive.
    (3) All seismic waves except surface waves are destructive.
    (4) All seismic waves are equally destructive.
    (5) Seismic waves are not destructive.

Read the following passage.

    Erosion is the moving and wearing away of rock by natural causes. A dramatic cause of erosion is a glacier.

    A **glacier** is a large mass of moving ice. Most glaciers form in mountains where snow builds up faster than it can melt. As snow falls upon snow, year after year, the snow changes into ice. When the ice becomes heavy enough, the pull of gravity causes it to move slowly down the mountain. As the glacier moves, it picks up blocks of rock. As the rocks become frozen into the bottom of the glacier, they carve away more rock. Some of this rock is left behind at the edges of the glacier.

    Sometimes, after flowing down a mountain, a glacier will enter a river valley that is narrower than the glacier. As the glacier squeezes through the valley, it erodes both the floor and sides of the valley. As a result, the shape of the valley changes from a V-shaped valley to a broad U-shaped valley.

Questions 3 and 4 refer to the passage. Circle the best answer for each question.

3. Which of the following is <u>not</u> an effect of a moving glacier?

    (1) the formation of tall mountain peaks
    (2) the movement of rocks
    (3) deposits of rock
    (4) the change in shape of a valley
    (5) the carving away of rock

4. What would be the effect of several unusually long, hot summers on mountain glaciers?

    (1) Glaciers would move down the mountain more rapidly.
    (2) Glaciers would reach farther down the mountain.
    (3) The edges of the glaciers would melt, making them smaller.
    (4) The glaciers would get thicker.
    (5) The glaciers would carve river valleys into U-shapes.

To check your answers, turn to page 173.

GED Mini-Test

Directions: Choose the best answer to each item.

Items 1 to 4 refer to the following article.

Deep within the Earth, the rock is a hot liquid called magma. In some places, magma works its way toward the Earth's surface by melting solid rock or by moving through cracks in rock. When magma reaches the Earth's surface, it is called lava. The place where lava reaches the Earth's surface is called a volcano.

In every volcano there is at least one opening called a vent. It is through the vent that the volcano erupts. You may think of a volcanic eruption as being a violent, dramatic event. Sometimes it is, but a volcanic eruption can also be a quiet flow of lava.

Volcanoes can be classified according to the type of eruptions that form them. There are three main types of volcanoes.

Cinder cone volcanoes are formed from explosive eruptions. Explosive eruptions are caused when lava in vents hardens into rocks. Steam and new lava build up under the rocks, causing pressure. Eventually the pressure becomes great enough to cause a violent explosion. The volcano is formed out of cinders and other rock particles that are blown into the air. A cinder cone volcano has a narrow base and steep sides.

Shield volcanoes result from quiet lava flows. The lava from shield volcanoes flows over a large area because it is thin and runny. A shield volcano, which forms after several quiet eruptions, is a gently sloping, dome-shaped mountain.

Composite volcanoes build up when explosive and quiet eruptions alternate. First an explosive eruption spews rock and cinders onto the Earth. Then a quiet eruption occurs, producing a lava flow that covers the cinders and rock particles. After many alternating eruptions, a cone-shaped mountain is formed. Two famous composite volcanoes are Mt. Vesuvius in Italy and Mt. St. Helens in the United States.

Volcanoes are like "windows" that let us see inside the Earth. By analyzing the chemical composition of lava, scientists are able to determine the chemical composition of the magma from which the lava formed.

1. Which of the following is a necessary condition for the formation of a volcano?

   (1) a violent eruption of lava
   (2) the melting of solid rock
   (3) the hardening of lava in vents
   (4) the release of cinders into the air
   (5) the movement of magma to the Earth's surface

2. According to the article, the classification of volcanoes is based on which of the following?

   (1) their shape
   (2) their size
   (3) the number of vents they have
   (4) the events that cause them to form
   (5) the effect they have on the environment

3. The formation of a gently sloping, dome-shaped volcano is the result of which of the following events?

   (1) several explosive eruptions
   (2) alternating explosive and quiet eruptions
   (3) several quiet eruptions
   (4) the release of cinders into the air
   (5) the formation of steam in vents

4. Which of the following statements about lava can be inferred from the article?

   (1) Lava produces steam.
   (2) Lava can be thin and runny.
   (3) Lava always flows over a wide area.
   (4) Lava produces magma.
   (5) Lava consists primarily of cinders.

<u>Items 5 and 6</u> refer to the following information.

     The dotted lines on this map show the world zones of earthquake and volcano activity. Most major earthquakes and volcanic eruptions occur along these lines.

EARTHQUAKE AND VOLCANIC ACTIVITY

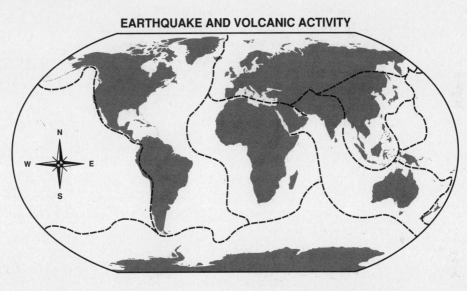

5. Which continent is generally free of major earthquakes?

   (1) North America
   (2) South America
   (3) Asia
   (4) Europe
   (5) Africa

6. Which of the following areas would provide the best opportunity to study volcanic activity?

   (1) northeastern Canada
   (2) western South America
   (3) western Africa
   (4) Australia
   (5) eastern South America

To check your answers, turn to page 173.

Answers and Explanations

Practicing Analysis (page 165)

1. (Analysis) **(3) The beach would get larger**. Since the action of winter storms, with high winds and waves, causes beaches to lose sand, it follows that several mild winters would slow this action. The deposit of sand during the summer would build up the beach. This sand would not be entirely removed during the next winter, so the beach would grow over the years.

2. (Analysis) **(2) Flooding and wind damage are likely to occur.** Barrier islands take the main force of a storm approaching from the sea. Since they are low and made of sand, they are very likely to be damaged by wind and flooding. Although the other options may or may not occur during a hurricane, they are not the major reasons for evacuating people from barrier islands when a bad storm is coming.

3. (Analysis) **(2) A steep face of rock is exposed.** Both normal and reverse faults move rock in such a way that a steep face of rock is created where previously the land had been flat. Option (1) is incorrect because it describes the movement of rock in horizontal faults. Option (3) is not indicated in any of the diagrams. Option (4) describes the land before faulting. Option (5) does not describe what happens during a fault. During a fault, rocks break and slide; they do not curve.

4. (Analysis) **(3) formation of mountains** Normal and reverse faults result in the creation of steep rock faces. If this was widespread, the likely result would be a mountainous landscape. Options (1), (2), and (4) do not describe the results of faulting. Option (5) is incorrect because it refers to the climate of an area, not the landscape features.

Exercises (pages 166–167)

1. (Analysis) **(4) wind** Option (1) is incorrect because it can cause all types of mass wasting. Option (2) is incorrect because it can cause soil creep. Options (3) and (5) are incorrect because they can cause landslides.

2. (Analysis) **(2) soil creep** Option (1) is incorrect because it is the <u>result</u> of mass wasting. Options (3), (4), and (5) are all incorrect because they <u>cause</u> rapid, not gradual, changes in the land.

3. (Analysis) **(5) gravity** No matter what other causes may be involved in mass wasting, gravity is always the basic cause. The other options are also causes, but they are not involved in each case of mass wasting, as gravity is.

4. (Application) **(1) The Great Plains** The Great Plains are least likely to suffer the effects of mass wasting because the landscape is generally flat. Options (2), (3), and (5) are mountainous areas where mass wasting occurs. Option (4), the Grand Canyon, would also be likely to have mass wasting because of the steep slopes.

5. (Comprehension) **(3) heavy rain** The third paragraph states that a mudflow is usually caused by heavy rain. The other options may or may not occur, but they are not necessary to cause a mudflow.

6. (Analysis) **(4) landslide** Since it is a form of rapid mass wasting, a landslide is likely to cause the most damage to things in its path. Options (1), (2), and (3) are slow changes that cause less severe results. Option (5) is a landscape feature that results from mass wasting.

7. (Evaluation) **(3) The general movement in mass wasting is from high ground to low ground**. Since the force of gravity pulls things from high places to lower places, all forms of mass wasting follow this pattern. Option (1) is only partially correct; there are other causes of mass wasting. Option (2) is incorrect because rapid mass wasting takes place in mountainous areas. Option (4) is incorrect because heavy rains are not necessary for some forms of mass wasting. Option (5) is incorrect because mass wasting tends to lower mountains, not form them.

8. (Comprehension) **(3) sediments piled at the bottom of a slope** According to the passage, a talus slope forms as a result of sediments piling up due to mass wasting. The other options do not describe a talus slope as defined in the passage.

1. (Comprehension) **(2) amount of energy released** The last paragraph of the passage states that the more energy an earthquake releases, the stronger it is. Options (1), (3), and (4) are incorrect because, while they may be important in other ways, they do not determine the strength of an earthquake. Option (5) may be related to the strength of an earthquake, but this is not mentioned in the article.

2. (Comprehension) **(1) Surface waves are the most destructive.** This is correct since it is the surface waves that cause the upheavals in the ground that are so destructive. Options (2) and (3) are incorrect because they contradict the correct answer. Options (4) and (5) are incorrect because the article describes the destruction that results from surface waves, but not from any other waves.

3. (Analysis) **(1) the formation of tall mountain peaks** A glacier wears away rocks, so it would not make a mountain peak taller. All other options are effects of a moving glacier.

4. (Analysis) **(3) The edges of the glaciers would melt, making them smaller.** Since glaciers form from the excess snow and ice that does not melt from season to season, it follows that unusually warm weather would cause more melting. This would make the glacier smaller. Options (1), (2), and (4) describe what happens when glaciers build up. Option (5) occurs whether or not the glacier is getting smaller.

GED Mini-Test (pages 170–171)

1. (Comprehension) **(5) the movement of magma to the Earth's surface** The key word in this question is <u>necessary</u>. All the other options can be associated with the formation of certain volcanoes, but only option (5) is a necessary condition for the formation of <u>all</u> volcanoes.

2. (Comprehension) **(4) the events that cause them to form** The article states that volcanoes are classified according to the type of eruptions that form them. The other options are incorrect because, although these factors may differ in each type of volcano, they are not the basis for classification.

3. (Comprehension) **(3) several quiet eruptions** Options (1) and (2) are incorrect because they form cone-shaped volcanoes. Options (4) and (5) are incorrect because they are associated with the formation of cone-shaped volcanoes.

4. (Comprehension) **(2) Lava can be thin and runny.** Options (1) and (5) are incorrect because the article does not link these factors to lava in cause and effect relationships. Option (3) is incorrect because lava does not flow over a wide area in an explosive eruption. Option (4) is incorrect because lava is formed from magma.

5. (Comprehension) **(5) Africa** The map shows that no earthquake zone passes through Africa; therefore it is free of major earthquakes. The other continents all have areas in an earthquake zone.

6. (Application) **(2) western South America** Options (1), (3), (4), and (5) are incorrect because the zones shown on the map do not pass through these areas.

Review: Earth Science

In Lessons 9–12 you have read about the Earth's structure and history. You have learned about the Earth's atmosphere, oceans, and natural resources. You have also studied some of the ways in which the Earth changes, both above and below the surface.

You have also learned some skills for understanding and thinking about what you read. The following exercises make use of the comprehension, application, analysis, and evaluation skills you have been practicing. These exercises also expand on the Earth science topics you have read about so far.

Directions: Choose the best answer for each item.

Items 1 and 2 refer to the following diagram.

The diagram below shows the characteristics of the four main layers of the atmosphere.

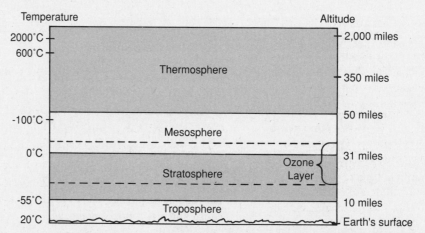

1. According to the diagram, temperature in the Earth's atmosphere

   (1) remains the same as altitude increases
   (2) increases as altitude increases
   (3) decreases as altitude increases
   (4) increases, then decreases as altitude increases
   (5) decreases, then increases, then decreases, then increases as altitude increases

2. The upper thermosphere is called the exosphere, where artificial satellites orbit the Earth. According to the diagram, which statement must be true about these satellites?

   (1) They orbit the Earth at an altitude of 62 to 124 miles.
   (2) They are protected from ultraviolet radiation by the ozone layer.
   (3) They can withstand extremely high temperatures.
   (4) They pass through thin clouds of ice.
   (5) They are used for television transmission.

Scientists studying ice in Antarctica have discovered evidence for the theory that the amount of methane gas in the atmosphere is related to changes in the Earth's climate. The scientists estimated the amount of methane in the atmosphere over a period of 160,000 years by measuring the amounts of different gases in air bubbles trapped in the polar ice. There was less methane in the atmosphere during ice ages and much more methane in the atmosphere during periods of global warming.

Other studies have shown that the amount of methane in the atmosphere nearly doubled — from 350 parts per billion during the ice ages to 650 parts per billion during warmer periods. Today, the methane level is 1,700 parts per billion. In recent years, some scientists have warned that methane can contribute to global warming because it traps heat in the atmosphere. Although the studies do not prove that increased methane levels cause the Earth's temperatures to rise, they do suggest that this is possible.

3. Which of the following is an opinion?

(1) In ice ages, the amount of methane was about 350 parts per billion.
(2) There is more methane in the atmosphere during periods of warming than during ice ages.
(3) There is a relationship between the level of methane in the atmosphere and the Earth's climate.
(4) Increased levels of methane cause global warming.
(5) Methane levels in the atmosphere change over long periods of time.

4. Which of the following supports the conclusion that Earth may be entering a warming period?

(1) There is methane in air bubbles trapped in polar ice.
(2) The methane level of the atmosphere changes over a long period of time.
(3) The methane level is much higher now than during previous warm periods.
(4) The methane level should drop sharply over the next 100 years.
(5) Methane levels over Antarctica are high.

To check your answers, turn to page 178.

5. What best describes the current scientific understanding of the relationship between methane and climate?

(1) Higher methane levels cause the Earth's climate to warm.
(2) Lower methane levels cause the Earth's climate to warm.
(3) In periods of warmer climate, more methane is in the atmosphere.
(4) In periods of warmer climate, less methane is in the atmosphere.
(5) There is no relationship between the level of methane in the atmosphere and the Earth's climate.

6. Which of the following is <u>not</u> true?

(1) The present level of methane in the atmosphere is 1,700 parts per billion.
(2) During ice ages, the level of methane in the atmosphere increases.
(3) In periods of global warming, the level of methane in the atmosphere increases.
(4) The Antarctica study provided measurements of methane levels in air samples from earlier periods.
(5) The present level of methane in the atmosphere is higher than at any known period in the past.

Over long periods of time, rocks change from one kind of rock to another. These changes occur again and again. This process is called the rock cycle.

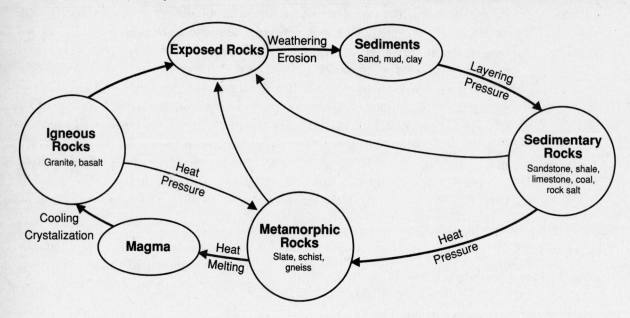

7. Which of the following statements is most accurate?

   (1) Metamorphic rocks are formed only from igneous rocks.
   (2) Sedimentary rocks are formed by heat.
   (3) Some igneous rocks are directly formed from sediments.
   (4) Weathering and erosion can affect all types of rocks.
   (5) Heat and pressure cause sedimentary rocks to form igneous rocks.

8. A hiker found a rock containing large and small crystals. What type of rock is this likely to be?

   (1) mud, clay, or sand
   (2) sedimentary
   (3) magma
   (4) metamorphic
   (5) igneous

9. What is a good title for the information and diagram?

   (1) The Rock Cycle
   (2) The Formation of Metamorphic Rocks
   (3) Uses of Rock
   (4) The Effect of Pressure on Rocks
   (5) Erosion and Weathering of Rocks

10. What kind of rocks would you be most likely to find at a place where there had once been a river?
   A. igneous
   B. metamorphic
   C. sedimentary

   (1) A only
   (2) B only
   (3) C only
   (4) A and B only
   (5) A and C only

Items 11 to 14 refer to the following article.

The oil shortages of the 1970s set off a search for new forms of energy. At the Natural Energy Laboratory of Hawaii, scientists are producing energy without burning fossil fuels and without causing pollution. They are using cold seawater.

In one process, called ocean thermal energy conversion, the difference in temperature between surface and deep seawater is used to generate electricity. In experiments conducted so far, it has taken more electricity to pump up deep seawater than the process has been able to produce. Scientists and engineers are planning to use much wider pipelines in order to pump enough seawater to produce more electricity than the process uses.

A more successful test used cold seawater for air conditioning and industrial cooling. Cold seawater was circulated through air-conditioning systems at the lab, and the electrical costs were cut by about 75 percent.

11. What does the author assume about sources of energy?

(1) Seawater will never be a practical source of energy.
(2) Solar energy is more efficient than energy from seawater.
(3) Seawater will eventually be used up.
(4) It is not wise to depend only on fossil fuels for energy.
(5) All air-conditioning systems can use cold seawater.

12. What is ocean thermal energy conversion?

(1) a process for removing salt and other minerals from seawater
(2) a process for farming sea plants and animals
(3) a process that uses the difference in temperature between warm and cold seawater to produce electricity
(4) a process to locate fossil fuel deposits under the ocean
(5) a process for air-conditioning buildings using seawater

13. In which of the following locations would using cold seawater to produce electricity be impractical?

(1) Hawaii
(2) California
(3) Florida
(4) Puerto Rico
(5) Kansas

14. Which of the following can be inferred from the article?

(1) Seawater will replace fossil fuels as the main source of energy within the next ten years.
(2) Until less electricity is used to pump up seawater than is produced by using seawater, the process will not be practical.
(3) Using fossil fuels to produce electricity costs more than using seawater.
(4) Seawater has been found to be more economical for producing electricity than for air-conditioning buildings.
(5) Using seawater to produce electricity causes a lot of pollution.

To check your answers, turn to pages 178–179.

Answers and Explanations

1. (Comprehension) **(5) decreases, then increases, then decreases, then increases as altitude increases** This is correct because the temperature decreases to -55°C at the border between the troposphere and stratosphere; then it increases to 0°C at the border between the stratosphere and mesosphere; then it decreases to -100°C at the top of the mesosphere; finally it reaches 2,000°C at the top of the thermosphere. The other options are incorrect because they do not describe the series of increases and decreases that occur as altitude increases.

2. (Analysis) **(3) They can withstand extremely high temperatures.** This is correct because the temperature in the upper thermosphere ranges from 600°C to 2,000°C. Option (1) is incorrect because the exosphere is 620 miles or higher above the Earth. Option (2) is incorrect because the ozone layer is located well below the thermosphere in the stratosphere and mesosphere. Option (4) is incorrect because ice clouds do not form where temperatures are very high. Option (5) is actually true in some cases, but this cannot be inferred from the information given in the diagram.

3. (Analysis) **(4) Increased levels of methane cause global warming.** Although there seems to be a relationship between methane levels and climate changes, it has not been proved that increased methane causes global warming. Options (1), (2), (3), and (5) are all provable facts.

4. (Analysis) **(3) The methane level is much higher now than during previous warm periods.** Since there seems to be a relationship between high methane levels and warmer climates, it is reasonable to conclude that the present high methane levels indicate that the Earth is in a warming period. Option (1) refers to methane levels in the past. Option (2) is true, but does not indicate the relationship of methane and climate. Option (4), if it were true, would indicate a cooling period rather than a warming period. Option (5) is true, but does not indicate a relationship between methane and climate.

5. (Analysis) **(3) In periods of warmer climate, more methane is in the atmosphere.** This is indicated by the results of the studies mentioned in the passage. Options (1) and (2) are incorrect because no cause and effect relationship has been proved. Option (4) is the opposite of what occurs. Option (5) is incorrect because there is a relationship between methane levels and climate.

6. (Evaluation) **(2) During ice ages, the level of methane in the atmosphere increases.** This is not true because the level of methane decreases during ice ages. The other options are all true according to the passage.

7. (Evaluation) **(4) Weathering and erosion can affect all types of rocks.** The diagram shows that when igneous, sedimentary, and metamorphic rocks are exposed, they are all subject to weathering and erosion. Option (1) is incorrect because metamorphic rocks are also formed from sedimentary rocks. Option (2) is incorrect because sedimentary rocks are formed by pressure. Option (3) is incorrect because igneous rocks are formed from magma. Option (5) is incorrect because heat and pressure cause sedimentary rocks to form metamorphic rock.

8. (Application) **(5) igneous** The cooling of magma produces the crystals in igneous rocks. Option (1) refers to sediments, not rocks. Option (2) is formed from particles of other rocks and would not be likely to contain large crystals. Option (3) is not a type of rock. Option (4) is incorrect because metamorphic rocks do not contain large crystals.

9. (Comprehension) **(1) The Rock Cycle** The diagram and information describe the steps in the process by which one type of rock turns into another, which is called the rock cycle. Options (2), (4), and (5) are too specific. Option (3) does not describe the information shown.

10. (Analysis) **(3) C only** An old river bed would contain many sediments that had been deposited by water. Over time, these sediments would form sedimentary rocks. Options (1), (4), and (5) are incorrect because sediments do not form igneous rocks. Option (2) is incorrect because sediments do not form metamorphic rocks. The area might contain some metamorphic rocks that formed from sedimentary rocks, but the sedimentary rocks would be more likely to be seen.

11. (Analysis) **(4) It is not wise to depend only on fossil fuels for energy.** Both short-term

shortages, as in the 1970s, and long-term limits on fossil fuel supply indicate a need to develop alternate sources of energy. Option (1) is incorrect because the author indicates that seawater may eventually be a useful source of energy. Option (2) is incorrect because nothing in the article indicates the author's opinions about solar energy. Option (3) is incorrect because seawater will never be used up. Option (5) is incorrect because only air-conditioning systems near a supply of cold seawater would be candidates for seawater as a coolant.

12. (Comprehension) **(3) a process that uses the differ-** ence in temperature between warm and cold seawater to produce electricity Ocean thermal energy conversion is defined in the second paragraph of the article. The other options do not explain this process. Options (1), (2), and (4) are not mentioned in the article. Option (5) describes a different process that uses seawater.

13. (Application) **(5) Kansas** Of the five options, only Kansas is located far from the ocean. It would thus be impractical to use seawater in such a location.

14. (Comprehension) **(2) Until less electricity is used to pump up seawater than is** produced by using seawater, the process will not be practical. At present, more electricity is used to pump seawater through the generator than the generator itself produces, causing a net loss in electricity. This obviously is not practical. Option (1) is not likely since the small-scale experiment has yet to produce electricity economically. Option (3) is not true. Option (4) is not true because the article indicates that cooling bills were cut by 75 percent, but that production of electricity was not economical. Option (5) is incorrect because there is no indication that generating electricity from seawater causes pollution.

Performance Analysis
Earth Science Review

Use the chart below to identify your strengths and weaknesses in each thinking skill area for the subject of Earth Science.

Circle the number of each item that you answered correctly on the Earth Science Review.

Thinking Skill Area	Earth Science	Lessons for Review
Comprehension	1, **9**, 12, 14	1, 2, 4, 9
Application	8, 13	10
Analysis	2, 3, 4, 5, **10**, 11	3, 5, 6, 8, 11, 12
Evaluation	6, **7**	7

Boldfaced numbers indicate items based on diagrams, charts, graphs, and illustrations.

If you answered 11 or more of the 14 items correctly, congratulations! You are ready to go on to the next section.

If you answered 10 or fewer items correctly, determine which skill areas are most difficult for you. Then go back and review the Earth Science lessons for those areas.

CHEMISTRY

Fireworks are the result of chemical reactions.

- **matter**
  anything that has mass and takes up space

- **atom**
  the smallest particle of an element that has all the properties of that element

- **ion**
  a charged particle formed from an atom

- **molecule**
  the smallest particle of a compound that has all the properties of that compound

Chemistry is the study of matter and changes in matter. You may think that chemistry exists only in a laboratory, but chemistry is all around you and within you. In fact, a chemical reaction within your brain enables you to understand and remember these words as you read them.

In Lesson 13 you will learn about matter. **Matter** is what everything in the world is made up of. Matter can exist in any one of three common physical states — solid, liquid, or gas.

The building blocks of matter are tiny particles called **atoms**. Atoms contain three particles called protons, neutrons, and electrons. Atoms can form charged particles called **ions**. Atoms can also bond together chemically to form molecules. A **molecule** is the smallest particle of a compound that still has all the properties of that compound.

☞ *See Also: GED Exercise Book Science, pages 22–27*

element
a substance that cannot be broken down into simpler substances by chemical means

periodic table
an arrangement of the elements according to their properties

solution
a mixture in which one substance is dissolved in another

acid
a substance that releases hydrogen ions in a water solution

base
a substance that releases hydroxide ions in a water solution

salt
a compound that results from a chemical reaction between an acid and a base

In this lesson you will also learn about the elements. An **element** is the simplest of substances, for it cannot be broken down into other substances by chemical means. Scientists have listed the elements in an arrangement called the **periodic table**. As you study the periodic table, you will find out how useful it is in predicting the properties of elements.

In chemical reactions substances are changed into new and different substances. In Lesson 14 you will discover how chemical changes take place. You will learn that during chemical reactions, bonds between atoms are broken and new bonds are formed. You will also learn that energy changes are part of chemical reactions.

In this lesson you will also learn about a familiar type of mixture called a **solution**. A solution consists of a substance called a solute dissolved in another substance called the solvent. Solutions can be made up of solids, liquids, or gases. The most familiar solutions are solid, liquid, or gas solutes dissolved in a liquid solvent.

You will also learn about an important group of chemical compounds known as acids, bases, and salts. Some of these compounds are very familiar to you. You can find an **acid** in an orange, a **base** in soap and cleaning solutions, and a **salt** right on your dinner table. And speaking of dinner, the last part of this lesson will teach you how to become a "kitchen chemist," as you learn what really happens when baking powder makes a cake rise.

LESSON 13 Application Skill: Other Contexts

When you use information you know in another situation, you are applying your knowledge in a new context.

Many of the topics you read about in science are very general. For example, you might read about the properties, or characteristics, of gases. You may read that some gases are lighter than others, such as the mixture of gases in the air. This is easier to grasp if you think about how helium-filled balloons float in the air. Often, it helps you to understand general science ideas when you can apply them to specific examples.

Read this paragraph and see how the ideas can be applied.

When you think about anything on Earth, you must be thinking about matter or energy. Everything on Earth can be classified as matter or energy. Matter has mass and takes up space. Energy has the ability to move matter or change it.

According to the passage, everything on Earth is either matter or energy. You can apply this general idea to specific things. Think of this book. It has mass and takes up space — it must be an example of matter. Now lift up the book. To move the book, you used energy. What other specific applications of the ideas of matter and energy can you think of?

Remember, when you read about general ideas it can be helpful to apply them to specific situations. Ask yourself:

♦ What is being described or explained?

♦ What situations might this information apply to?

♦ How would this information be used in those situations?

Practicing Application

Read the following paragraph.

> At sea level, where the air is at standard pressure, pure water boils at 100°C. When water is heated in a closed container so that the steam cannot escape, the boiling point of the water is raised above 100°C.

Question 1 refers to the paragraph. Circle the best answer.

1. Which of the following appliances is based on the principles described above?

   (1) a microwave oven
   (2) a washing machine
   (3) a toaster
   (4) a refrigerator
   (5) a pressure cooker

Read the following paragraph.

> Minerals dissolved in water can be removed from water by the process of distillation. Water is added to a distilling flask and boiled. The water vapor released then passes through a cooled tube, and the condensed water flows down into a receiving container. The minerals are left behind in the distilling flask.

Question 2 refers to the paragraph. Circle the best answer.

2. Which of the following is an example of distillation?

   (1) obtaining drinking water from ocean water
   (2) putting ice cubes in a cold drink
   (3) boiling water for tea
   (4) adding antifreeze to a car engine
   (5) bottling mineral water

To check your answers, turn to page 190.

Topic 13: Matter

Read the following passage.

Matter can exist in any of three physical states. These states are solid, liquid, and gas.

♦ A **solid** is any form of matter that has a definite shape and volume. A bar of gold is an example of a solid. If you try to put a square bar of gold into a round hole, it will not fit. The gold has a definite shape. If you try to put the bar of gold into a space that is too small, it also will not fit. The bar of gold has a definite volume.

♦ A **liquid** has a definite volume, but it does not have a definite shape. If you pour a quart of milk into a gallon jug, the milk will fill only one-fourth of the jug. If you pour the same quart of milk into an eight-ounce glass, the milk will overflow. The volume of the milk does not change. However, its shape changes each time you pour it into a different container.

♦ A **gas** has neither definite shape nor definite volume. A gas will spread out to fill the volume of a container that it is placed in. You can understand this property of a gas if you think of how quickly the smell of an apple pie baking in the oven fills the whole kitchen.

Most matter can change from one state to another. If enough heat is removed from a liquid, it will freeze into a solid. Similarly, if enough heat is added to a solid, it will melt into a liquid. The temperature at which these changes of state occur is called the **freezing point** or **melting point** of substances. If enough heat is added to a liquid, the liquid will change into a gas. This process is called **vaporization**. The temperature at which vaporization occurs is called the **boiling point** of a substance. When a gas is cooled enough, it changes into a liquid in a process called **condensation**.

Exercises

Questions 1 to 8 refer to the previous passage. Circle the best answer for each question.

1. Which of the following situations would produce a change in the state of matter?

    (1) A drop of food coloring is added to a glass of water.
    (2) A container of ice cream is left on top of a radiator.
    (3) A hole is pricked in a balloon.
    (4) A can of soup is placed on a scale.
    (5) A square block of wood is cut in half.

2. A heart-shaped cake is made by pouring cake batter into a heart-shaped mold. Which principle is illustrated by this situation?

    (1) A solid takes the shape of the container into which it is placed.
    (2) A solid has a definite shape but not a definite volume.
    (3) A liquid has a definite shape and volume.
    (4) A liquid has no definite shape.
    (5) A gas has a definite shape and volume.

3. What causes matter to change state?

    (1) the addition or removal of heat
    (2) its mass
    (3) the space it occupies
    (4) changes in volume
    (5) changes in shape

4. Which of the following is not an example of a change of state of matter?

    (1) steam condensing on a cold window
    (2) filling a balloon with helium
    (3) icicles melting
    (4) freezing leftover soup
    (5) boiling water until the pot is empty

5. Which of the following situations illustrates the process of condensation?

    (1) dew forming on the grass in early morning
    (2) ice cubes melting in a cold drink
    (3) rain changing to sleet
    (4) steaming vegetables
    (5) melting butter in a frying pan

6. Which of the following situations shows that gases have neither definite shape nor definite volume?

    (1) pouring a pint of orange juice into a quart container
    (2) lighting a match
    (3) adding salt to boiling water
    (4) smelling pollutants released by a nearby factory
    (5) pouring concrete into a mold

7. What are the properties that characterize a liquid?

    (1) definite shape; definite volume
    (2) definite shape; no definite volume
    (3) no definite shape; no definite volume
    (4) no definite shape; definite volume
    (5) none of the above

8. Which of the following statements is most accurate?

    (1) All matter can be found naturally in all three states.
    (2) Liquids spread out to completely fill their containers.
    (3) All matter is liquid below 100°C.
    (4) Water freezes and vaporizes at different temperatures.
    (5) Liquids melt and vaporize at different temperatures.

To check your answers, turn to page 190.

Reviewing Lesson 13

Read the following passage and look at the diagram.

All matter is made up of elements. An **element** is a substance that cannot be broken down into simpler substances by chemical means. Oxygen, carbon, iron, and copper are examples of elements.

Can an element be broken down into smaller and smaller pieces forever and still be an element? For example, can a piece of copper be cut into smaller and smaller pieces and still be copper? The answer is <u>no</u>. Eventually, a tiny piece would be obtained that could not be divided and still be copper. This smallest piece of an element is called an atom.

An **atom** is the smallest particle of an element that still has the properties of that element. All elements are made up of atoms. The element copper is made only of copper atoms, the element iron is made only of iron atoms, and so on.

What is an atom made of? Each atom has a small, dense core called a **nucleus**. The nucleus contains particles called **protons**, which have a positive electric charge, and **neutrons**, which have no charge. Moving in orbits around the nucleus are tiny, negatively charged particles called **electrons**. A neutron and a proton have the same mass, which is about 1,800 times greater than the mass of an electron.

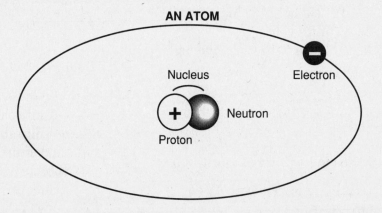

Questions 1 to 4 refer to the passage and diagram. Circle the best answer for each question.

1. The structure of an atom is most similar to which of the following?

   (1) a salt crystal
   (2) a chair
   (3) an egg
   (4) the Moon
   (5) the solar system

2. Which of the following statements about substance X would indicate that it is <u>not</u> an element?

   (1) It breaks down when heated, producing mercury and oxygen.
   (2) It is red-orange in color.
   (3) It reacts with water to form an acid.
   (4) It is a solid at room temperature.
   (5) It can react with certain substances to form salts.

3. The total charge on an atom is zero, which means an atom is electrically neutral. Which statement explains why this is so?

(1) The number of protons in an atom is greater than the number of electrons.

(2) The number of protons in an atom is less than the number of electrons.

(3) The number of protons in an atom is equal to the number of electrons.

(4) The number of protons in an atom is equal to the number of neutrons.

(5) The number of neutrons in an atom is equal to the number of electrons.

4. Atom X has five more protons than atom Y. Both atoms have the same mass. Which statement explains why this is so?

(1) Atom X has more electrons than atom Y.

(2) Atom X has fewer electrons than atom Y.

(3) Atom X has more neutrons than atom Y.

(4) Atom X has fewer neutrons than atom Y.

(5) Atom X has the same number of neutrons as atom Y.

Read the following passage.

When two or more elements combine chemically, they form a **compound**. Water is a compound because it consists of the elements oxygen and hydrogen.

The smallest particle of a compound that still has the properties of that compound is called a **molecule**. A molecule is made up of two or more atoms chemically bonded together. When atoms bond together, they share or transfer electrons. If electrons are shared, the bond is said to be a **covalent bond**. If electrons are transferred from one atom to another, charged particles, called **ions**, form. A bond between two ions is said to be an **ionic bond**. Some atoms form bonds easily with other atoms, while some atoms hardly ever form bonds.

Questions 5 and 6 refer to the passage. Circle the best answer for each question.

5. Which of the following combinations represents a compound?

(1) two carbon atoms side-by-side in a diamond crystal

(2) sugar and water mixed together

(3) a hydrogen atom and a chlorine atom sharing a pair of electrons

(4) hydrogen and nitrogen existing together in air

(5) water and carbon dioxide mixed together in carbonated water

6. It can be inferred from the passage that atoms that do not bond easily with other atoms

(1) have no electrons

(2) form only ionic bonds

(3) are present in very few compounds

(4) are found only in water

(5) form only covalent bonds

To check your answers, turn to pages 190–191.

GED Mini-Test

Directions: Choose the best answer to each item.

Items 1 and 2 refer to the following passage and table.

The atoms of each element are different from the atoms of every other element. What makes them different? The atoms of each element have a certain number of protons in the nucleus. The number of protons in the nucleus of an atom of an element is called its atomic number. Thus, each element has its own atomic number. The number of protons plus the number of neutrons in the atom of an element is called the atomic mass.

Scientists found that elements could be organized into groups with similar physical and chemical properties. The periodic table shows elements arranged according to their atomic number and their properties. In the horizontal rows, called periods, the elements are arranged in order of increasing atomic number. Elements with similar properties are lined up in vertical columns called groups or families. For convenience, two subgroups are shown below the main part of the table.

Because the properties of elements vary in a regular pattern, you can tell a lot about an element by where it appears in the periodic table. For example, all the elements on the left side and in the center of the table are metals. All the elements on the right side are nonmetals. A heavy zigzag line separates the metals and nonmetals.

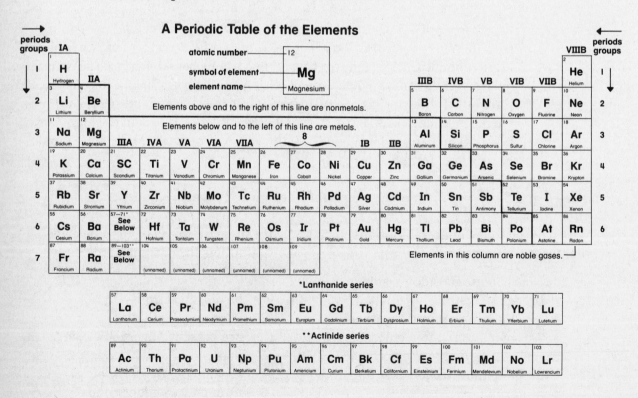

A Periodic Table of the Elements

1. According to the passage, the periodic table is based on a relationship between which two factors?

   (1) atomic number and atomic size
   (2) atomic number and number of electrons
   (3) atomic number and properties
   (4) number of electrons and properties
   (5) atomic size and properties

2. According to the periodic table, which of the following groups of elements have similar properties?

   (1) sodium (Na), chlorine (Cl), hydrogen (H)
   (2) neon (Ne), argon (Ar), krypton (Kr)
   (3) lithium (Li), magnesium (Mg), sulfur (S)
   (4) oxygen (O), carbon (C), chlorine(Cl)
   (5) phosphorus (P), sulfur (S), chlorine (Cl)

Items 3 and 4 refer to the following information.

Elements in the same group in the periodic table have many similar properties. Five groups of elements are described below.

- group I = very reactive metals; react violently with water

- group II = fairly reactive metals; often found in salts

- transition elements = metals, many of which can react with oxygen to form more than one compound

- group VII = very reactive nonmetals that combine with hydrogen to form acids

- group VIII = very inactive elements that rarely take part in chemical reactions

Each of the following questions describes an element that belongs to one of the groups listed above. Classify each element into one of the groups. More than one element may belong to the same group.

3. Element X is a silvery-gray solid that can cause an explosion when it is dropped into water.

   (1) group I
   (2) group II
   (3) transition elements
   (4) group VII
   (5) group VIII

4. Element Z combines with sulfur and oxygen to form a compound known as Epsom salts.

   (1) group I
   (2) group II
   (3) transition elements
   (4) group VII
   (5) group VIII

To check your answers, turn to page 191.

Answers and Explanations

Practicing Application (page 183)

1. (Application) **(5) a pressure cooker** A pressure cooker uses the buildup of steam to raise the boiling point of water. The hotter water cooks the foods faster. Options (1) and (3) are appliances that do not need water to cook foods. Options (2) and (4) involve water at temperatures lower than the boiling point.

2. (Application) **(1) obtaining drinking water from ocean water** Ocean water, which contains salts and other minerals, can be distilled to yield pure water for drinking. Options (2) and (3) involve uses of pure water. Although water is boiled for making tea, no water vapor is collected. Option (4) lowers the freezing point; it has nothing to do with removing minerals or other impurities. Option (5) does not involve distillation, since mineral water is sold because of its mineral content.

Exercises (page 185)

1. (Application) **(2) A container of ice cream is left on top of a radiator.** This is correct because the solid ice cream would melt into a liquid. Options (1), (3), and (5) are incorrect because changing color, volume, or size does not involve a change of state. Option (4) is incorrect because no change of state occurs when the can of soup is weighed.

2. (Application) **(4) A liquid has no definite shape.** This is correct because liquid cake batter would take the shape of the mold. After the batter is baked, the solid that forms keeps that shape. Options (1), (2), (3), and (5) are not true.

3. (Comprehension) **(1) the addition or removal of heat** According to the passage, heating substances causes them to melt or vaporize. Cooling substances causes them to freeze or condense. All of these changes are changes of state. Options (2), (3), (4), and (5) describe properties of types of matter, not changes of state.

4. (Application) **(2) filling a balloon with helium** This is an example of one of the properties of gases — expanding to the available space in a container. Options (1), (3), (4), and (5) are all examples of matter changing from one state to another.

5. (Application) **(1) dew forming on the grass in early morning** Water vapor in the air condenses as the air cools during the night. Options (2) and (5) are incorrect because they involve melting. Option (3) is incorrect because it involves freezing. Option (4) is incorrect because it involves vaporization.

6. (Application) **(4) smelling pollutants released by a nearby factory** The gases spread out in all directions and can be smelled from a distance. Options (1) and (5) are incorrect because they involve liquids. Options (2) and (3) are incorrect because they involve solids.

7. (Comprehension) **(4) no definite shape; definite volume** Liquids change shape and fill the container they are in, but their volume remains the same regardless of shape. Option (1) describes solids. Option (2) does not describe any of the types of matter. Option (3) describes gases. Option (5), none of the above, is incorrect because there is a correct description, option (4).

8. (Evaluation) **(4) Water freezes and vaporizes at different temperatures.** Water freezes at 0°C and vaporizes at 100°C. Option (1) is incorrect because few substances are found naturally in all three states. Option (2) is incorrect because gases, not liquids, spread out to completely fill their containers. Option (3) is incorrect because some matter is solid and some matter is gas below 100°C. Option (5) is incorrect because liquids are already melted.

Reviewing Lesson 13 (pages 186–187)

1. (Application) **(5) the solar system** The structure of the atom is most similar to the solar system, with the nucleus in the position of the Sun and the electrons like the planets orbiting the Sun. None of the other options involves objects in orbit around another object.

2. (Comprehension) **(1) It breaks down when heated, producing mercury and oxygen.** This statement shows that substance X is not an element because elements cannot be broken down into other elements. Option (2) is incorrect because nothing stated in the passage refers to the color of an element. Options (3) and (5) are incorrect because elements can react with other substances.

Option (4) is incorrect because many elements are solids.

3. (Analysis) **(3) The number of protons in an atom is equal to the number of electrons.** Option (1) is incorrect because the atom would have a positive charge. Option (2) is incorrect because the atom would have a negative charge. Options (4) and (5) are incorrect because neutrons do not have a charge, so they do not affect the charge of an atom.

4. (Analysis) **(4) Atom X has fewer neutrons than atom Y.** This is correct because neutrons and protons have the same mass — thus extra neutrons in Y could balance the extra protons in X. Options (1) and (2) are incorrect because electrons make up so little of an atom's mass that they could not possibly balance the extra protons. Option (3) would make atom X even heavier. Option (5) is incorrect because atom X would still be heavier by five protons.

5. (Application) **(3) a hydrogen atom and a chlorine atom sharing a pair of electrons** This is correct because it describes a covalent bond between hydrogen and chlorine atoms. Option (1) is incorrect because the carbon is still just one element. Options (2), (4), and (5) are incorrect because they describe mixtures.

6. (Comprehension) **(3) are present in very few compounds** If an atom does not bond easily with other atoms, it cannot form many compounds. Option (1) is incorrect because it is clear from the information given that all atoms have electrons. Options (2) and (5) are incorrect because the passage does not indicate that one kind of bond is more likely to form than the other. Option (4) is incorrect because water molecules contain bonds.

GED Mini-Test (page 189)

1. (Comprehension) **(3) atomic number and properties** Options (1) and (5) are incorrect because atomic size is not the basis of the periodic table. Options (2) and (4) are incorrect because the number of electrons is not the basis of the periodic table.

2. (Comprehension) **(2) neon (Ne), argon (Ar), krypton (Kr)** This is correct because these elements are all in the same family, group VIII. The elements in options (1), (3), and (4) are not all members of the same family. If you selected option (5), you confused elements in the same period with elements in the same group.

3. (Application) **(1) group I** This element is silvery, like most metals. It causes an explosion when it is dropped into water. This is a violent reaction, which is typical of elements in group I. Option (2) is incorrect because there is no mention of a salt. Option (3) is incorrect because there is no description of oxygen-containing compounds. Option (4) is incorrect because there is no mention of acids. Option (5) is incorrect because group VIII elements are inactive and would not cause an explosion.

4. (Application) **(2) group II** This element forms Epsom salts. Salts are only mentioned in reference to group II elements. The other options are incorrect because they are not described as forming salts. If you answered option (3), you were misled by the mention of oxygen. However, only one compound was described in the question, and transition elements form more than one compound with oxygen.

LESSON 14 Evaluation Skill: Drawing Conclusions

When you draw a conclusion, you use details and facts to prove that a broad statement is logical.

In Lesson 8, you learned how to tell the difference between a conclusion and a supporting statement. A conclusion is a logical generalization. Supporting statements are details, facts, measurements, or other kinds of information that help prove that the conclusion is true.

In science, conclusions must be drawn from observable, measurable facts. When you read about a science topic, be alert for the information the author provides to prove that an idea or conclusion is correct. The information may be in the form of words, diagrams, or mathematical equations.

Read this passage and see what conclusions can be drawn based on the information that is given.

Two or more elements may either be mixed mechanically or combined chemically. If they are mixed mechanically, the result is a **mixture**. In a mixture, each element keeps its own properties. The elements may be mixed in any amount, and they may be separated from the mixture by mechanical means. Examples of mixtures are seawater, air, and mayonnaise.

If the elements are combined chemically, the result is a **compound**. In a compound, the elements lose their identity and become another substance with different properties. The elements that form a compound are always combined in the same proportions. Water is a compound formed from the elements hydrogen and oxygen.

What conclusion can we draw from this information? Can we conclude, for example, that chicken vegetable soup is a mixture? We must see if the characteristics of a mixture apply to chicken vegetable soup. Does each ingredient keep its own properties? Yes, the carrots, peas, potatoes, beans, and broth remain themselves. Can the vegetables and the broth be separated by mechanical means? Yes, a strainer can be used to remove the solids from the broth. Using the information in this passage, we can reasonably conclude that chicken vegetable soup is a mixture.

Remember, when you are asked what conclusions can be drawn from the information you are given, pay close attention to the information that is actually there. A statement may be true, yet the information you are given may not be enough to prove that it is true. Ask yourself:

1. Does this statement make sense? Is it logical and accurate?
2. If the answer is yes, then ask: What information presented here proves that this statement is true?

If there is not enough information to support a conclusion, then either the conclusion is inaccurate or there is not enough data.

Practicing Evaluation

Read the following passage.

A chemical equation uses formulas and symbols to show what happens during a chemical reaction. In chemical reactions, the substances you start out with are called the **reactants**. The substances that are formed are called the **products**. The general form of a chemical equation is:

$$\text{reactants} \longrightarrow \text{products}$$

Question 1 refers to the passage. Circle the best answer.

1. Which of the following conclusions about the chemical equation shown below can be supported by the information provided?

$$C + O_2 \longrightarrow CO_2$$

(1) C is a product of the chemical reaction.
(2) CO_2 is present in the atmosphere.
(3) CO_2 is formed when coal burns.
(4) CO_2 is a product of the chemical reaction.
(5) CO_2 is a reactant of the chemical reaction.

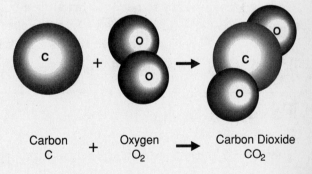

Carbon   +   Oxygen   →   Carbon Dioxide
C   O₂   CO₂

To check your answer, turn to page 200.

Topic 14: Chemical Reactions

Read the following passage and look at the illustrations.

In a **chemical reaction**, one or more substances change, forming one or more new substances. In some chemical reactions, molecules split apart into atoms. In other reactions, atoms join together to form molecules. In yet other reactions, atoms change places with other atoms to form new molecules.

Chemical reactions either release or absorb energy. A chemical reaction that releases energy is called an **exothermic reaction**. For example, when wood or oil is burned, large amounts of heat energy are released. A chemical reaction that absorbs energy is called an **endothermic reaction**. Cooking an egg is an endothermic reaction, because the egg absorbs heat energy as it sets.

EXOTHERMIC REACTION

(change from liquid to solid)

ENDOTHERMIC REACTION

(change from liquid to gas)

Energy must be added to start many reactions. The energy is necessary to begin to break the bonds in the reactant molecules. The energy that must be added to start a chemical reaction is called the **activation energy**. When you light a match to start a charcoal grill, you are supplying activation energy.

Exercises

Questions 1 to 8 refer to the passage. Circle the best answer for each question.

1. What is activation energy?

    (1) energy absorbed by a reaction
    (2) energy released by a reaction
    (3) electricity
    (4) energy needed to start a reaction
    (5) energy needed to stop a reaction

2. Which of the following is an example of an exothermic reaction?

    (1) condensing water vapor
    (2) cooking a custard
    (3) melting ice
    (4) exploding dynamite
    (5) boiling water

---

3. Which of the following conclusions can be supported by the information provided?

   (1) Exothermic reactions can be more useful for the energy they release than for their products.

   (2) Electricity is produced in endothermic reactions.

   (3) New chemical bonds are formed only during exothermic reactions.

   (4) Boiling water is an endothermic reaction.

   (5) Activation energy is needed for all reactions.

4. Which of the following statements supports the conclusion that rusting is an exothermic reaction?

   (1) During the process of rusting, small amounts of heat energy are released.

   (2) Activation energy produces rust.

   (3) Rusting affects only certain substances.

   (4) During rusting, metal absorbs heat.

   (5) Rusting occurs in humid conditions.

5. In the chemical reaction of photosynthesis, plants take in water and carbon dioxide and produce sugar, oxygen, and water in the presence of sunlight. What role does sunlight play in this reaction?

   (1) It is a reactant.

   (2) It is a product.

   (3) It is activation energy.

   (4) It is endothermic.

   (5) It stops the reaction.

6. Which of the following best describes what happens when hydrogen and oxygen combine to form water?

   (1) Molecules are breaking apart to form atoms.

   (2) Atoms are combining to form molecules.

   (3) Atoms are changing place with other atoms.

   (4) Hydrogen and oxygen are undergoing a change of state.

   (5) Oxygen is a product of the reaction.

7. What is a chemical reaction?

   (1) a change in the state of matter

   (2) the wearing away of a substance

   (3) the change of one or more substances into a new substance with different properties

   (4) a change in the number of molecules

   (5) a change in the size of a substance

8. What is an endothermic reaction?

   (1) a reaction in which energy is released

   (2) a reaction in which energy is absorbed

   (3) a reaction in which energy is neither released nor absorbed

   (4) a reaction that requires activation energy to start

   (5) a reaction that changes chemical bonds

To check your answers, turn to page 200.

Reviewing Lesson 14

Read the following passage.

A **solution** is a mixture in which two or more substances are dissolved in one another. Solutions can involve solids, liquids, or gases. The substance being dissolved is called the **solute**. The substance doing the dissolving is called the **solvent**. In a sugar-water solution, sugar is the solute and water is the solvent. In carbonated water, carbon dioxide is a gas solute dissolved in water, a liquid solvent.

The greatest amount of a solute that will dissolve in a given amount of solvent at a certain temperature is called **solubility**. Usually, solubility increases with temperature. For gases dissolved in a liquid, however, the reverse is true — solubility increases as the temperature of the solvent decreases. The rate of solution is also affected by temperature. An increase in temperature causes molecules to move and spread apart more quickly.

Questions 1 and 2 refer to the passage. Circle the best answer for each question.

1. A glass of carbonated beverage left in a warm room goes flat, while a glass of the same beverage stored in the refrigerator does not. According to the information provided, which of the following statements could explain why this happens?

(1) Water molecules evaporate more rapidly at high temperatures.
(2) The solubility of a gas in liquid decreases as temperature increases.
(3) Molecules move more slowly at low temperatures.
(4) The solubility of a gas in liquid decreases as temperature decreases.
(5) The rate of solution is slower at low temperatures than at high temperatures.

2. If two identical mixtures of salt and water are prepared, and only one is stirred, the one that is stirred dissolves the salt faster. Which of the following conclusions can be supported by the information provided?

(1) Stirring decreases the temperature of the water.
(2) Stirring increases solubility.
(3) Stirring increases the amount of solute.
(4) Stirring increases the amount of solvent.
(5) Stirring causes molecules to move and spread apart more quickly.

Read the following passage.

Three important groups of chemical compounds are acids, bases, and salts. When dissolved in water, these compounds produce ions. Ions are atoms or molecules with an electric charge. In water, **acids** produce hydrogen, or H^+ ions. In water, **bases** produce hydroxide, or OH^- ions. Citric acid is found in citrus fruits. Magnesium hydroxide is the base that is the active ingredient in many stomach remedies.

A strong acid, such as sulfuric acid, or a strong base, such as sodium hydroxide, is poisonous and can burn the skin. Yet a weak acid, such as citric acid, or a weak base, such as magnesium hydroxide, can be safely handled. The strength of an acid or base is measured on a scale called the pH scale. The **pH scale** ranges from 0 to 14. The number 7 is the neutral point. Substances with a pH below 7 are acidic, and substances with a pH above 7 are basic. The strongest acid would have a pH of 0, and the strongest base would have a pH of 14.

When an acid and a base combine chemically, the result is a neutral compound called a **salt**. Water is also a product of this reaction. A familiar salt is sodium chloride, table salt.

Questions 3 to 6 refer to the passage. Circle the best answer for each question.

3. Which of the following statements would indicate that substance X is <u>not</u> an acid?

   (1) It combines chemically with certain other compounds to produce salts.
   (2) It dissolves in water to produce OH^- ions.
   (3) It can be poisonous if swallowed.
   (4) Its pH is lower than that of water.
   (5) It can be corrosive to skin.

4. An antacid relieves indigestion by neutralizing excess acid in the stomach. Based on this information, which of the following statements must be true about an antacid?

   (1) It contains salt.
   (2) It has a low pH.
   (3) It contains a base.
   (4) It produces H^+ ions.
   (5) It dissolves very rapidly.

5. Which of the following represents a correct ordering of substances from lowest to highest pH?

   (1) magnesium hydroxide, pure water, sulfuric acid, sodium hydroxide
   (2) pure water, sulfuric acid, citric acid, sodium hydroxide
   (3) sulfuric acid, sodium hydroxide, citric acid, magnesium hydroxide
   (4) sulfuric acid, citric acid, pure water, sodium hydroxide
   (5) sodium hydroxide, pure water, citric acid, sulfuric acid

6. It can be inferred from the passage that a solution of table salt and water would have a pH of approximately

   (1) 0
   (2) 4
   (3) 7
   (4) 10
   (5) 14

To check your answers, turn to pages 200–201.

GED Mini-Test

Directions: Choose the best answer to each item.

Items 1 to 4 refer to the following passage.

You may not think of the kitchen as being a chemistry lab, but many chemicals can be found right on your kitchen shelf. One substance that contains several interesting chemicals is baking powder, which is used to make cake batter rise.

The principal ingredient in baking powder is sodium bicarbonate, $NaHCO_3$. When sodium bicarbonate reacts with an acid, it produces the gas carbon dioxide (CO_2) and water. When sodium bicarbonate is heated strongly, it breaks down to form carbon dioxide and sodium carbonate (Na_2CO_3). Baking powder also contains a substance that will react with water to form acids. This substance is usually a type of compound called a tartrate.

---

1. It can be inferred from the passage that the purpose of the tartrate in baking powder is to

   (1) provide an acid for sodium bicarbonate to react with
   (2) break down to form carbon dioxide
   (3) react with carbon dioxide
   (4) provide a salt that will make dough rise
   (5) react with water to form sodium bicarbonate

2. Sodium bicarbonate reacts with vinegar to produce carbon dioxide and water. Which of the following must be true about vinegar?

   (1) It contains sodium carbonate.
   (2) It makes bread dough rise.
   (3) It decomposes when heated.
   (4) It contains an acid.
   (5) It contains a salt.

3. When baking powder is added to cake batter, the substance that actually makes the cake rise is

   (1) salt
   (2) oxygen
   (3) water
   (4) tartrate
   (5) carbon dioxide

4. When baking powder is left uncovered in damp or humid weather, it quickly loses its effectiveness. Based on the passage, which of the following statements could explain why this happens?

   (1) Moisture in the air causes sodium bicarbonate to break down.
   (2) Oxygen in the air causes the tartrate to break down.
   (3) Moisture in the air reacts with the tartrate.
   (4) Oxygen in the air reacts with carbon dioxide.
   (5) Oxygen in the air reacts with sodium bicarbonate.

Items 5 and 6 refer to the following information.

Iron (Fe) is one of the most plentiful elements on Earth, but it usually occurs combined with other elements in rock called ores. The major ore of iron is hematite (Fe_2O_3). Iron is separated from hematite in a blast furnace. Coke is burned with air, producing carbon dioxide. The carbon dioxide reacts with more coke to form carbon monoxide. Carbon monoxide removes the oxygen from the hematite, leaving the iron. The steps in the process are shown in the flow chart.

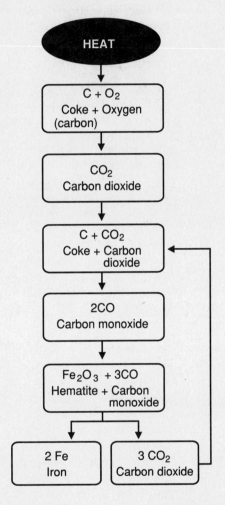

5. What substance must continually be added to the blast furnace to keep the process of separating iron going?

(1) coke
(2) carbon dioxide
(3) carbon monoxide
(4) pure iron
(5) pure oxygen

6. According to the diagram, what substances react to form carbon monoxide?

(1) coke and oxygen
(2) hematite and coke
(3) hematite and oxygen
(4) carbon dioxide and iron
(5) coke and carbon dioxide

To check your answers, turn to page 201.

Answers and Explanations

1. (Evaluation) **(4) CO_2 is a product of the chemical reaction.** Since CO_2 is on the right side of the arrow, it is a product of the chemical reaction. Option (1) is incorrect because C is a reactant. Options (2) and (3) may be true but are not supported by the information provided. Option (5) is the opposite of the correct answer.

Exercises (pages 194–195)

1. (Comprehension) **(4) energy needed to start a reaction** According to the passage, many reactions are started by the application of energy from an outside source, called activation energy. The other options are therefore incorrect.

2. (Application) **(4) exploding dynamite** Exploding dynamite is a reaction in which energy is released, an exothermic reaction. Options (1), (3), and (5) are changes of state, not chemical reactions. Option (2) is an endothermic reaction since heat is absorbed when something is cooked.

3. (Evaluation) **(1) Exothermic reactions can be more useful for the energy they release rather than for their products.** The purpose of burning fuels, an exothermic reaction, is to release heat energy for human use. Option (2) is incorrect because endothermic reactions absorb, not release, energy. Option (3) is not true; new bonds can form in exothermic reactions. Option (4) is a change of state, not a chemical reaction. Option (5) is not true, according to the passage.

4. (Analysis) **(1) During the process of rusting, small amounts of heat energy are released.** Since the release of energy is what characterizes an exothermic reaction, this means that rusting is an exothermic reaction. Option (2) is not related only to exothermic reactions. Options (3) and (5) are true but do not support the conclusion. Option (4) would support the conclusion that rusting is an endothermic reaction.

5. (Application) **(3) It is activation energy.** Without energy from the Sun, photosynthesis would not occur. Options (1) and (2) are incorrect because sunlight is not among the reactants and products described in the item. Option (4) applies to chemical reactions, not to sunlight. Option (5) is not true.

6. (Comprehension) **(2) Atoms are combining to form molecules.** When hydrogen atoms and oxygen atoms combine, they form molecules of water, H_2O. Option (1) is incorrect because it describes the process of breaking down a molecule. Option (3) does not describe a process of combining. Option (4) is not true; they are undergoing a chemical change. Option (5) is not true; oxygen is a reactant.

7. (Comprehension) **(3) the change of one or more substances into a new substance with different properties** Option (1) is incorrect because changes of state do not result in new substances; they merely change the form of a substance from gas to liquid to solid.

Option (2) is also a physical rather than a chemical change. Option (4) may or may not be true of any given reaction. Option (5) is incorrect because size of a substance is not part of the definition of a chemical reaction.

8. (Comprehension) **(2) a reaction in which energy is absorbed** According to the passage, the absorption of energy is part of an endothermic reaction. Options (1) and (3) are incorrect because they contradict the correct definition. Options (4) and (5) are incorrect because they apply to exothermic reactions as well as endothermic reactions.

Reviewing Lesson 14 (pages 196–197)

1. (Analysis) **(2) The solubility of a gas in liquid decreases as temperature increases.** Options (1), (3), and (5) are true statements, but they do not explain the situation, nor are they all covered in the passage. Option (4) is incorrect because it is an untrue statement.

2. (Evaluation) **(5) Stirring causes molecules to move and spread apart more quickly.** Stirring moves molecules around, causing the rate of solution to increase. Option (1) is incorrect because stirring will not affect the water temperature. Option (2) is incorrect because stirring does not increase the amount of salt that eventually dissolves; it increases the rate at which the salt dissolves. Options (3) and (4) are incorrect because stirring does not change the amount of matter in the container.

3. (Comprehension) **(2) It dissolves in water to produce OH⁻ions.** Option (1) is incorrect because acids combine with bases to produce salts. Options (3) and (5) are incorrect because these properties are true if the acid is strong. Option (4) is incorrect because all acids have a pH lower than that of water.

4. (Evaluation) **(3) It contains a base.** This is correct because bases neutralize acids. Option (1) is incorrect because a neutralization reaction produces a salt. Options (2) and (4) are incorrect because they are properties of acids. Option (5) is incorrect because how fast the antacid dissolves does not depend on its ability to neutralize acids.

5. (Comprehension) **(4) sulfuric acid, citric acid, pure water, sodium hydroxide** This is correct because these compounds have been identified as strong acid, weak acid, neutral substance, and strong base, in that order. If you answered option (5), you probably did not read the question carefully — the correct order is exactly reversed.

6. (Comprehension) **(3) 7** This is correct because both salt and water are neutral substances. Options (1) and (2) are incorrect because they are pH values of acids, and options (4) and (5) are incorrect because they are pH values of bases.

GED Mini-Test (pages 198–199)

1. (Comprehension) **(1) provide an acid for sodium bicarbonate to react with** Option (2) is incorrect because sodium bicarbonate breaks down to form carbon dioxide. Options (3) and (4) are incorrect because carbon dioxide is a product of the reaction between sodium bicarbonate and an acid. Option (5) is incorrect because sodium bicarbonate is one of the substances in baking powder to begin with.

2. (Evaluation) **(4) It contains an acid.** This is correct because the passage states that sodium bicarbonate reacts with acids to form carbon dioxide and water. The other options are incorrect because the passage includes no other conditions that would cause carbon dioxide and water to be produced from sodium bicarbonate.

3. (Analysis) **(5) carbon dioxide** This is correct because it is logical that a gas produced would cause the cake to rise. Options (1), (3), and (4) are incorrect because they are not gases, and options (1) and (4) are not produced when sodium bicarbonate reacts. Option (2) is a gas, but there is no mention of oxygen in the passage.

4. (Analysis) **(3) Moisture in the air reacts with the tartrate.** This is correct because tartrate plus water produces acid, which in turn reacts with the sodium bicarbonate in baking powder to release carbon dioxide gas. Option (1) is incorrect because heat causes sodium bicarbonate to break down. Options (2), (4), and (5) are incorrect because no mention is made of reactions involving oxygen.

5. (Comprehension) **(1) coke** Since coke is used up in the process, more must be added to keep it going. Options (2) and (3) are produced again and again, and need not be added. Option (4) makes no sense, since pure iron is the desired product of the reactions. Option (5) is not needed since oxygen is available in the air.

6. (Comprehension) **(5) coke and carbon dioxide** The flow chart shows that coke and carbon dioxide react to produce carbon monoxide. No other steps produce carbon monoxide.

Review: Chemistry

In Lessons 13 and 14 you have learned about matter and the different states of matter. You have learned how changes in matter are produced by chemical reactions. In the material that follows, you will learn more about matter.

Directions: Choose the best answer for each item.

Items 1 to 4 refer to the following passage.

All molecules are constantly in motion. The motion depends upon whether the molecules are in a solid, liquid, or gas.

Molecules in a solid move the least. They only vibrate. Forces of attraction hold the molecules of a solid very tightly in place, which is why solids have both definite volume and definite shape.

The molecules in a liquid move more freely and have more energy than the molecules of a solid. Forces of attraction between liquid molecules hold them together. That is why, when you pour a liquid, it stays together.

Molecules of a gas move freely and randomly. There are almost no forces of attraction between gas molecules. A gas will escape from an open container and "disappear."

---

1. In a carbonated beverage you can see bubbles of gas rising to the surface. What causes these bubbles?

   (1) vibrations of gas molecules
   (2) collisions between liquid and gas molecules
   (3) gas molecules moving rapidly through less freely-moving liquid molecules
   (4) force of attraction between liquid and gas molecules
   (5) liquid molecules pushing against less freely-moving gas molecules

2. Which of the following conclusions can be supported by the passage?

   (1) Large molecules move faster than small molecules.
   (2) Molecules in gases have less energy if the gas is in a small container.
   (3) You can feel the molecules vibrating when you hold a solid.
   (4) Molecules in solids stay in one place, while molecules in liquids and gases are free to move from place to place.
   (5) Molecules of a gas will spread out in a regular pattern as they move to fill a container.

3. A block of ice is heated, and it changes into a liquid. This shows that as temperature increases,

(1) movement of molecules decreases
(2) movement of molecules increases
(3) movement of molecules becomes less random
(4) molecules collide less often
(5) changes of state take place more slowly

4. The passage discusses a relationship between which two factors?

(1) states of matter and movement of molecules
(2) temperature and energy of molecules
(3) states of matter and sizes of molecules
(4) sizes of molecules and movement of molecules
(5) temperature and attractions between molecules

Items 5 and 6 refer to the following information.

If the numbers of protons and neutrons in the nucleus of an atom are very different, the atom will be radioactive. A radioactive atom decays and gives off particles until the nucleus is balanced. The time needed for half the nuclei in a sample of radioactive material to decay is called its half-life. The half-life of carbon-14, for example, is 5,700 years. This means that after 5,700 years, half the carbon-14 in a given sample will have decayed into another substance.

HALF-LIFE OF CARBON 14

1 gram
carbon-14

0 years

1/2 gram
carbon-14

5,770 years

1/4 gram
carbon-14

11,540 years

1/8 gram
carbon-14

17,310 years

5. Using information about the half-life of radioactive substances would be of most interest to which scientist?

(1) a metallurgist looking for ways to remove metals from ores
(2) a geologist interested in estimating the age of rock samples
(3) a chemist developing new products from organic compounds
(4) an oceanographer studying wave motion
(5) a biochemist studying photosynthesis

To check your answers, turn to page 204.

6. Which of the following conclusions can be supported by the information provided?

(1) After 23,080 years, none of the carbon-14 will remain.
(2) After 23,080 years, 1/8 gram of carbon-14 will remain of the original 1 gram.
(3) After 23,080 years, 1/16 gram of carbon-14 will remain of the original 1 gram.
(4) Radioactive elements can be made in the laboratory.
(5) The number of electrons affects the radioactivity of an atom.

Answers and Explanations

1. (Analysis) **(3) gas molecules moving rapidly through less freely-moving liquid molecules** This is correct because gas molecules have more freedom of motion than liquid molecules. Option (1) is incorrect because solid molecules, not gas molecules, move by vibrating. Option (2) is not a good choice because, although gas molecules could collide with liquid molecules, this would not account for the gas bubbles rising to the top of the liquid. Option (4) is incorrect because no mention is made of forces of attraction between liquids and gases. Option (5) is incorrect because it is the reverse of the correct answer.

2. (Evaluation) **(4) Molecules in solids stay in one place, while molecules in liquids and gases are free to move from place to place.** This is correct because molecules of a solid can only vibrate, while molecules of a gas are moving freely and have no forces of attraction between them.

Nothing in the passage supports options (1), (2), and (3). Option (5) contradicts the statement that gas molecules are in constant random motion.

3. (Analysis) **(2) movement of molecules increases** Because the passage states that molecular motion is greatest for a gas and least for a solid, you can conclude that the movement of molecules must increase as temperature increases. Option (1) is incorrect because it is opposite to the correct answer. Options (3) and (4) are opposite to what can be inferred from the passage. Option (5) is incorrect because the passage says nothing about how quickly matter changes state.

4. (Comprehension) **(1) states of matter and movement of molecules** This is correct because the passage describes how molecular motion is different for gases, liquids, and solids. Options (2) and (5) are incorrect because no mention of temperature is made in the passage. Options (3) and (4) are incorrect because no mention of molecular size is made.

5. (Application) **(2) a geologist interested in estimating the age of rock samples** By finding out the amount of radioactive material compared to decayed material in a rock sample, geologists can estimate the age of the rock. The other activities would not make use of radioactive half-life.

6. (Evaluation) **(3) After 23,080 years, 1/16 gram of carbon-14 will remain of the original 1 gram.** The diagram shows that every 5,770 years, half of the carbon-14 has decayed. At 17,310 years, 1/8 gram is left. After another 5,770 years, or 23,080 years altogether, half of 1/8, or 1/16, gram would be left. Option (1) is incorrect because 1/16 gram would be left. Option (2) is incorrect because 1/8 gram remains after 17,310 years. Option (4) is true, but it is not supported by the information provided. Option (5) is incorrect because the passage indicates that the number of protons and neutrons affects the radioactivity of an element.

Performance Analysis
Chemistry Review

Use the chart below to help you identify your strengths and weaknesses in each thinking skill area for the subject of chemistry.

Circle the number of each item that you answered correctly on the Chemistry Review.

Thinking Skill Area	Chemistry	Lessons for Review
Comprehension	4	1, 2, 4, 9
Application	**5**	10, 13
Analysis	1, 3	3, 5, 6, 8, 11, 12
Evaluation	2, **6**	7, 14

Boldfaced numbers indicate items based on diagrams, charts, graphs, and illustrations.

If you answered 4 or more of the 8 items correctly, congratulations! You are ready to go on to the next section.

If you answered 3 or fewer questions correctly, determine which skill areas are most difficult for you. Then go back and review the chemistry lessons for those areas.

PHYSICS

All the energy needed for a roller coaster ride comes from the first big hill.

Physics is the study of matter and energy. It is the branch of science that looks for answers to these questions: What is matter? What is energy? How are energy and matter related?

◆ **force**
a push or pull acting on an object

In Lesson 15 you will learn some scientific laws that describe the behavior of matter and energy. You will read about two sets of laws — Newton's laws of motion and the laws of thermodynamics.

You will learn how motion is described by Newton's first and second laws. You will discover that motion is changed by **forces**. You will also learn how machines can change the size and direction of a force to make work easier.

☞ *See Also: GED Exercise Book Science, pages 28–34*

thermodynamics
the study of the relationship between heat energy and the energy of motion

electric current
the flow of electrons through a wire

electric circuit
a continuous, unbroken pathway over which electric current can flow

wave
a disturbance that travels through space or matter

Lesson 15 also describes the laws of **thermodynamics**. These laws explain the relationship between the energy of heat and the energy of motion. Because heat is a form of energy, it can be converted into other forms of energy. When heat is converted into the energy of motion, it is able to do work. You will discover how this takes place in the gasoline engine of an automobile.

In Lesson 16 you will discover what scientists mean when they say "opposites attract," as you learn about electricity and magnetism. You will discover how electricity is related to the tiny, charged particles inside an atom. You will also find out how an **electric current** is produced and how an **electric circuit** makes the flow of current possible.

In Lesson 16 you will also read about the discovery of the first magnet over 2,000 years ago. You will learn about the properties of magnets, and how an important relationship exists between electricity and magnetism.

In Lesson 17 you will read about the properties of waves. A **wave** is a disturbance that transfers energy from one place to another. You will find out how a sound wave transfers energy by disturbing the molecules of the matter through which it travels. You will also learn about light waves, and how they are able to travel through space where there is no matter to carry the waves.

In this lesson you will discover what happens to a light wave when it strikes a substance and bounces back, or when it passes from one material into another.

LESSON 15 Evaluation Skill: Making Judgments

When you make a judgment, you choose the best alternative from a group of possibilities.

You make judgments all the time. Whenever you are faced with a problem, you think about the possible solutions, and you select the solution you decide is best. For example, suppose you have to make a weekend trip to visit your brother in a city two hundred miles away. You can take a plane, train, or bus, or you can drive your own car. If you have a limited amount of money to spend, you would have to compare the costs of these options to decide which is best.

People also make judgments about science-related matters. Scientists have to decide the best way to set up an experiment in order to get reliable results. Engineers have to decide the best way to solve problems of energy, force, and motion when designing roads, buildings, and machines.

Read this passage and decide on the best way to solve a problem.

Microwaves have high heating power. In a microwave oven, microwaves are produced by a device called a magnetron located in the top or side of the oven. The beam of microwaves strikes a spinning fan located in the center of the oven. The fan reflects the microwaves in all directions. In addition, the microwaves bounce off the inner surface of the oven and are reflected in still more directions. Microwaves pass through the container holding the food and heat it throughout.

Suppose you are having trouble heating food evenly in a microwave oven. You are placing the food on the side of the oven where the magnetron is located. How can you improve the heating action? You could heat the food longer so the cool spots also heat up. This would work, but the hot spots might overheat. Another possibility is to interrupt the cooking to stir the food. That might work, but you would have to pay attention to the cooking time. Perhaps you can solve the problem by placing the food in the center of the oven. The food will cook evenly, because more microwaves will reach the food in the center of the oven than at the sides.

When you are deciding the best way to solve a problem or thinking about the advantages and disadvantages of various actions, first decide if each possibility is logical and would indeed solve the problem. Then select the solution with the most advantages.

Practicing Evaluation

Read the following passage.

Climbing to the top of a mountain by the steepest slope requires a great deal of effort, or force. However, the distance that has to be covered is short. On the other hand, climbing the same mountain by the gentlest slope requires less force, but the distance to be covered is greater. In both cases, the amount of work required is the same. Work is the result of a force moving an object. This relationship can be expressed as:

$$\text{work} = \text{force} \times \text{distance}$$

Work is measured in units called joules; force is measured in units called newtons, which measures the pull of gravity on an object, or its weight; and distance is measured in meters. For example, the amount of work involved in lifting an object that weighs 10 newtons a distance of 0.5 meters is

$$\text{work} = \text{force} \times \text{distance} = 10 \text{ newtons} \times 0.5 \text{ meters} = 5 \text{ joules}$$

Question 1 refers to the passage. Circle the best answer.

1. A child is asked to help clean a room by picking up one toy and putting it away. Putting away which toy would require the least work?

    (1) a truck weighing 12 newtons that goes on a shelf 0.5 meters high
    (2) a ball weighing 6 newtons that goes in a box 1 meter high
    (3) a catcher's mitt weighing 8 newtons that goes on a hook 1.5 meters high
    (4) a plastic dinosaur weighing 10 newtons that goes in a basket 1.5 meters high
    (5) a bucket of blocks weighing 20 newtons that goes on a shelf 0.1 meter high

To check your answer, turn to page 216.

Topic 15: Motion

Read the following passage.

> A **force** is a push or pull that acts on matter, causing it to speed up, slow down, or change direction. A change in the speed or direction of motion is called **acceleration**.
>
> The English scientist Sir Isaac Newton formulated laws about motion. The **first law of motion** states that an object at rest tends to remain at rest, and an object in motion tends to keep moving in a straight line at the same speed, until they are acted upon by outside forces. For example, when you throw a ball across a field it eventually falls to the ground and stops rolling. It stops moving because the force of gravity has pulled it down, and the friction from the air and the ground has slowed it down. Without gravity and friction, the ball would have continued moving in a straight line.
>
> The tendency of an object to keep moving or remain at rest is called **inertia**. You become aware of inertia when you are riding in a car that suddenly stops. Because of inertia, you will keep moving forward until a force stops you. If you are wearing a seat belt, the seat belt will exert a force to slow and stop you. Otherwise, the dashboard or windshield will stop you.
>
> Newton's **second law of motion** states that an object will accelerate in the direction of the force that acts upon it. The mass of an object and the force that acts upon it will affect how the object accelerates. For example, a ten-ton truck requires more force than a one-ton compact car to accelerate away from a stoplight at the same rate. Newton's second law can be expressed by the formula:
>
> $$force = mass \times acceleration$$

Exercises

Questions 1 to 8 refer to the passage. Circle the best answer for each question.

1. Which of the following designs would be best for a racing car?

    (1) small engine and lightweight body
    (2) large engine and lightweight body
    (3) small engine and heavy body
    (4) large engine and heavy body
    (5) small engine and large gas tank

2. Once a spacecraft reaches outer space, its inertia would keep it moving in a straight line at a constant speed, even if its engine is not used. What is most likely to cause the spacecraft to change direction?

    (1) running out of fuel
    (2) energy from the Sun
    (3) the force of friction
    (4) the force of gravity
    (5) Nothing can cause it to change direction.

3. If a force of 8 newtons is applied to a 2-kilogram ball, and a force of 6 newtons is applied to a 3-kilogram ball, how will the accelerations of the balls compare?

(1) Both accelerations will be the same.
(2) The acceleration of the second ball will be twice that of the first.
(3) The acceleration of the first ball will be twice that of the second.
(4) The acceleration of the first ball will be four times that of the second.
(5) The acceleration of the second ball will be four times that of the first.

4. According to Newton's laws, an outside force would be required to make which of the following situations occur?

(1) A cyclist traveling at 15 mph continues to travel at the same speed in the same direction.
(2) A person who stood through the first hour of a sold-out lecture stands through the second hour.
(3) A passenger on the subway sits reading while the train travels four miles in four minutes.
(4) Two children forehead-to-forehead stare into each other's eyes.
(5) A car traveling at 40 mph goes around a curve at the same speed.

5. Which of the following is an example of Newton's second law of motion?
   A. a pitcher throwing a fast ball
   B. a seat belt preventing a person from hitting the windshield
   C. a parked car

(1) A only
(2) B only
(3) C only
(4) A and B
(5) B and C

6. Which of the following situations is an example of Newton's first law?

(1) A man finds that pushing a full wheelbarrow takes more force than pushing an empty one.
(2) A package on the seat of a car going 60 mph slides forward when the car stops suddenly.
(3) Passengers notice that a bus rides more smoothly at 50 mph than at 25 mph.
(4) A car stuck on an ice patch is able to move when a blanket is placed under the back wheels.
(5) A football player intercepts a pass and runs in the opposite direction from which the ball was thrown.

7. Football teams use large players in the defense lines and smaller players in the backfield to run and catch passes. What is an advantage of this strategy?

(1) Small players can accelerate quickly, while large players can apply force to stop the motion of opponents.
(2) Large players remain at rest, while light players remain in motion.
(3) Light players can catch passes, while large players can tackle.
(4) Large players tend to move in a straight line, while light players can change direction easily.
(5) The force needed to stop a large player is much greater than the force needed to stop a small one.

8. Newton's second law of motion describes the relationships among which factors?

(1) mass and acceleration
(2) direction and force
(3) direction, force, and acceleration
(4) mass, force, and acceleration
(5) mass, direction, and acceleration

To check your answers, turn to page 216.

Reviewing Lesson 15

Read the following paragraph.

Energy is the ability to move matter from one place to another or to change matter from one substance to another. Energy is never used up. It just changes from one form to another. For example, you use the energy stored in your muscles to lift an object such as a pen. At first, some of the energy is changed into the motion of the pen. This energy is called **kinetic energy**, or the energy of motion. But when the pen stops moving, what happens to the kinetic energy? It is stored in the form of **potential energy**, or the energy of position. Potential energy is released when matter moves. Drop the pen and you are releasing potential energy.

Questions 1 and 2 refer to the paragraph. Circle the best answer for each question.

1. Which of the following is an example of potential energy?

   (1) a woman jogging
   (2) a moving bicycle
   (3) a car parked on a hill
   (4) wind
   (5) a swiftly flowing stream

2. Which of the following is an example of kinetic energy?

   (1) a boy sitting still
   (2) a turning ferris wheel
   (3) a sleeping animal
   (4) a book on a table
   (5) a coat hanging in a closet

Read the following passage.

Thermodynamics is the study of the relationship of heat energy and kinetic energy. Heat energy can be changed into other forms of energy, and other forms of energy can be changed into heat. You can feel this change when you rub your hands together briskly. Your hands begin to feel warm because the kinetic energy of rubbing them together is being changed into heat.

When one form of energy is changed into another, energy is conserved. This means that no energy is lost in the process of changing from one form to another. The conservation of energy is the first law of thermodynamics.

Questions 3 and 4 refer to the previous passage. Circle the best answer for each question.

3. Why is the water at the bottom of Niagara Falls slightly warmer than the water at the top of the falls?

(1) The rocks at the base of the falls give off heat energy.

(2) More heat energy from the Sun reaches the bottom of the falls.

(3) The kinetic energy of the falling water changes to heat energy when it hits the bottom of the falls.

(4) The heat energy of the falling water changes to kinetic energy when it hits the bottom of the falls.

(5) Heat energy is lost when water hits the bottom of the falls.

4. On a chilly day you go out without a jacket and become cold. What is the best way to warm up?

(1) Sit down and wrap your arms around your legs to conserve heat energy.

(2) Lie down on the ground to conserve heat energy.

(3) Rub your hands briskly to change kinetic energy into heat energy.

(4) Wave your arms to change heat energy into kinetic energy.

(5) Exercise your whole body to change kinetic energy into heat energy.

Read the following passage.

The second law of thermodynamics describes the way heat moves from one object to another. Heat always flows from hotter objects to colder ones. The transfer of heat from one object to another can be explained in terms of moving molecules. The molecules in hot objects have more energy and move faster than the molecules in cold objects. When two objects come in contact with each other, the higher-energy molecules in the warmer object begin to collide with the molecules in the cooler object. Energy is transferred in the process. This process shows that there is no such thing as "coldness" — coldness is the absence of heat. As heat energy leaves an object, the object becomes less warm, or cold.

Questions 5 and 6 refer to the passage. Circle the best answer for each question.

5. Which of the following is most similar to the action of molecules in a heated substance?

(1) a car traveling at a steady speed in a straight line

(2) popcorn popping

(3) a ball rolling

(4) the Earth rotating

(5) a truck coming to a sudden stop

6. As an ice cube in your hand melts, your hand begins to feel cold. Which of the following is a conclusion?

(1) Coldness from the ice flows into your hand.

(2) Heat leaves your hand and is absorbed by the ice.

(3) The ice loses energy and becomes water.

(4) Molecules in your hand gain energy from the ice.

(5) Energy changes from one form to another as it goes from the ice to your hand.

To check your answers, turn to pages 216–217.

GED Mini-Test

Directions: Choose the best answer to each item.

Items 1 to 4 refer to the following passage.

Energy is the ability to do work. Because heat is a form of energy, it also can do work. Scientists define work as a force acting on an object, causing it to move.

One way in which heat is used to do work is in a heat engine. The gasoline engine in a car is a heat engine. In a gasoline engine, hot gases move into cylinders that contain pistons. When an electric spark ignites the gases, they explode and push against the pistons. The movement of the pistons eventually transfers energy to the wheels through a series of shafts and gears. The wheels exert a force against the ground, moving the car.

Most of the energy produced by gasoline engines is wasted. Only about 12 percent of the energy provided by the fuel is used to power the car. As a result, much of the heat produced by a gasoline engine is released through the exhaust pipe into the atmosphere.

---

1. Which of the following is the force that begins to set a car in motion?

   (1) the burning of gasoline
   (2) the push of hot gases on a piston
   (3) the push of hot gases on shafts and gears
   (4) the push of pistons on shafts and gears
   (5) the push of the wheels against the ground

2. According to the information provided, in which of the following situations would you be doing work?

   (1) studying for a physics test
   (2) standing in line for an hour
   (3) holding a heavy bag of groceries for an hour
   (4) hitting a tennis ball across the net
   (5) dropping a ball out of a second-story window

3. In a gasoline engine, hot gases release energy to power a car by

   (1) releasing oxygen
   (2) exerting pressure
   (3) transferring heat to other substances
   (4) releasing heat to the atmosphere
   (5) changing water into steam

4. In the same town, why is the temperature on heavily traveled streets higher than on streets with less traffic?

   (1) The heavily traveled streets absorb more sunlight than the less traveled streets.
   (2) Heat released from many cars raises the temperature of heavily traveled streets.
   (3) The wearing away of road surfaces raises the temperature of heavily traveled streets.
   (4) Heavily traveled streets have fewer trees to provide shade.
   (5) More heat-generating industries are near heavily traveled streets.

A machine is a device that helps you do work. A machine helps you by changing the force or the distance, or both. The amount of help a particular machine gives is called its mechanical advantage. The bigger the mechanical advantage, the less force you must provide.

The mechanical advantage of a machine can be found in two ways. The first way is to divide the force of the resistance (or the weight of the object to be moved) by the force of the effort. For example, if you apply 25 pounds of effort force to raise a 200-pound weight using a pulley, you can figure out that the mechanical advantage of the pulley is 8:

$$\text{mechanical advantage} = \frac{\text{resistance force}}{\text{effort force}} = \frac{200 \text{ pounds}}{25 \text{ pounds}} = 8$$

Another way to figure out mechanical advantage is to use the distances involved. Divide the effort distance by the resistance distance. The mechanical advantage of the inclined plane shown in the diagram is the length of the slope divided by its height, or 2:

$$\text{mechanical advantage} = \frac{\text{effort distance}}{\text{resistance distance}} = \frac{4 \text{ m}}{2 \text{ m}} = 2$$

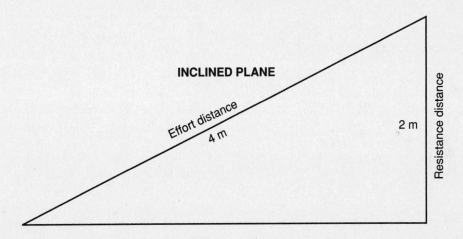

INCLINED PLANE

Effort distance
4 m

2 m

Resistance distance

---

5. The pulley with the greatest mechanical advantage is the one that requires

(1) 40 pounds of effort to raise a 480-pound rock
(2) 10 pounds of effort to raise a resistance of 50 pounds
(3) 80 pounds of effort to raise an 800-pound boulder
(4) 15 pounds of effort to raise a 300-pound box
(5) 50 pounds of effort to raise a resistance of 650 pounds

To check your answers, turn to page 217.

6. You must raise a box four meters off the ground. Which inclined plane would give you the greatest mechanical advantage?

(1) 3 meters high and 9 meters long
(2) 4 meters high and 6 meters long
(3) 4 meters high and 8 meters long
(4) 4 meters high and 12 meters long
(5) 3 meters high and 4 meters long

Answers and Explanations

Practicing Evaluation (page 209)

1. (Evaluation) **(5) a bucket of blocks weighing 20 newtons that goes on a shelf 0.1 meter high** To answer this question, you must figure out the amount of work involved in putting away each toy. Do this by multiplying the weight of the object by the distance it must be lifted. Thus the amount of work involved is 6 joules for option (1), 6 joules for option (2), 12 joules for option (3), 15 joules for option (4), and 2 joules for option (5). Putting away the bucket of blocks requires the least amount of work (2 joules).

Exercises (pages 210–211)

1. (Evaluation) **(2) large engine and lightweight body** The large engine provides a lot of force, and the lightweight body has little mass. This combination makes possible rapid acceleration, which is desirable for a racing car. Options (1), (3), and (5) are incorrect because a small engine would not provide enough force for rapid acceleration. Option (4) is incorrect because the heavy body would decrease the amount of acceleration the large engine could provide.

2. (Analysis) **(4) the force of gravity** Gravity from a celestial body is the most likely force to affect a spacecraft in outer space. Option (1) is incorrect because the spacecraft is not using fuel in outer space. Option (2) might warm the spacecraft, but will not cause a change of direction. Option (3) is incorrect because the force of friction is absent in outer space where there is no matter. Option (5) is incorrect because something can cause a change in direction.

3. (Analysis) **(3) The acceleration of the first ball will be twice that of the second.** According to the equation, force equals mass times acceleration, the acceleration of the first ball will be 4 meters per second² (the way acceleration is expressed) and the second ball's acceleration will be 2 meters per second². The remaining options indicate a misuse of the formula.

4. (Application) **(5) A car traveling at 40 mph goes around a curve at the same speed.** This is correct because force is required to change the direction of motion. All the other options are incorrect because they describe either objects at rest or objects continuing in motion in a straight line.

5. (Application) **(1) A only** By changing the force applied to the ball, a pitcher can alter its acceleration, which demonstrates Newton's second law of motion. The other options include examples B and C, which show inertia, Newton's first law of motion.

6. (Application) **(2) A package on the seat of a car going 60 mph slides forward when the car stops suddenly.** This is correct because the package keeps moving forward even though the car stops. Option (1) is incorrect because it is an example of Newton's second law. Options (3), (4), and (5) are incorrect because they describe types of motion that are not related to the information provided.

7. (Evaluation) **(1) Small players can accelerate quickly, while large players can apply force to stop the motion of opponents.** This explanation is based on Newton's second law. Option (5) is a true statement according to Newton's second law, but it does not explain the strategy. Option (2) is not necessarily true. Options (3) and (4) may or may not be true, but they do not explain the strategy.

8. (Comprehension) **(4) mass, force, and acceleration** Newton's second law of motion can be expressed as force equals mass times acceleration. The other options list only two of the three factors or list an incorrect factor.

Reviewing Lesson 15 (pages 212–213)

1. (Application) **(3) a car parked on a hill** This is an example of potential energy because, if the brake were released, the car would roll down the hill. The other options are all examples of things in motion, which have kinetic energy rather than potential energy.

2. (Application) **(2) a turning ferris wheel** Since the ferris wheel is in motion, it is an example of kinetic energy. The other options are examples of things at rest, which have potential energy rather than kinetic energy.

3. (Analysis) **(3) The kinetic energy of the falling water changes to heat energy when it hits the bottom of the falls.** The falling water strikes rocks at the base of the falls, causing some of its

kinetic energy to be converted into heat energy and raising the temperature of the water at the bottom slightly. Option (1) is incorrect because it is not likely that underwater rocks would give off heat energy. Option (2) is incorrect because the amount of energy from the Sun reaching the base of the falls would be about the same as the amount reaching the top. Option (4) is the opposite of what happens. Option (5) is incorrect because energy cannot be lost; it just changes into another form.

4. (Evaluation) **(5) Exercise your whole body to change kinetic energy into heat energy.** The idea here is to use motion to generate heat. Options (1) and (2) do not generate heat, so they can be eliminated. Options (3), (4), and (5) all involve changing kinetic energy into heat energy, so all are possible. However, the more kinetic energy involved, the more heat energy will be generated. Therefore, it follows that exercising your whole body will generate the most heat energy, option (5).

5. (Application) **(2) popcorn popping** As popcorn pops, the kernels move and collide with one another. This is most similar to the movement of molecules in a heated substance. Options (1), (3), and (4) are steady, even movements not typical of heated molecules. Option (5) is the opposite of what happens to a heated molecule.

6. (Analysis) **(2) Heat leaves your hand and is absorbed by the ice.** Option (1) is incorrect because there is no such thing as "coldness." Options (3) and (4) are incorrect because the heat transfer

from your hand to the ice causes molecules in your hand to lose energy and molecules in the ice to gain energy. Option (5) is not true.

GED Mini-Test (pages 214–215)

1. (Comprehension) **(2) the push of hot gases on a piston** Option (1) is incorrect because burning is a chemical reaction, not a force, although heat from burning gasoline can produce a force. Option (3) is incorrect because hot gases do not act directly on shafts and gears. Options (4) and (5) are incorrect because they are forces that occur later in the process of moving the car.

2. (Application) **(4) hitting a tennis ball across the net** This option is correct because it is the only situation in which you are producing a force that acts on an object and causes it to move. Options (1), (2), and (3) do not involve motion. In option (5), which is tricky, you are not producing the force that is moving the ball — you are just allowing gravity to pull it to Earth.

3. (Comprehension) **(2) exerting pressure** This is correct because the gases push against the pistons as a result of exploding upon ignition. Option (1) is incorrect because burning the gases would require oxygen rather than release it. Option (3) is incorrect because, although heat is transferred to surrounding substances, this is not what powers the car. Option (4) is incorrect because this is how heat energy not used to power the car is wasted. Option (5) has nothing to do with the gasoline engine in a car.

4. (Analysis) **(2) Heat released from many cars raises the temperature of heavily traveled streets.** All of the other options could be true — although option (1) sounds a little far-fetched — but they do not explain the information provided.

5. (Evaluation) **(4) 15 pounds of effort to raise a 300-pound box** This is correct because the mechanical advantage is 300/15, or 20. The mechanical advantages of the other options are lower. Option (1) is 12, option (2) is 5, option (3) is 10, and option (5) is 13.

6. (Evaluation) **(4) 4 meters high and 12 meters long** Since you want to raise the box 4 meters, you can immediately eliminate options (1) and (5), because the inclined planes are only 3 meters high. To find the best answer among the remaining options, calculate the mechanical advantage of each inclined plane by dividing the length of the slope by its height. The mechanical advantage of option (2) is 1.5. The mechanical advantage of option (3) is 2. The mechanical advantage of option (4) is 3; thus option (4) is correct.

Lesson 16 Analysis Skill: Cause and Effect

When you think about cause and effect, you are thinking about how one thing influences another. A cause is what makes something happen. An effect is what happens as a result of the cause.

When you flip on a light switch, you cause the lights to come on. This is a simple example of cause and effect. An obvious action, or cause, produces an obvious result, or effect.

Sometimes cause and effect relationships are not so obvious in science. In the light switch example, for instance, you can go further and think about what is really happening. When you flip the switch, you are completing an electric circuit, allowing current to flow into the bulb and produce light. That's another way to think of cause and effect in regard to turning on a light.

Now take this example further still. You can think about the nature of electricity, which you will learn about in this lesson. When you flip the switch, you are providing electrons with a path to flow through. Thus you can think about the cause and effect of one situation in many different ways.

Remember, when you are looking for cause and effect relationships, watch for words and phrases like thus, therefore, as a result, what happens is, because, since, effect, and consequently.

Practicing Analysis

Read the following paragraph.

A photocopier works by using static electricity to produce copies of documents. Inside the photocopier is a metal drum that is given a negative charge. Lenses then project an image of the document onto the drum. Where light strikes the metal surface, the electric charge disappears. Only the dark parts of the image remain negatively charged. The copier contains a dark powder called toner, which has a positive charge. The toner is attracted to the negatively charged dark parts of the image. The toner is transferred to a piece of paper and is sealed by heat. A warm copy of the document comes out of the machine.

Questions 1 and 2 refer to the previous paragraph. Circle the best answer for each question.

1. What causes the toner to form the image of the document on the drum?

(1) The positively charged toner is attracted to the negatively charged areas of the drum.

(2) The toner fills in the raised image of the document on the drum.

(3) The document is negatively charged, and the positively charged toner is attracted to it.

(4) Heat from the machine causes the toner to form the image of the document.

(5) Light from the lenses causes the image of the document to be traced on the drum.

2. The lenses failed to project an image of the document on the negatively charged drum. Rather, just light was projected. What would happen as a result?

(1) The toner would be attracted to the drum.

(2) The toner would be attracted to the document rather than the drum.

(3) A grayish-black sheet of copy paper would result.

(4) A blank sheet of copy paper would result.

(5) An acceptable copy of the document would result.

Read the following information.

The friction created when objects move in air is called **drag**. Drag acts to slow down a moving object. The amount of drag depends on the shape of the object. Air flows more smoothly around objects with a tapered shape.

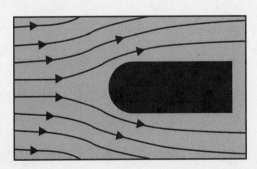 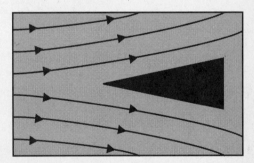

Question 3 refers to the information. Circle the best answer.

3. Which of the following would be an effect of streamlining the shape of an automobile?

(1) increased drag

(2) more speed for the same amount of fuel energy

(3) less speed for the same amount of fuel energy

(4) more speed for more fuel energy

(5) There would be no effect.

To check your answers, turn to page 226.

Topic 16: Electricity and Magnetism

Read the following passage.

Have you ever heard the saying "opposites attract"? People usually use this phrase to describe opposite personalities. But this phrase can also describe the forces in two important areas of physics — electricity and magnetism.

You may think of electricity as turning on the lights or starting an electric motor. Certainly these are important uses of electricity — but what exactly <u>is</u> electricity? Where does it come from? And how does it power the appliances that you use every day?

To answer these questions, you must first go back to the atom that you studied in Lesson 13. You will recall that atoms are made up of protons, electrons, and neutrons. Protons and electrons have a property called **electric charge**. Protons are positively charged (+1) and electrons are negatively charged (–1). Neutrons have no charge — as their name implies, they are neutral (0).

If protons and electrons did not have opposite charges, an atom would simply fly apart. The force of attraction between the positively charged protons and the negatively charged electrons holds an atom together. The structure of an atom shows the basic rule of electric charge: Unlike charges attract each other, while like charges repel each other. You can remember this rule easily if you just think of the old saying "opposites attract."

If one charged particle comes near another charged particle, it will experience a force. The **force** will be one of attraction if the particles are of unlike charge, and repulsion if the particles are of like charge. The area of force that surrounds a charged particle is called an **electric field**. The strength of an electric field depends on the distance from the charged particle — as the distance increases, the strength of the field decreases.

THE HELIUM ATOM

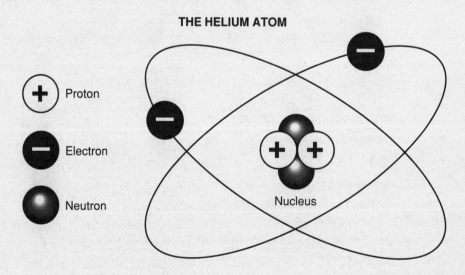

Exercises

Questions 1 to 6 refer to the previous passage. Circle the best answer for each question.

1. A negatively charged strip of metal is suspended from the ceiling by a string. When another strip of metal is brought close to it, the first strip bends away. Which statement explains why this happens?

   (1) The second strip is positively charged.
   (2) The second strip is negatively charged.
   (3) The second strip contains fewer electrons than the first strip.
   (4) The second strip has a larger electric field than the first strip.
   (5) The atoms in the second strip repel the atoms in the first strip.

2. Which of the following particles always has a negative charge?

   (1) proton
   (2) neutron
   (3) electron
   (4) atom
   (5) field

3. What is an electric field?

   (1) a group of electrons
   (2) a group of protons
   (3) an area of force
   (4) a charged particle
   (5) positively charged atoms

4. If an atom contains 3 protons, 4 neutrons, and 2 electrons, what is the charge on the atom?

   (1) +2
   (2) +1
   (3)   0 (neutral)
   (4) −1
   (5) −2

5. Atoms and molecules with an electric charge are called ions. What is likely to happen when positively charged ions are mixed with negatively charged ions?

   (1) The negative and positive ions will attract each other.
   (2) The negative ions and positive ions will remain separate.
   (3) The negative ions produce a larger electric field than the positive ions.
   (4) The negative and positive ions will repel each other.
   (5) Positive ions will remain grouped together.

6. What holds an atom together?

   (1) force of gravity
   (2) chemical bonds
   (3) its nucleus
   (4) force of attraction between protons and electrons
   (5) force of attraction between neutrons and electrons

To check your answers, turn to page 226.

Reviewing Lesson 16

Read the following passage.

Have you ever rubbed a balloon against your sleeve and stuck it to the wall? You were able to do this because the balloon became electrically charged.

Electrons in atoms are free to move. As you rub a balloon on your sleeve, electrons from the cloth move onto the balloon. The extra electrons give the balloon a negative charge. When the balloon comes near the wall, it repels the electrons in the wall, leaving the wall positively charged. Thus, the negatively charged balloon and the positively charged wall attract each other.

The ability of electrons to move from one place to another makes electric current possible. Electric current is the flow of electrons through a wire, and it is what powers your appliances.

Electrons flow through a wire in much the same way as water flows through a hose. Just as water pressure pushes water through a hose, a source of energy pushes electrons through a wire. This source of energy is called **voltage**. Voltage is measured in units called **volts**. A battery marked "9V" means that the battery supplies nine volts of energy to move electrons through a wire.

Questions 1 to 4 refer to the passage. Circle the best answer for each question.

1. If you scuff your feet across a wool carpet, you may feel a shock when you touch a metal object. Which statement explains why this happens?

   (1) An electric charge builds up on the rug and then is transferred to the metal object.
   (2) Electrons from the carpet move onto your feet, causing you to temporarily have an electric charge.
   (3) Electrons flow from your feet to the carpet, causing the carpet to have an electric charge.
   (4) Electrons in the metal object are repelled by electrons on the carpet.
   (5) Scuffing across the carpet causes an increase in voltage, which then produces a shock.

2. Which of the following statements cannot be supported by the information provided?

   (1) Protons flowing onto an object cause the object to become positively charged.
   (2) Low voltage could cause a power "brownout" in a town or city.
   (3) If two balloons that had been rubbed on cloth were brought close together, they would probably repel each other.
   (4) A torn wire could cause a loss of electricity, just as a leaky hose would cause a loss of water.
   (5) Current flow increases as voltage increases.

3. Extra electrons on an electrically charged object eventually leave the object, returning it to its neutral state. A person who wishes to stick balloons to the wall as party decorations should

(1) place the balloons on the wall just before the party begins
(2) place the balloons far apart
(3) place the balloons on the wall at least a day ahead of time
(4) rub the balloons on several different types of cloth
(5) place the balloons as close to the ceiling as possible

4. In the passage, water pressure is compared to voltage because

(1) both provide energy to move electrons
(2) both make possible the flow of a substance from one place to another
(3) both are measured in units called volts
(4) both involve the transfer of electrically charged particles
(5) both can be used to power a battery

Read the following information.

In order for electric current to flow, electrons must have a closed, continuous pathway over which to travel. Such a pathway is provided by an electric circuit. An **electric circuit** consists of a source of electrons, a load or resistance, and a switch. In the diagram the source of electrons is an eight-volt battery. The wavy lines you see are resistances. A resistance can be a light bulb, an appliance, or a motor — anything that uses electrical energy.

Electric current (I) is measured in amperes ("amps" for short). Resistance (R) is measured in ohms. Current, resistance, and voltage are related by the equation $V = I \times R$.

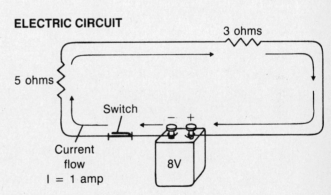

ELECTRIC CIRCUIT

Questions 5 and 6 refer to the information. Circle the best answer for each question.

5. If the voltage in the circuit above were reduced to four volts, what would be the effect on the current?

(1) It would stay the same.
(2) It would be cut in half.
(3) It would double.
(4) It would increase four times.
(5) It would be equal to the voltage.

6. If more resistance were added to the circuit, what would be the result?

(1) an increase in voltage
(2) a decrease in current
(3) an increase in current flow
(4) a decrease in voltage
(5) a greater number of electrons passing through each resistor

To check your answers, turn to pages 226–227.

GED Mini-Test

Directions: Choose the best answer to each item.

Items 1 to 4 refer to the following passage.

A magnetic field is formed around a wire conducting an electric current. This relationship between electricity and magnetism is known as electromagnetism.

Because of electromagnetism, powerful temporary magnets called electromagnets can be made by wrapping coils of wire around soft iron and passing an electric current through the wire. When the current passing through the wire is turned off, the magnet loses its magnetic properties. When the current is turned back on, the magnet regains its magnetic properties. The strength of the electromagnet depends on the number of loops of wire and the size of the current.

1. A heavy crane used at construction sites is equipped with a large electromagnet. The electromagnet will make the machine most useful for which of the following tasks?

    (1) transporting pieces of iron to distant locations
    (2) picking up pieces of scrap metal on the site and depositing them elsewhere on the site
    (3) generating electricity for the area surrounding the site
    (4) making magnets out of pieces of scrap metal found on the site
    (5) lifting objects too heavy for other types of machines on the site

2. The relationship between electricity and magnetism is best expressed by which of the following statements?

    (1) Electromagnets form magnetic fields.
    (2) Magnets cause electricity.
    (3) Wrapping wire around iron causes magnetism.
    (4) An electric current produces a magnetic field.
    (5) An electric field can be produced by an iron magnet.

3. The needle of a compass placed near the wire of an electric circuit turns away from north when the circuit is turned on. Why does this happen?

    (1) The needle has lost its ability to point north.
    (2) The wire in the circuit must be pointing north.
    (3) The needle is responding to the magnetic field produced by the current in the wire.
    (4) Electrons in the needle are repelled by electrons in the wire.
    (5) The compass needle has become an electromagnet.

4. Which hypothesis can be supported by the information given?

    (1) Magnets make electric current strong.
    (2) Electromagnets are not as strong as natural magnets.
    (3) A magnet can reverse the direction of an electric current.
    (4) The strength of an electromagnet depends upon the material used to make the magnet.
    (5) Magnetism is related to the movement of electrons.

Over 2,000 years ago, a mysterious stone was discovered that could attract bits of material containing iron. The Greeks who found this stone named it "magnetite." They had discovered an interesting property that we now call magnetism.

Magnetite is an example of a natural magnet. Most of the magnets you have probably used are artificial magnets. The simplest artificial magnet is an iron bar magnet.

If you suspend a bar magnet horizontally on a string and allow it to swing freely, one end of the magnet will always point north. This end of the magnet is called the north magnetic pole. The other end of the magnet, which points south, is called the south magnetic pole.

The area in which magnetic forces can act is called a magnetic field. The magnetic field of a bar magnet is strongest around the poles. If a north pole of one magnet and a south pole of another magnet are brought together, they will attract each other. If two north poles or two south poles are brought together, they will repel each other. Thus the rule for magnetic poles is: Unlike poles attract each other while like poles repel each other. Once again, "opposites attract."

---

5. The discovery of magnetite was important because it

   (1) made artificial magnets possible
   (2) was the first natural iron magnet
   (3) gave scientists information about the properties of iron
   (4) illustrated the property of magnetism
   (5) explained why one pole of a magnet always points north

6. The needle of a compass always points to the north. Which of the following must be true about compass needles?

   (1) Compass needles are made of magnetite.
   (2) Compass needles are made of iron.
   (3) Compass needles are magnets.
   (4) A compass needle has a north magnetic pole but no south magnetic pole.
   (5) Two compass needles will repel each other.

7. The fact that one end of a freely suspended magnet always points north suggests that

   (1) Climate influences magnetism.
   (2) The Sun has a magnetic field.
   (3) A deposit of magnetite is located north of the magnet.
   (4) The Earth has a magnetic field.
   (5) The polar ice caps are magnetic.

To check your answers, turn to page 227.

Answers and Explanations

1. (Analysis) **(1) The positively charged toner is attracted to the negatively charged areas of the drum.** Because the negatively charged areas of the drum correspond to the document image, this causes the toner to form the pattern of the image on the drum. Option (2) is incorrect because the image on the drum is an electrical image, not a raised image. Option (3) is incorrect because the document and toner never come in contact with one another. Option (4) is incorrect because electricity, not heat, causes the toner to form the image on the drum. Option (5) is incorrect because the image is not traced; it consists of negatively charged areas.

2. (Analysis) **(4) A blank sheet of copy paper would result.** Since light would reach all areas of the drum, the negative charge would disappear, leaving nothing for the toner to be attracted to. As a result, the copy would be blank. Option (1) is incorrect because the negative charge would be gone. Option (2) is incorrect because the toner and the document never come in contact. Option (3) is incorrect because no toner has been attracted to the drum; therefore no toner appears on the copy. Option (5) is incorrect because the image of the document did not reach the drum.

3. (Analysis) **(2) more speed for the same amount of fuel energy** If the automobile were streamlined, it would have less drag. This means that it would take less force (fuel energy) to

move at a particular speed or distance. Options (1), (3), and (4) are the opposite of what would happen. Option (5) is incorrect because there is an effect, option (2).

Exercises (page 221)

1. (Analysis) **(2) The second strip is negatively charged.** This is correct because like charges repel each other. Options (1) and (3) are not true. Option (5) is incorrect because it is positively and negatively charged particles, not atoms, that attract and repel each other. Option (4) may or may not be true, but it does not explain the effect being described.

2. (Comprehension) **(3) electron** According to the passage, electrons have a negative charge. Option (1) is incorrect because protons have a positive charge. Option (2) is incorrect because neutrons have a neutral charge. Option (4) is incorrect because atoms can be positive, negative, or neutral. Option (5) is incorrect because a field is not a particle.

3. (Comprehension) **(3) an area of force** According to the passage, an electric field is an area of force. The other options are all examples of matter, not force.

4. (Analysis) **(2) +1** There is one more proton than electrons, so the atom would have a positive charge. The number of neutrons does not matter because neutrons have no charge.

5. (Analysis) **(1) The negative and positive ions will attract each other.** This is true because like charges repel

and unlike charges attract. Options (2), (4), and (5) are incorrect because they state the opposite. Option (3) may or may not be true, but it does not answer the question.

6. (Comprehension) **(4) force of attraction between protons and electrons** According to the passage, the attraction between protons and electrons keeps atoms from flying apart. The other options do not explain what holds an atom together.

Reviewing Lesson 16 (pages 222–223)

1. (Analysis) **(2) Electrons from the carpet move onto your feet, causing you to temporarily have an electric charge.** This is correct because the situation is similar to the rubbing of a balloon on a sleeve. Options (1) and (3) are incorrect because you, not the carpet, receive the shock — so you must have an electric charge as a result of acquired electrons. Option (4) leaves you out of the action entirely, and option (5) takes the concept of voltage out of context to create a senseless answer.

2. (Evaluation) **(1) Protons flowing onto an object cause the object to become positively charged.** This is correct because no mention is made in the passage of protons being able to move and, in fact, they cannot. Options (2) and (5) are supported by the passage because electric current is pushed through wires by voltage, and a change in voltage would cause a change in current. Option (3) could be supported because

both balloons would have a negative charge, and like charges repel each other. Option (4) is supported by the analogy between the flow of electricity and the flow of water.

3. (Analysis) **(1) place the balloons on the wall just before the party begins** This option is correct because this would leave little time for the balloons to lose their negative charge and come loose from the wall. Option (3) would work exactly against this idea. Options (2), (4), and (5) would not help the balloons stick to the wall.

4. (Comprehension) **(2) both make possible the flow of a substance from one place to another** Options (1), (3), and (4) are incorrect because they apply only to voltage. Option (5) is incorrect because voltage is supplied by a battery — and water pressure has nothing to do with powering a battery.

5. (Analysis) **(2) It would be cut in half.** This is correct because the formula $V = I \times R$ shows that lowering voltage would lower current as long as resistance stays the same. The other options contradict the correct answer.

6. (Analysis) **(2) a decrease in current** This is correct because as resistance increases, current decreases. Options (1) and (4) are incorrect (this is tricky) because a battery made to deliver eight volts can only deliver eight volts. Options (3) and (5) both refer to an increase in current, and thus contradict the correct answer.

GED Mini-Test (pages 224–225)

1. (Evaluation) **(2) picking up pieces of scrap metal on the site and depositing them elsewhere on the site** This is correct because the electromagnet will gain and lose its magnetism as the current is turned on and off — thus it is ideal for picking up metal objects, and then dropping them again. Option (1) is incorrect because there is no way of knowing whether this machine is equipped to travel distances. Option (5) may or may not be true, depending on the relative strengths of this and other machines on the site. Options (3) and (4) are incorrect because the crane's electromagnet would not cause these effects.

2. (Comprehension) **(4) An electric current produces a magnetic field.** Option (1) is true, but it does not answer the question because it does not say anything specific about electricity. Options (2) and (5) are true, but this passage discusses only the production of a magnetic field from electric current, not the reverse. Option (3) fails to specify that the wire must be conducting an electric current.

3. (Analysis) **(3) The needle is responding to the magnetic field produced by the current in the wire.** Option (1) is incorrect because it does not adequately relate cause and effect. Also, once the current is turned off, the needle will again point northward. Option (5) is incorrect because the magnetism of the compass needle was not produced by the current in the wire. Options (2) and (4) have nothing to do with the situation being described.

4. (Evaluation) **(5) Magnetism is related to the movement of electrons.** The correct option is (5) because a flow of electrons (electric current) produces a magnetic field. Options (2) and (4) are not discussed in the passage. There is nothing in the passage to support options (1) and (3).

5. (Comprehension) **(4) illustrated the property of magnetism** Option (3) is possible, but the discovery of magnetite did much more than just reveal the properties of one metal. Option (2) is incorrect because magnetite is not made of pure iron — it is made of an iron-containing compound. Options (1) and (5) are incorrect because the understanding of poles and artificial magnets came much later.

6. (Evaluation) **(3) Compass needles are magnets.** Options (1) and (2) are possible, but not necessarily true — thus they are not the best choices. Option (4) is incorrect because any magnet must have a north and south pole. Option (5) is incorrect because it would depend upon how the compass needles were lined up — if the north pole of one needle were near the south pole of the other, they would attract each other.

7. (Comprehension) **(4) The Earth has a magnetic field.** The magnets must be turning so that one of their poles is pointing to the Earth's magnetic pole. Option (1) is unlikely. Option (2) is true, but does not explain the behavior of magnets on Earth. Option (3) cannot be true of all locations north of a magnet. Option (5) is incorrect because the ice caps are not magnetic.

LESSON 17 Evaluation Skill: Drawing Conclusions

When you draw a conclusion, you use examples, details, facts, and observations to prove that the conclusion is true.

You draw conclusions all the time. For example, when you look out the window to see what the weather is, you observe whether it's sunny or cloudy, windy or still. From these observations you conclude what kind of day it will be.

Scientists also draw conclusions. From measurable, observable data and from carefully designed experiments, scientists draw conclusions about the nature of the universe. If the conclusions are correct, they can be used to predict other events in the physical world.

When you read science, be on the lookout for rules, laws, principles, and equations. These are usually generalizations that can be used to evaluate specific information. For example, in an earlier lesson you learned about two of Newton's laws of motion. In answering questions about these laws, you evaluated specific events to see if they followed Newton's laws.

Thinking about conclusions also involves recognizing when conclusions are wrong. In addition, you must recognize when a conclusion is true, but it cannot be proved using the information at hand. You must examine the passage, diagram, or equation carefully to determine whether a conclusion follows logically from the information that is there.

Practicing Evaluation

Read the following passage.

A **particle accelerator** is an extremely long, narrow tunnel, either straight or circular. It is charged with powerful electric and magnetic fields. When physicists fire a tiny particle such as a proton or electron into the tunnel, the fields cause the particle to accelerate. The longer the tunnel, the greater the acceleration and energy of the particle.

Physicists can guide the particles along a path that makes them collide with each other or with another target. When the impact is great enough, the energy released as a result of the collision is converted into new particles. By studying these new particles, scientists can learn about the nature of matter.

Question 1 refers to the previous passage. Circle the best answer.

1. Which statement can be supported by the information provided?

   (1) A straight accelerator is better than a round one.

   (2) A round accelerator can achieve higher speeds than a straight one.

   (3) More valuable data is likely to be provided by an accelerator with a long tunnel than by one with a short tunnel.

   (4) Electrons move faster than protons in an accelerator.

   (5) Accelerators can produce large amounts of electricity.

Read the following passage.

Some materials carry electric current better than others. **Resistance** is a material's opposition to the flow of current. Resistance is measured in **ohms**. Resistance, voltage, and current are related in a formula called Ohm's law:

$$\text{Resistance} = \frac{\text{Voltage}}{\text{Current}}$$

Question 2 refers to the passage. Circle the best answer.

2. Which statement can be supported by the information provided?

   (1) The resistance of a lamp that uses 4 amps of current in a 12-volt circuit is 48 ohms.

   (2) The resistance of a doorbell that uses 6 amps of current in a 9-volt circuit is 0.66 ohms.

   (3) The resistance of a toaster that uses 5 amps of current in a 120-volt circuit is 24 ohms.

   (4) The resistance of an electric range that uses 5 amps of current in a 230-volt circuit is 40 ohms.

   (5) The resistance of a light bulb that uses 0.2 amps of current in a 14-volt circuit is 7 ohms.

To check your answers, turn to page 236.

Topic 17: Waves

Read the following information.

A **wave** is a disturbance that travels through space or matter. Waves transfer energy from one place to another. Some familiar types of waves include sound waves, light waves, radio waves, microwaves, and water waves. All waves, no matter what their type, have certain basic characteristics.

The **amplitude** of a wave is how high it rises from its rest position. When the sea is absolutely calm, the water is at its rest position. As a wave rises, the water reaches a high point, and then falls back down again. This high point is called the **crest**. The distance from the rest position to the crest is the amplitude. The greater the amplitude, the greater the energy of the wave.

The **shape** of a wave is created as the wave moves from its rest position to a high point, or crest, then back through the rest position to a low point. The low point of a wave is called the **trough**, and it is the same distance from the rest position as is the crest.

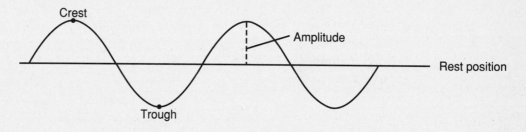

Exercises

Questions 1 to 3 refer to the passage and diagram. Circle the best answer for each question.

1. Which of the following statements <u>cannot</u> be supported by the information provided?

    (1) The amplitude of an ocean wave is equal to the distance from the rest position to the trough.
    (2) A three-foot high wave has a trough three feet deep.
    (3) The depth of the trough of a wave is double the distance from the rest position to the crest.
    (4) The higher the crest of a wave, the deeper its trough.
    (5) Crests and troughs alternate in a wave.

2. What is a wave?

    (1) a movement of water
    (2) a disturbance that travels through matter
    (3) energy in space
    (4) a disturbance that travels through space or matter
    (5) energy in space or matter

3. Which characteristic of an ocean wave would be of most interest to a surfer?

    (1) amplitude
    (2) rest position
    (3) trough
    (4) disturbance
    (5) spray

Read the following passage.

If you have ever watched the ocean, you have probably noticed that sometimes the waves seem to come very close together, while at other times they seem very far apart. What you were noticing was a difference in wavelength. **Wavelength** is the distance between the crests of two consecutive waves. Waves that appear very close together have a shorter wavelength, while those that appear far apart have a longer wavelength.

Frequency is the number of waves that pass by a given point in a specific unit of time. For example, if you watched an object in the ocean bob up and down ten times in one minute, the frequency of the wave would be ten cycles per minute. In order to be counted as one complete cycle, a wave must pass through both its crest and its trough.

If you know the wavelength and frequency of a wave, you can find its velocity, or speed. **Velocity** is the distance the wave covers in a specific unit of time. If the frequency of the wave is measured in Hertz, and the wavelength is measured in meters, then the velocity in meters per second is given by this equation:

$$velocity = wavelength \times frequency$$

Questions 4 to 7 refer to the passage. Circle the best answer for each question.

4. A wave's frequency is the relationship between

   (1) height and weight
   (2) height and distance between crests
   (3) distance between crests and amplitude
   (4) number of waves that pass a given point and unit of time
   (5) unit of time and distance

5. To calculate the frequency of a wave, you must

   (1) multiply wavelength times frequency
   (2) divide velocity by wavelength
   (3) multiply velocity times wavelength
   (4) divide wavelength by velocity
   (5) multiply velocity times frequency

6. What is the velocity of a wave with a wavelength of 3 meters and a frequency of 6 Hertz?

   (1) 2 meters per second
   (2) 3 meters per second
   (3) 9 meters per second
   (4) 18 meters per second
   (5) 36 meters per second

7. What is the result of decreased wavelength and frequency?

   (1) decreased velocity
   (2) the same velocity
   (3) increased velocity
   (4) increased distance
   (5) decreased distance

To check your answers, turn to page 236.

Reviewing Lesson 17

Read the following passage.

> Sounds travel as waves. For a sound wave to travel, it must have a **medium**, a substance capable of transmitting the wave. The medium can be a solid, a liquid, or a gas.
>
> The wave pushes the molecules of the medium back and forth parallel to its line of motion. During one complete cycle of a sound wave, the molecules are pushed together in a **compression**, then spread out in a **rarefaction**. You can think of the wave's motion as push forward-pull back. Such waves are called **longitudinal waves**.

Questions 1 and 2 refer to the passage. Circle the best answer for each question.

1. The motion of a longitudinal wave most resembles the motion of

  (1) an ocean wave
  (2) a pulsating rope held between two people
  (3) an accordion being played
  (4) a bouncing ball
  (5) a bicycle on a bumpy road

2. The Moon has no atmosphere. Which items would be <u>useless</u> to take along on a trip to the Moon?

  (1) a flashlight
  (2) thermal underwear
  (3) oxygen supply
  (4) cassette player with earphones
  (5) tape deck and speaker system

Read the following passage.

> Sound waves travel best through solids, because the molecules are packed tightly together. Elastic solids, such as nickel, steel, and iron, carry sound especially well; inelastic solids, such as sound-proofing materials, carry sound less well. Liquids are second-best to solids in carrying sound, and gases are the least effective carriers of sound.
>
> Sound travels through air at room temperature at about 1,140 feet per second. The speed of sound averages about 4,950 feet per second in water and about 19,700 feet per second in stone.

Questions 3 to 6 refer to the passage. Circle the best answer for each question.

3. It can be inferred from the passage that a gas is the <u>least</u> effective medium for the transmission of sound waves because

  (1) its molecules are too close together
  (2) its molecules are too far apart
  (3) it is too dense
  (4) it is inelastic
  (5) its temperature is too low

4. "Solid state" on audio equipment probably refers to which statement?

  (1) Sound molecules travel fastest in solids.
  (2) Solid components are more durable than other types of components.
  (3) Solids transmit sounds best.
  (4) Solids make the best amplifiers.
  (5) Solids are more elastic than gases.

5. On a very cold day, a baseball fan notices the ball in the air before hearing the crack of the bat. Which statement offers the best explanation?

(1) The person's ears have become stopped up by the cold.
(2) The players are hitting the ball slower because of the cold.
(3) Some people receive sound waves slower than other people.
(4) The sound waves have farther to travel than usual.
(5) Sound waves travel slower in cold air than in warm air.

6. The expression "Put your ear to the ground" probably originated because

(1) sound that travels through ground will reach your ears before sound that travels through air
(2) sound waves travel horizontally
(3) sound waves travel faster at low altitudes than at high altitudes
(4) some types of sounds travel only through solids
(5) sound waves are loudest near the ground

Read the following passage.

Light waves are **electromagnetic waves**. Unlike sound waves, electromagnetic waves do not need a medium through which to travel.

Light is the part of the electromagnetic spectrum that we can see. It includes those electromagnetic waves with frequencies between 400 trillion and 750 trillion Hertz. Below these frequencies are invisible waves such as infrared, radio waves, microwaves, and radar. Above the frequencies of visible light are other invisible waves, including ultraviolet, X-rays, and gamma rays. All electromagnetic waves travel at the same speed in a given medium. The speed of light in a vacuum is 186,282 miles per second.

Questions 7 and 8 refer to the passage. Circle the best answer for each question.

7. Compared to radio waves, ultraviolet waves are

(1) visible, while radio waves are invisible
(2) lower in energy than radio waves
(3) lower in frequency than radio waves
(4) higher in frequency than radio waves
(5) of longer wavelength than radio waves

8. Which statement can be <u>disproved</u> by the information provided?

(1) Air is the only medium through which light can travel.
(2) Light waves can travel through space.
(3) Gamma rays travel just as fast as microwaves.
(4) While sound cannot be heard on the Moon, light can be seen there.
(5) No one can see radio waves.

To check your answers, turn to pages 236–237.

GED Mini-Test

Directions: Choose the best answer to each item.

Items 1 and 2 refer to the following information.

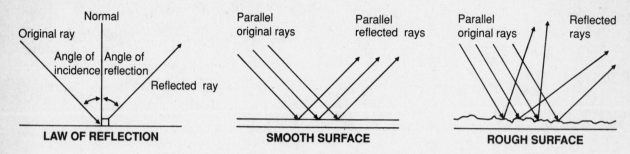

When a light ray strikes a surface and bounces back, we say that it is reflected. The law of reflection states that the angle at which the original ray strikes the surface (the angle of incidence) will be equal to the angle at which the ray is reflected (angle of reflection). Both angles are measured in relation to the normal line which is at a right angle to the surface.

Because light waves travel in straight lines, they can be represented by lines that show direction, called rays. A beam of light, such as that produced by a flashlight, contains many parallel rays. If a beam of light strikes a smooth surface such as a mirror, all of the rays will be reflected parallel to one another in the same order in which they originated. If, however, a beam of light strikes a rough surface, each ray will strike the surface at a different angle of incidence, and the light will be scattered in many different directions.

1. A woman wishes to use a mirror to send a signal by flashlight to a man. The angle formed by the woman, the mirror, and the man is 60°. When the woman aims the flashlight toward the mirror, what should the angle of incidence be?

   (1) 60°
   (2) 15°
   (3) 30°
   (4) 90°
   (5) 120°

2. Dust and other foreign particles in the atmosphere scatter light. An airport control tower located near a city with polluted air would find that its light signals

   (1) appear sharper and clearer than usual
   (2) travel longer distances than usual
   (3) take longer to reach their destinations than usual
   (4) appear fuzzy and blurred
   (5) are all reflected at the same angle

Where do the colors of the rainbow come from? Seven colors are seen in a rainbow: red, orange, yellow, green, blue, indigo, and violet. Each color is made up of electromagnetic waves of a different wavelength. Violet has the shortest wavelength, and red has the longest wavelength. When all the colors are mixed together, you see white light, which is colorless. A rainbow occurs when white light passes through drops of water in the air and is separated into its different colors. The water bends, or refracts, the white light at different angles. Light waves of different colors are bent to different angles and become separated into the colors of the rainbow.

3. What is white light?

(1) one of the colors of visible light
(2) one of the colors of the rainbow
(3) the absence of color
(4) a mix of all the visible colors
(5) a type of rainbow

4. Scientists estimate that people can see about 17,000 different colors. What causes these colors?

(1) Each rainbow has different colors.
(2) The order of colors in the rainbow is always the same.
(3) There are 17,000 different mixtures of the colors in white light.
(4) White light is invisible.
(5) The colors are mixtures of black and the seven colors of the rainbow.

To check your answers, turn to page 237.

Answers and Explanations

1. (Evaluation) **(3) More valuable data is likely to be provided by an accelerator with a long tunnel than by one with a short tunnel.** The passage indicates that the longer the tunnel, the greater the acceleration and energy of the particle. The greater the acceleration, the more energy is released upon impact, and the more likely that new particles will form as a result. Options (1) and (2) cannot be supported by the passage since the advantages and disadvantages of the two shapes of accelerators are not discussed. Option (4) is incorrect because the passage does not compare the speeds of protons and electrons. Option (5) is incorrect because the passage says nothing about producing electricity.

2. (Evaluation) **(3) The resistance of a toaster that uses 5 amps of current in a 120-volt circuit is 24 ohms.** To answer this question, you must substitute the numbers given in each option in the formula for Ohm's law, and see if the equation is properly solved. Using option (3) as an example:

$$24 \text{ ohms} = \frac{120 \text{ volts}}{5 \text{ amps}}$$

Since 120 divided by 5 is 24, option (3) is correct. When you do this calculation for the other options, you will find that they are incorrect. The resistance for option (1) should be 3 ohms; for option (2), 1.5 ohms; for option (4), 46 ohms; and for option (5), 70 ohms.

1. (Evaluation) **(3) The depth of the trough of a wave is double the distance from the rest position to the crest.** The diagram shows, and the passage states, that the depth of the trough of a wave is equal to the distance between the rest position and the crest. The other options are all true and are based on information in the passage.

2. (Comprehension) **(4) a disturbance that travels through space or matter** This definition appears in the first paragraph of the passage. Option (2) is incorrect because it leaves out space.

3. (Application) **(1) amplitude** A surfer would be most interested in the amount of energy a wave has to carry him or her to shore. The passage indicates that the greater the amplitude, the greater the energy. Options (2) and (3) refer to properties of waves that would not be a surfer's main interest. Option (4) is incorrect because all waves are a disturbance. Option (5) is not an important factor in surfing.

4. (Comprehension) **(4) number of waves that pass a given point and unit of time** Frequency is defined in the second paragraph of the passage. The other options do not describe frequency.

5. (Analysis) **(2) divide velocity by wavelength** Since velocity equals wavelength times frequency, to find frequency, you must divide velocity by wavelength. Options (1) and (5) are incorrect because you don't know the frequency and are trying to solve for it.

Options (3) and (4) are incorrect because they do not calculate frequency.

6. (Application) **(4) 18 meters per second** To find velocity, multiply wavelength by frequency (3 x 6 = 18).

7. (Analysis) **(1) decreased velocity** If both wavelength and frequency decrease, it follows that when you multiply them, the resulting velocity will also decrease. Options (4) and (5) are incorrect because distance is not related to these two factors.

1. (Application) **(3) an accordion being played** This is correct because it represents a back-and-forth motion along a straight line. The other options are incorrect because they represent up-and-down motion.

2. (Application) **(5) tape deck and speaker system** This is correct because the sound from the speakers could not travel without air. Option (4) is incorrect because the earphones and wires attached to the cassette player would serve as a medium to transmit sound to your ears. The other options are incorrect because they are not affected by a lack of air.

3. (Comprehension) **(2) its molecules are too far apart** Options (1) and (3) are incorrect because substances with molecules close together carry sound better than substances with molecules far apart. Option (4) is incorrect because inelasticity was described only for solids. Option (5) is

incorrect because a gas can be any temperature.

4. (Application) **(3) Solids transmit sounds best.** If you answered option (1), you read the statement too quickly — because there is no such thing as a "sound molecule." Option (2) is incorrect because it does not relate to the information in the passage. Option (4) is not discussed in the passage, and the validity of option (5) cannot be determined from the passage.

5. (Evaluation) **(5) Sound waves travel slower in cold air than in warm air.** Options (1) and (2), while plausible, are not related to the subject of the passage. Options (3) and (4) are not true.

6. (Analysis) **(1) sound that travels through ground will reach your ears before sound that travels through air** This is correct because sound travels faster through solids than through gases. Option (2) does not make sense with respect to the question. Options (3), (4), and (5) are not supported by the information provided.

7. (Comprehension) **(4) higher in frequency than radio waves** Option (1) is incorrect because both types of waves are invisible. Option (2) cannot be determined from the information given. (Actually, ultraviolet waves are higher in energy.) Option (3) contradicts the correct answer. Option (5) can be disproved by applying the wave equation given in the introductory passage of the lesson. Even without recalling this information, option (5) would not be the best answer because it cannot be determined directly from the information given.

8. (Evaluation) **(1) Air is the only medium through which light can travel.** This is correct because electromagnetic waves do not need any medium through which to travel. Options (2) and (4) are supported by the fact that light can travel without a medium. Option (3) is true because all electromagnetic waves travel at the same speed in a given medium. Option (5) is supported by the fact that radio waves are invisible.

GED Mini-Test (pages 234–235)

1. (Analysis) **(3) 30°** This is correct because light aimed at an angle of incidence of 30° would be reflected at an angle of 30°, totaling the required 60°. Option (1) is incorrect because the light would be reflected at an angle of 60°, thus hitting a target that would form an angle of 120°. Option (2) is incorrect because the light would hit a target forming an angle of 30°. The light in option (4) would be reflected straight back to the sender. Option (5) is incorrect because an incident angle must be between 0° and 90°.

2. (Application) **(4) appear fuzzy and blurred** Option (1) is incorrect because the light would be reflected randomly in many directions, resulting in a less focused signal. Options (2) and (3) are not supported by the information provided. Option (5) is incorrect because the angle of reflection for each ray will depend upon its angle of incidence.

3. (Comprehension) **(4) a mix of all the visible colors** According to the paragraph, white light contains all the colors. Options (1) and (2) are incorrect because white light contains all the colors, not just one color. Option (3) defines

black, not white. Option (5) makes no sense — a rainbow is not white.

4. (Analysis) **(3) There are 17,000 different mixtures of the colors in white light.** Although the rainbow shows seven basic colors, these can be mixed in many combinations, resulting in the huge number of colors we can see. Option (1) is contradicted by the paragraph. Option (2) is true, but it does not explain why we can see so many colors. Options (4) and (5) are not true.

Review: Physics

In Lessons 15 and 16 you have read about some of the laws that describe the behavior of matter and energy. You have learned about electricity and magnetism. You have also learned about waves, including sound and light waves.

The following exercises review the comprehension, application, analysis, and evaluation skills you have been learning. They also expand upon the physics subjects you have read about so far.

Directions: Choose the best answer to each item.

Items 1 and 2 refer to the following passage.

Atoms with unstable nuclei are radioactive. Scientists believe that unstable nuclei are caused by unequal numbers of protons and neutrons. Radioactive elements can change into other elements by a spontaneous process known as radioactive decay. In radioactive decay, the unstable nucleus of a radioactive atom breaks down until it becomes the stable nucleus of another element. For example, an atom of uranium goes through thirteen changes until it becomes a stable atom of lead.

As it breaks down, an atom gives off radiation. There are three types of radiation. Alpha radiation consists of two positively charged protons and two neutrons released together in what is known as an alpha particle. Beta radiation consists of negatively charged beta particles that are actually electrons. Gamma radiation is made up of high-energy electromagnetic waves called gamma rays.

Some radioactive elements undergo alpha decay, while others undergo beta decay. Both alpha and beta decay are nearly always accompanied by the release of gamma rays. Of the three types of radiation, gamma rays are the most harmful. With tremendous penetrating power, gamma rays have the ability to destroy the cells of living things.

1. According to the passage, an atom of a radioactive element does all of the following except

   (1) release energy
   (2) increase in size
   (3) change its identity
   (4) release subatomic particles
   (5) give off gamma rays

2. Which of the following would be most useful in separating alpha particles from beta particles?

   (1) a microscope
   (2) an electric field
   (3) a powerful lamp
   (4) a mirror
   (5) a block of lead

Items 3 and 4 refer to the following passage.

Large amounts of energy are released by atoms in a process called nuclear fission. Nuclear fission is the splitting of an atomic nucleus into two smaller nuclei. Unlike radioactive decay, a fission reaction must be made to happen. The first nuclear fission reaction was engineered in 1938.

The rapid splitting of many nuclei is called a nuclear chain reaction. This process produces the energy generated at a nuclear power plant. If uncontrolled, a chain reaction can result in a nuclear explosion. For this reason, fission reactions take place in a device called a nuclear reactor. The purpose of a reactor is to control the speed of the reaction and to prevent the escape of radioactive materials.

3. Which of the following statements about nuclear fission is <u>not</u> true?

(1) They occur spontaneously in nature.
(2) They release energy and matter.
(3) They can be controlled.
(4) They involve changes in the nucleus of an atom.
(5) They never took place before the twentieth century.

4. What would be the most likely long-term effect of an accident at a nuclear power plant?

(1) a nuclear explosion
(2) contaminating the environment
(3) speeding up of the nuclear chain reaction
(4) stopping of the nuclear chain reaction
(5) There would be no long-term effect.

Items 5 and 6 refer to the following diagrams.

SERIES CIRCUIT

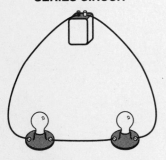

A series circuit has only one path for the electric current. When the circuit is broken, the current stops.

PARALLEL CIRCUIT

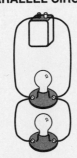

In a parallel circuit, the current flows in two or more separate paths. If the current in one path stops, it still flows in the other branches.

6. One light bulb in a kitchen circuit burns out, but the other lights still work. Which statement supports the conclusion that the kitchen uses a parallel circuit?

(1) The current has stopped in the entire circuit.
(2) The current continues in all but one path of the circuit.
(3) The current is automatically shut off by a fuse.
(4) The burned-out light bulb was on its own series circuit.
(5) Only a power shortage would cause the light to go out.

5. Which of the following is likely to use a series circuit?

(1) the wiring for a house
(2) a string of decorating lights
(3) heavy duty transmission lines
(4) the wiring in a car
(5) a billboard with light bulbs

To check your answers, turn to page 240.

Answers and Explanations

1. (Comprehension) **(2) increase in size** This is correct because, if a radioactive atom were to change size, it would get smaller as it releases sub-atomic particles and energy. Options (1), (3), and (4) are incorrect because during the process of radioactive decay, energy and subatomic particles are released as a radioactive element changes into another element. Option (5) is incorrect because all radioactive activity is accompanied by the release of gamma rays.

2. (Evaluation) **(2) an electric field** This is correct because alpha particles are positively charged and beta particles are negatively charged. There is no evidence from the information given that any of the other options would have an effect on these particles.

3. (Evaluation) **(1) They occur spontaneously in nature.** This is correct because the passage states that a fission reaction must be made to happen. Options (2) and (4) are incorrect because a fission reaction is defined as the splitting of an atomic nucleus in which two smaller nuclei (matter) and energy are released. Option (3) is incorrect because fission reactions are controlled in a nuclear reactor. Option (5) is incorrect because the first fission reaction took place in 1938.

4. (Analysis) **(2) contaminating the environment** Options (1), (3), and (4), if they happened, would be short-term effects of an accident at a nuclear power plant. Option (2) is correct because the radioactive material released into the environment would continue to decay for a long time. Option (5) is incorrect because there would be a long-term effect, option (2).

5. (Application) **(2) a string of decorating lights** These are often made such that if one bulb is removed, the whole string goes out. In more important electricity uses such as options (1), (3), (4), and (5), parallel circuits are used to ensure that the whole circuit does not fail if one item on it burns out or is removed.

6. (Analysis) **(2) The current continues in all but one path of the circuit.** Because the other lights and appliances still work after the bulb burns out, this indicates that the circuit must be a parallel circuit. Option (1) would happen if it were a series circuit. Options (3) and (5) are incorrect because blown-out fuses and power shortages have the same effect on both kinds of circuits. Option (4) is contradicted by information in the question.

Performance Analysis
Physics Review

Use the chart below to help you identify your strengths and weaknesses in each thinking skill area for the subject of physics.

Circle the number of each item that you answered correctly on the Physics Review.

Thinking Skill Area	Physics	Lessons for Review
Comprehension	1	1, 2, 4, 9
Application	**5**	10, 13
Analysis	4, **6**	3, 5, 6, 8, 11, 12, 16
Evaluation	2, 3	7, 14, 15, 17

Boldfaced numbers indicate items based on diagrams, charts, graphs, and illustrations.

If you answered 4 or more items correctly, congratulations! You are ready to go on to the GED Posttest.

If you answered 3 or fewer questions correctly, determine which skill areas are most difficult for you. Then go back and review the physics lessons for those areas.

POSTTEST

Science

Directions

The Science Posttest consists of multiple-choice questions intended to measure your understanding of the general concepts in science. The questions are based on short readings that often include a graph, chart, or figure. Study the information given and then answer the question(s) following it. Refer to the information as often as necessary in answering the questions.

You should spend no more than 95 minutes answering the 66 questions on the Science Posttest. Work carefully, but do not spend too much time on any one question. Be sure you answer every question. You will not be penalized for incorrect answers.

Record your answers to the questions on the answer sheet provided on page 271. To record your answers, mark one numbered space on the answer sheet beside the number that corresponds to the question on the Science Posttest.

Example   Which of the following is the smallest unit in a living thing?

(1) tissue
(2) organ
(3) cell
(4) muscle
(5) capillary   ① ② ● ④ ⑤

The correct answer is "cell"; therefore, answer space 3 should be marked on the answer sheet.

Do not rest the point of your pencil on the answer sheet while you are considering your answer. Make no stray or unnecessary marks. If you change an answer, erase your first mark completely. Mark only one answer space for each question; multiple answers will be scored as incorrect. Do not fold or crease your answer sheet.

Directions: Choose the best answer to each item.

Items 1 and 2 refer to the following map.

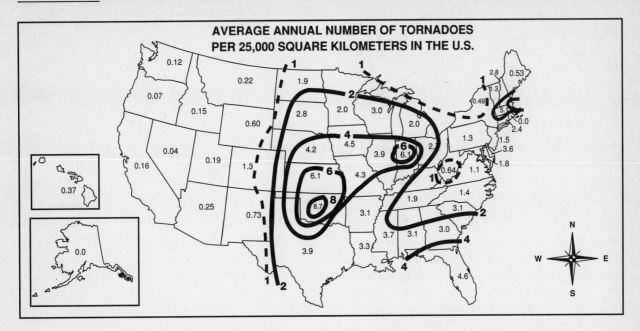

AVERAGE ANNUAL NUMBER OF TORNADOES PER 25,000 SQUARE KILOMETERS IN THE U.S.

1. According to the map, tornadoes in the United States can be found most often

   (1) in the mountains
   (2) in coastal areas
   (3) in the North
   (4) in large, flat areas
   (5) over the ocean

2. A person with a fear of tornadoes would be happiest living in

   (1) New England
   (2) Florida
   (3) Texas
   (4) the Midwest
   (5) Alaska

Items 3 and 4 refer to the following information.

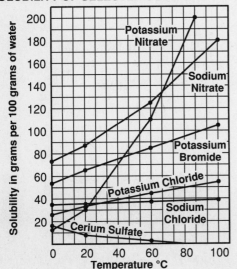

SOLUBILITY OF SELECTED SOLIDS IN WATER

solves in 100 grams of water at a particular temperature. The temperature is important because the solubility of some substances changes if the water temperature changes. The solid whose solubility changes most as the temperature of water changes is

(1) potassium nitrate
(2) potassium bromide
(3) potassium chloride
(4) sodium chloride
(5) sodium nitrate

3. The term solubility means the ability to dissolve. Solubility is usually measured in terms of the amount of a solid that dis-

4. Many people think that increasing the temperature of the water will increase the solubility of any solid. Which line on the graph disproves this idea?

   (1) potassium nitrate
   (2) potassium bromide
   (3) potassium chloride
   (4) cerium sulfate
   (5) sodium nitrate

Items 5 to 10 refer to the following article.

Many insects have a wormlike stage called a larva before becoming adults. The screwworm is the larva of a fly that lays its eggs in the open sores of cattle and other animals. Hundreds of worms can develop in a single sore and kill the host animal. In warm climates, the screwworm does millions of dollars of damage each year.

Controlling the screwworm has long been a problem. Chemical control using insecticides could kill the insects. However, applying insecticides to individual animals is too expensive. A biological control method was first tried in 1958 on the island of Curaçao in the West Indies. Large numbers of sterile male flies raised in laboratories were released on the island. The theory was that if there were more sterile males that could not fertilize female flies than fertile males, many females would mate with the sterile males. Since the females mate only once, most of the eggs would not be fertilized. The number of young produced would eventually reach zero.

The Curaçao project was successful, and screwworm flies were eliminated from the island. Since then, sterile-male release has been used to control the screwworm in the southwestern United States. However, the screwworm also lives in Mexico, where it is not well controlled. Therefore, release of sterile males must be continued along the border.

5. According to the article, which method of controlling screwworms continues to be used today?

(1) spraying insecticides
(2) treating wounded cattle
(3) sterilizing the males in the wild
(4) releasing laboratory-bred sterile males
(5) sterilizing the eggs laid by the females

6. Why was the decision made to try to control screwworm damage on cows in Curaçao by biological rather than chemical methods?

(1) Chemical insecticides were hard to obtain.
(2) Treating cattle individually with chemical insecticides was too expensive.
(3) The chemical methods were less effective than biological control.
(4) Biological methods do not harm the environment.
(5) Biological methods can be easily developed by individual farmers.

7. Which of the following is an unstated assumption made by the author of this article?

A. A change in the ecology of an area is allowed if the results can be helpful.
B. Getting rid of an agricultural pest has more advantages than disadvantages.
C. Chemical pesticides may be suitable for other insect-control problems.

(1) A only
(2) B only
(3) C only
(4) A and B only
(5) A, B, and C

8. According to the article, you can conclude that biological control

(1) works best on insect populations that cover a wide geographical area
(2) works best on isolated insect populations
(3) is always better than chemical insecticides
(4) does not work because insects evolve different ways of reproducing
(5) is more expensive than chemical pesticides in the long run

9. Another type of biological control is to release males that have been exposed to radiation. The sperm of these males carry deadly genetic defects. In such a situation, you would expect that the population would

(1) die out in one generation because none of the eggs would be fertilized
(2) die out in several generations because most of the eggs would not be fertilized
(3) decrease in several generations because many fertilized eggs would be seriously defective and would die
(4) stay the same because the females would continue to mate
(5) increase because more eggs would be fertilized

10. Which of the following is most similar to the sterile-male release method of controlling insects?

(1) releasing chemical insecticides in areas such as swamps
(2) spaying animals to prevent their reproduction
(3) introducing predators to feed on the pest
(4) introducing viruses or bacteria that are harmful to the pest
(5) setting traps for animal pests such as rats and mice

Items 11 and 12 refer to the following information.

A blood pressure reading is made up of two numbers such as 125/85. The first number is the greatest pressure exerted by the blood in the arteries when the heart beats. This is called the systolic pressure. The second number is the lowest pressure of the blood in the arteries. This is called the diastolic pressure. The chart shows how blood pressure varies as the heart beats, forcing blood through the arteries. Three beats would have occurred in the period shown.

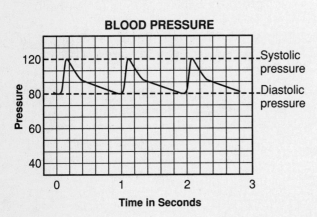

BLOOD PRESSURE

11. What is the blood pressure reading of the patient whose blood pressure is charted above?

(1) 120/80
(2) 80/120
(3) 130/90
(4) 90/130
(5) 1.5

12. What is the effect of the pumping action of the heart on blood pressure?

(1) When the heart pumps, blood pressure goes down.
(2) When the heart pumps, blood pressure goes up.
(3) When the heart pumps, blood pressure remains the same.
(4) When the heart relaxes, blood pressure goes up.
(5) When the heart relaxes, blood pressure remains the same.

Items 13 to 16 refer to the following article.

Earthquakes start deep in the Earth. The point where an earthquake starts is called its focus. From the focus, energy moves outward in all directions, causing the Earth to vibrate. These vibrations are called seismic waves.

In 1935, the American scientist Charles F. Richter developed a scale to measure how much energy there is at the focus of an earthquake. On the Richter scale, the amount of energy increases ten times between each whole number on the scale. For example, an earthquake that measures 5 has ten times more energy than one that measures 4.

Earthquakes that measure 4.5 or greater on the Richter scale cause damage. Earthquakes that measure 7 or greater are considered severe. The famous 1906 San Francisco earthquake measured 8.3, and the 1989 earthquake measured 7.1.

The Mercalli scale is another scale related to earthquakes. This scale rates earthquakes according to their effects on various places where the quake is felt. This scale uses Roman numerals. For example, an earthquake felt only by scientific monitors would have a rating of I. An earthquake that destroyed all the buildings in a wide area would have the highest rating of XII. Since the effects of an earthquake differ from place to place, the Mercalli scale ratings in each place can differ.

13. What does the Richter scale measure?

    (1) the amount of energy on the Earth's surface during an earthquake
    (2) the amount of energy at the focus of an earthquake
    (3) the seismic wave patterns of an earthquake
    (4) the amount of damage done by an earthquake
    (5) the distance between the focus of an earthquake and Earth's surface

14. What does the Mercalli scale measure?

    (1) the amount of energy on the Earth's surface during an earthquake
    (2) the amount of energy at the focus of an earthquake
    (3) the effect of an earthquake on the area precisely over the focus
    (4) the different effects of an earthquake on various places
    (5) the seismic wave patterns of an earthquake

15. When an earthquake first occurred, seismologists estimated its rating as 6 on the Richter scale. Later, they checked their data and gave the earthquake a rating of 7. What caused them to revise the Richter scale rating? Their data showed that the earthquake

    (1) was not considered severe
    (2) had 10 times less energy than they had first estimated
    (3) had 10 times more energy than they had first estimated
    (4) had 100 times less energy than they had first estimated
    (5) had 100 times more energy than they had first estimated

16. An earthquake whose focus was located near San Jose, California was rated 5.2 on the Richter scale and VII on the Mercalli scale. In nearby Oakland, the earthquake was rated 5.2 on the Richter scale and V on the Mercalli scale. Why were the two Mercalli ratings different?

    (1) The Mercalli scale measures energy at the focus of the earthquake.
    (2) The Mercalli scale measures energy at various points affected by the earthquake.
    (3) Each earthquake has two Mercalli ratings.
    (4) The earthquake caused less damage in Oakland than in San Jose, so the Mercalli rating in Oakland was lower.
    (5) The earthquake caused more damage in Oakland than in San Jose, so the Mercalli rating in Oakland was lower.

Items 17 to 20 refer to the following chart.

EMERGENCY ACTION FOR POISONING	
Type of Poison	**First Aid***
Inhaled poison	Provide fresh air. If victim is not breathing, begin artificial respiration.
Poison on the skin	Remove contaminated clothing. Flood skin with water for 10 minutes. Wash gently with soap and rinse.
Poison in the eye	Flood eye with lukewarm water. Repeat for 15 minutes.
Swallowed poison	Medicine: Call for advice. Chemical or household products: Unless victim is unconscious, convulsing, or cannot swallow, give milk or water. Then call for advice on whether to induce vomiting.

*After emergency first-aid actions, call the local poison control center, a hospital, or a doctor.

17. A three-year-old girl ate the contents of a bottle of children's aspirin tablets. What is the first thing her caregiver should do?

    (1) Flood with water.
    (2) Induce vomiting.
    (3) Give milk.
    (4) Give water.
    (5) Call the local poison control center.

18. Because of a malfunction in the exhaust system of his car, a man was overcome by breathing carbon monoxide gas when he started the car in a closed garage. What emergency action should the person who finds him take first?

    (1) Open the doors and windows.
    (2) Flood the skin with water.
    (3) Call the local poison control center.
    (4) Induce vomiting.
    (5) Give artificial respiration.

19. A gardener working in her shed spilled a solution of insecticide, and it splashed on her blue jeans. She continued to mix the chemical. Then her vision started to blur and her head began to ache. What should she have done immediately after the accident?

    (1) opened the doors and windows
    (2) removed her blue jeans
    (3) called the local poison control center
    (4) applied an ointment
    (5) washed with soap and rinsed with water

20. A man who got some poison in his eye was given emergency first aid by a friend. After about 15 minutes of having his eye flooded with water, the man started to feel better. What should the friend do now?

    (1) Continue flooding the eye with water.
    (2) Flood the eye with a solution of baking soda and water.
    (3) Apply a bandage to the eye.
    (4) Call the local poison control center.
    (5) Do nothing.

The path taken by an electric current is called a circuit. For electricity to flow, the circuit must be complete. In a flashlight, the electric current flows from one end of a battery through a wire to the bulb. From the bulb, the current returns along another wire to the other end of the battery. A switch is a device that enables you to break or connect the circuit, which stops or restarts the flow of electric current.

Batteries used in toys and portable radios are called dry cells. These batteries are a common source of electric current for objects that must be portable. Dry cell batteries produce electricity by chemical action. This chemical action starts whenever the circuit is complete. When the circuit is broken, the current stops flowing and the chemical action continues, only more slowly. When the supply of chemicals is used up, the dry cell battery is dead.

A storage battery is often used to power larger devices. Chemical action is also the source of electric current in the storage battery. As long as the circuit is complete, the chemical action starts. When the chemicals are used up, the production of current stops. A storage battery does not need to be thrown away; its chemicals can be restored by recharging it with electric current from an outside source.

21. You pick up a flashlight that uses dry cell batteries and press the switch. Nothing happens. From your observations, which of the following is a valid conclusion?

    A. The light does not go on.
    B. The chemicals in the batteries may have been used up.
    C. Storage batteries are better than dry cell batteries.

(1) A only
(2) B only
(3) C only
(4) A and B only
(5) B and C only

22. A portable CD player stopped working. When its batteries were recharged, it worked again. Why did recharging help?

(1) The chemical action continued at a reduced rate when the circuit was broken, and the chemicals were used up.
(2) The chemical action continued at a steady rate when the circuit was broken, and the chemicals were used up.
(3) An outside electric current restored the chemicals in the battery so that they could produce current again.
(4) The CD player needed to be plugged in to operate.
(5) The batteries were new.

23. A boy left his toy flashlight on, but the father told him to turn it off. Why did the father think it should be off?

(1) The battery produces the same amount of current whether the light is on or off.
(2) The battery can produce electric current as long as it is being re-charged.
(3) The life of the battery is the same whether the flashlight is left on or off.
(4) The life of the battery is shorter when the flashlight is left on.
(5) The battery will not wear out when the flashlight is left off.

24. Which of the following items is best suited for getting electric current from a dry cell battery?

(1) a washing machine
(2) a car
(3) a motorized wheelchair
(4) a toy truck
(5) a doorbell

25. You need a large amount of electric current to start a car's engine. A car uses a large storage battery to produce this current. It does not take all this current to keep the car running. In fact, the battery is constantly being recharged while the car's engine is running. As a result, batteries do not have to be plugged into an electric outlet for recharging. Which of the following is a conclusion from this information?

(1) A car battery is not powerful enough to provide electric current to start the motor.
(2) Car batteries frequently need to be jump-started.
(3) The car's engine generates electricity that recharges the battery as the car runs.
(4) A car battery does not need external recharging for the first 25,000 miles.
(5) A car battery does not need external recharging for the first 100,000 miles.

Item 26 refers to the following diagram.

EARTH'S ROTATION

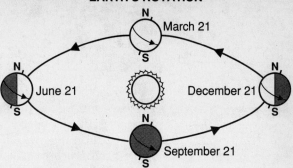

26. The Earth rotates on its axis, or spins around on a tilt, at the same time that it revolves around the Sun. Which of the following conclusions can be supported by the diagram?

(1) High tides occur when the Earth is closest to the Sun.
(2) High tides occur when the Earth is closest to the Moon.
(3) On June 21, the Northern Hemisphere is tilted toward the Sun.
(4) On June 21, the Northern Hemisphere is tilted away from the Sun.
(5) On June 21, all parts of the Earth have days and nights of equal length.

27. A freighter is scheduled to make the trip from New York to Liverpool, England in eight days. The trip back to New York is scheduled to take nine days. What accounts for the difference in the two trips?

   (1) Liverpool's time zone is six hours ahead of New York's time zone.
   (2) The ocean currents in the North Atlantic flow from west to east.
   (3) The ocean currents in the North Atlantic flow from east to west.
   (4) The ocean currents in the South Atlantic flow from east to west.
   (5) The ocean currents in the South Atlantic flow from west to east.

Items 28 and 29 refer to the following information and diagram.

Methane is a colorless, odorless gas that is found in nature. It is the product of the breaking down of organic matter in marshes. It is also found in coal mines and natural gas deposits. When methane burns, it combines with oxygen to form carbon dioxide and water vapor, as shown in the following equation:

$$CH_4 + 2O_2 \longrightarrow CO_2 + 2H_2O$$

In the reaction, atoms are rearranged to form different substances. The reaction begins with one molecule of methane (CH_4)—made up of one carbon and four hydrogen atoms—and one molecule of oxygen (O_2)—made up of two oxygen atoms. These recombine to form the new substances of carbon dioxide and water. The carbon dioxide (CO_2) molecule is made up of one carbon atom from the methane and two oxygen atoms from the oxygen molecules. The water (H_2O) is made up of two hydrogen atoms taken from the methane and another oxygen atom from the oxygen molecules. The atoms have formed new molecules, but the total number of each kind of atom remains the same. For example, there are four hydrogen atoms at the beginning of the reaction and four at the end.

The reaction of methane and oxygen can be hazardous to humans; the quantities of the gases and the speed of the reaction must be carefully controlled.

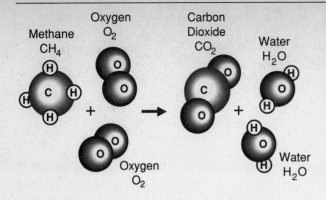

Methane CH₄ + Oxygen O₂ + Oxygen O₂ → Carbon Dioxide CO₂ + Water H₂O + Water H₂O

28. In order to write or express a chemical reaction in the form of an equation such as the one above, what must be true?

(1) The result of the reaction must be a gas.
(2) The atoms must be made of one or more molecules.
(3) There must be an equal number of atoms from each element on each side of the equation.
(4) The same molecules must appear on each side of the equation.
(5) Each molecule in the equation must be either divided or multiplied by two.

29. The chemical reaction shown above is hazardous. It is most dangerous to people in

(1) natural gas deposits deep within the Earth
(2) marshes
(3) mines
(4) laboratories
(5) factories where other chemicals are made

Items 30 to 32 refer to the following article.

Nearly a third of colds are caused by rhinoviruses. These viruses can survive for up to three hours outside the body—on skin, certain fabrics, and hard materials such as stainless steel and wood. Often people infect themselves. For example, a person picks up the live viruses from a contaminated surface. Later when the person rubs the nose or eyes, the viruses are transferred to the mucous membranes in the nose—a cold virus's favorite spot. Shaking hands, opening a door, and picking up a toy are frequent infection routes—more frequent than sneezes, coughs, and kisses. Contrary to what you were told as a child, walking in puddles will not cause a cold.

To prevent a cold, experts advise you to keep your hands away from your eyes and nose and to wash your hands frequently with hot water and soap. You should also clean contaminated objects and use disposable tissues. To stop the spread of viruses that are airborne, cover a cough or a sneeze. Teaching all these methods to children is especially important. Children have the most colds and share the most viruses.

30. Why should objects contaminated with cold viruses be washed?

(1) The viruses can ruin the objects.
(2) The viruses survive indefinitely on objects.
(3) A person with a cold may touch them.
(4) Another person may pick up the viruses by touching the objects.
(5) Even after the cold viruses die, they can infect someone.

31. Which of the following ways to avoid catching a cold is NOT based on scientific fact?

(1) Wash hands often with soap and water.
(2) Keep hands away from eyes and nose.
(3) Clean contaminated objects.
(4) Cover sneezes.
(5) Avoid going out in the rain.

32. What is the effect of touching your eyes or nose with unwashed hands that have touched contaminated objects?

(1) Your face gets soiled.
(2) Rhinoviruses may be transferred from your hands to the mucous membranes in your nose.
(3) Rhinoviruses die when they come in contact with your skin.
(4) Airborne germs come in contact with the mucous membranes in your nose.
(5) There is no effect.

Items 33 and 34 refer to the following diagram.

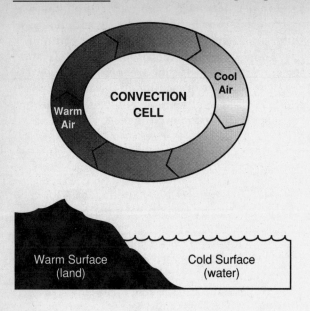

33. The convection cell in the diagram is an area of the atmosphere in which air movement is caused by variations in temperature. Which of the following conclusions is best supported by the diagram?

(1) Warm air rises and cool air sinks.
(2) Warm air sinks and cool air rises.
(3) Warm air pushes cold air upward.
(4) The temperature of the Earth's surface has no effect on air movement.
(5) Convection cells occur only where land meets water.

34. According to the diagram, in which direction should you face to keep the wind out of your eyes at this particular beach and at this time?

(1) Face the water.
(2) Face inland.
(3) Have your left side to the water.
(4) Have your right side to the water.
(5) Face in any direction.

Items 35 and 36 refer to the following diagram.

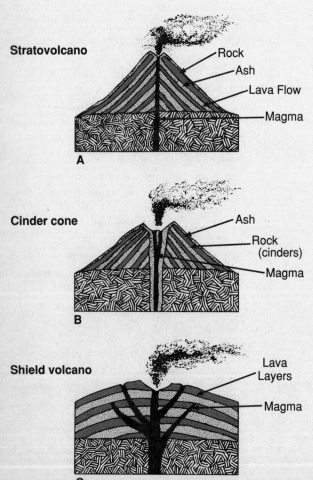

35. Mt. St. Helens erupted in 1980, spewing forth rock, ashes, and lava over a wide area. Mt. St. Helens is

(1) a stratovolcano
(2) a cinder cone
(3) a shield volcano
(4) a magma volcano
(5) an extinct volcano

36. Which of the following conclusions is best supported by the diagram of volcanoes?

(1) Volcanoes are steep mountains.
(2) Volcanic eruptions are always rapid and explosive.
(3) Volcanoes are mountains built from deposits of lava, rock, or ashes.
(4) Volcanoes are most likely to occur along the boundaries of the Earth's tectonic plates.
(5) Volcanoes may lie dormant for years between eruptions.

Items 37 and 38 refer to the following information.

Matter is anything that has mass and takes up space. Matter is found in three different states — solid, liquid, and gas.

1. A solid such as iron has a definite shape and takes up a definite amount of space.
2. A liquid such as milk does not have a definite shape since it can be poured. It takes up a definite amount of space.
3. A gas such as air has no definite shape and does not take up a definite amount of space. It expands to fill whatever space is available.

Most matter can exist in any one of the three states and can change from one state to another through changes in temperature.

37. Which of the following is an example of a single form of matter in three different states?

(1) lump of sugar, grains of sugar, sugar syrup
(2) rock salt, salt crystals, salt water
(3) ice, water, water vapor
(4) glass, broken glass, molten glass
(5) dry ice, carbon dioxide, carbonated beverage with carbon-dioxide bubbles

38. What causes a form of matter to change from one state to another?

(1) changes in chemical makeup
(2) changes in position
(3) changes in color
(4) changes in amount of heat
(5) changes in hardness

Items 39 and 40 refer to the following diagram.

THE HUMAN EYE

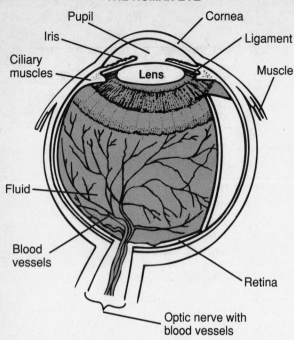

Items 41 and 42 refer to the following information.

One way matter can change is through a chemical reaction. When matter undergoes a chemical change, the old substances turn into new substances with new properties, or characteristics. The energy needed to start a chemical reaction is called activation energy. Once they have started, some chemical reactions release heat energy. These are called exothermic reactions. Other chemical reactions absorb energy. These are called endothermic reactions.

41. Lighter fluid is poured on charcoal in a barbecue grill, and a match is used to light the coals. In this chemical reaction, the lit match is

    (1) a property of the reaction
    (2) an endothermic reaction
    (3) a physical change
    (4) the resulting form of matter
    (5) the source of activation energy

39. The lens in the human eye must change shape in order for the eye to focus. What part of the eye is in the best position to control the lens shape?

    (1) the retina
    (2) the cornea
    (3) the ciliary muscles
    (4) the pupil
    (5) the optic nerve

42. What is an endothermic reaction?

    (1) a chemical reaction in which the properties of matter change
    (2) a chemical reaction in which heat energy is released
    (3) a chemical reaction in which heat energy is absorbed
    (4) the application of energy to start a reaction
    (5) a reaction that does not need activation energy to start

40. Light enters the eye through the pupil. The iris is the colored area around the pupil that makes the pupil larger or smaller. What is the effect of the adjustments made by the iris?

    (1) The pupil changes color.
    (2) The eye has a blind spot.
    (3) The amount of light entering the eye is changed.
    (4) The cornea bulges.
    (5) The ability to see color is changed.

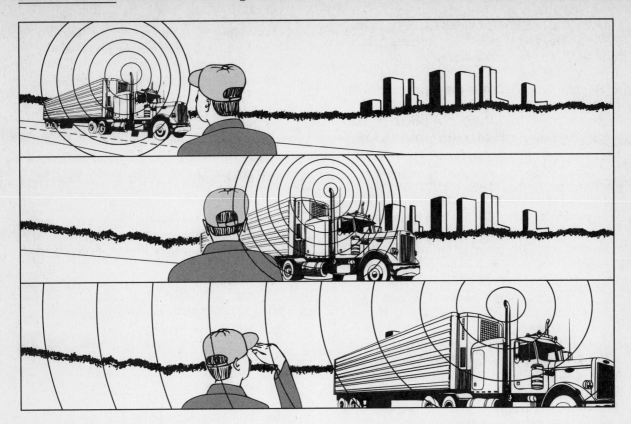

The Doppler effect describes the apparent changes that take place in waves, such as sound waves, as a result of the movement of the source or receiver of the waves. As the source of a sound and the receiver move closer together, the frequency seems to increase and the sound seems higher to the receiver. As the sound and the receiver move apart, the frequency seems to decrease and the sound seems lower.

43. Why does a truck's engine sound different when it is moving toward you and away from you?

    (1) The truck's engine makes more noise when it is moving toward the listener.
    (2) The frequency of the sound waves changes.
    (3) To the listener, the frequency of the sound waves seems to change.
    (4) The driver has shifted gears.
    (5) To the listener, the engine sounds the same at all times.

44. Which of the following conclusions can be supported by the diagram?

    (1) The Doppler effect works only with sound waves.
    (2) The Doppler effect can be used to determine whether a source of sound is moving away from or toward someone.
    (3) The Doppler effect works only when the source of sound is close.
    (4) The Doppler effect is a change in the actual frequency of sound waves as they are sent by the source.
    (5) The Doppler effect is used to overcome interference from other sources of sound.

Items 45 to 48 refer to the following article.

Cholesterol is a fatty material that is made by our bodies and provided by our diets. Cholesterol helps produce cell membranes and is part of many key hormones. When the body cannot use all its cholesterol, much of the excess builds up on artery walls. Over the years, the arteries become narrower, resulting in a condition called atherosclerosis. Blood clots can form in clogged arteries. When this happens in the arteries of the heart, the flow of blood to the heart is blocked, and a heart attack occurs.

Until recently, the relationship between cholesterol and heart disease was not clear. Then 3,800 men with high cholesterol levels were studied. They were given a drug that lowers cholesterol and then were monitored. For every 1 percent decrease in the cholesterol level, there was a 2 to 3 percent decrease in the number of heart attacks and sudden death. Other studies confirmed these results.

Cholesterol levels can usually be lowered without drugs by reducing the amount of fat in the diet. The main sources of cholesterol are saturated fats, which the body converts to cholesterol. Saturated fats generally come from animal products. Some plant foods, such as coconut oil, palm oil, and chocolate, also contain saturated fats.

45. A person who is trying to lower his or her cholesterol level should eat less of which of the following foods?

   (1) fruit
   (2) vegetables
   (3) steak
   (4) bread
   (5) sweetened juices

46. The effect of excess cholesterol in the body is that it

   (1) lodges in the heart
   (2) attacks the cell membranes of the heart
   (3) builds up on the artery walls and can block the flow of blood to the heart
   (4) causes uneven beating of the heart
   (5) lowers the blood's ability to carry oxygen

47. Atherosclerosis is a condition in which

   (1) the arteries become rigid
   (2) the arteries become clogged
   (3) the arteries become perforated
   (4) heart tissue dies when it is deprived of oxygen
   (5) heart tissue becomes less elastic

48. Which type of fat is the main source of cholesterol in our diets?

   (1) unsaturated fat
   (2) saturated fat
   (3) polyunsaturated fat
   (4) monounsaturated fat
   (5) essential fatty acids

Items 49 to 52 refer to the following article.

The basic particles of matter — atoms and molecules — are always moving. Random motion of particles is present in solids, liquids, and gases. The energy of this movement is called kinetic energy. When a substance is heated, the kinetic energy increases, and the atoms and molecules move faster and farther apart. The volume of the substance increases. When a substance is cooled, it loses kinetic energy, and its volume decreases.

For example, the mercury in a thermometer expands when the thermometer is warmed. The more the thermometer is heated, the more it expands, and the higher the mercury rises. When the temperature decreases, the mercury contracts and moves down the thermometer.

Different substances expand by different amounts when heated. Engineers must account for these differences when designing buildings, roads, and other structures. Steel bridges, for example, have special structures at each end to allow for expansion in warm weather. Concrete bridges have expansion joints between slabs. These joints prevent the concrete from buckling when it expands.

Water behaves unlike most other substances. When it cools and freezes, it does not contract; it expands.

49. Which of the following can be concluded from the information provided?

(1) All substances expand when heated and contract when cooled.
(2) All substances contract when heated and expand when cooled.
(3) Most substances expand when heated and contract when cooled.
(4) Most substances contract when heated and expand when cooled.
(5) Most substances do not change volume when heated or cooled.

50. Why does mercury rise in a thermometer when the temperature increases?

(1) Mercury is a liquid.
(2) Mercury expands when heated.
(3) Mercury contracts when heated.
(4) Heat replaces mercury at the base of the thermometer.
(5) The glass tube of the thermometer contracts, forcing the mercury upward.

51. Given the behavior of matter when heated and cooled, which of the following would be most likely to happen?

A. Telephone wires sag in the summertime.
B. A glass breaks when boiling water is poured into it.
C. Soup contracts when frozen.

(1) A only
(2) B only
(3) C only
(4) A and B only
(5) A, B, and C

52. When enough heat energy is added, substances change from solids to liquids and then to gases. What can be concluded about the molecules and atoms in solids, liquids, and gases?

(1) The molecules and atoms of gases are the farthest apart and the most mobile.
(2) The molecules and atoms of liquids are the farthest apart and the most mobile.
(3) The molecules and atoms of solids are the farthest apart and the most mobile.
(4) The molecules and atoms of matter behave in the same way in solid, liquid, and gas forms.
(5) The molecules and atoms of a solid do not move.

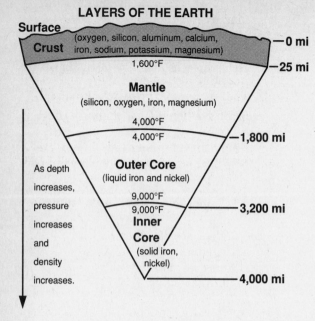

LAYERS OF THE EARTH

Surface

Crust (oxygen, silicon, aluminum, calcium, iron, sodium, potassium, magnesium) —0 mi

1,600°F —25 mi

Mantle
(silicon, oxygen, iron, magnesium)

4,000°F

4,000°F —1,800 mi

As depth increases, pressure increases and density increases.

Outer Core
(liquid iron and nickel)

9,000°F

9,000°F —3,200 mi

Inner Core
(solid iron, nickel)

—4,000 mi

53. Using information about the Moon's motion and distance from the Earth, scientists calculated that the average density of the Earth is more than 5 grams per cubic centimeter. However, the density of rocks in the Earth's crust alone is less than 3 grams per cubic centimeter. What statement would help explain the difference between the Earth's average density and the density of the crust alone?

(1) The Earth's interior is composed of material much more dense than the crust.
(2) The Earth's interior is composed of material much less dense than the crust.
(3) The gravitational pull of the Moon causes the crust material to be less dense.
(4) The gravitational pull of the Sun causes the crust to be less dense.
(5) The lower density of the oceans was left out of the calculations.

54. The temperature of the inner core of the Earth is hot enough to melt iron and nickel. However, iron and nickel are solid at the core. What factor in the diagram could be used to explain this?

(1) the presence in the core of other substances with high melting points
(2) the magnetic field of the Earth
(3) the gravitational pull of the Sun
(4) the great pressure on the inner core
(5) the chemical composition of the mantle

Items 55 and 56 refer to the following information.

A lack of vitamin C weakens some body tissues and can result in a disease called scurvy. Many foods contain little or no vitamin C. However, large amounts of vitamin C are found in citrus fruits, tomatoes, and members of the cabbage family. Stored or cooked food may contain less vitamin C than fresh food.

55. What foods are NOT good sources of vitamin C?

(1) dairy products
(2) citrus fruits
(3) mustard greens
(4) tomatoes
(5) red cabbage

56. A person concerned with getting the most vitamin C should eat which of the following foods?

(1) peas
(2) dried fruit
(3) cooked mustard greens
(4) a banana
(5) a fresh orange

Items 57 to 60 refer to the following article.

Test-tube conception, or *in vitro* fertilization, is a procedure for fertilizing eggs outside a woman's reproductive system. Although fertilization occurs outside the body, the egg and sperm are taken from the parents, and the child is genetically their own.

In a healthy female reproductive system, the fallopian tube transports an egg from the ovary, where the egg is produced, to the uterus. In some women, the tubes are blocked, and eggs cannot reach the uterus. If these women are otherwise healthy, they are prime candidates for *in vitro* fertilization.

Early in the woman's menstrual cycle, she can take fertility drugs to ensure that several eggs are produced. Without these drugs, only one egg would be produced. The eggs are surgically removed just before they are ready to be released by the ovaries. Each egg is placed in a culture dish, kept warm, and allowed to mature. Meanwhile, the man's sperm is processed. Each egg is then mixed with 100,000 to 200,000 of the processed sperm and allowed to incubate. After 12 hours, each egg is examined under a microscope to see if it has been fertilized. Usually 75 to 90 percent of the eggs will be fertilized.

One of the eggs is allowed to develop into an embryo. When it has grown into four or eight cells, the embryo is placed in the woman's uterus, and she is given hormones that help the uterus accept the egg. To succeed, the development of the embryo must be perfectly timed with developments in the uterus. Embryo transfer succeeds only about 20 percent of the time in any given cycle. During the next 12 to 16 weeks, almost half of the embryos that are transferred are lost by miscarriage. Ultimately, the couple has only a 10 percent chance that *in vitro* fertilization will result in the birth of a baby.

57. Why is it important to give fertility drugs early in the woman's menstrual cycle?

(1) The fallopian tubes will be cleared.
(2) More than one egg will be produced.
(3) The couple will conceive on their own.
(4) The man's sperm count will increase.
(5) The embryo is more likely to develop normally.

58. *In vitro* fertilization has an overall success rate of only 10 percent, and it can be stressful and very expensive. What information in the article would explain why a couple would choose this procedure instead of adoption?

(1) Adopting a child can take a long time.
(2) They want to be sure they have a healthy child.
(3) They want a child who is genetically their own.
(4) Adoption agencies tend to reject infertile couples.
(5) There are not enough children available for adoption.

59. What causes the high failure rate of transferring the embryo into the uterus?

(1) Many eggs are not fertilized.
(2) The sperm are defective.
(3) Many culture dishes are contaminated.
(4) The sperm are improperly processed.
(5) The timing of embryo development and uterus development is off.

60. Which of the following conditions would make a woman a likely candidate for *in vitro* fertilization?

(1) pelvic inflammatory disease
(2) non-functioning ovaries
(3) two damaged fallopian tubes
(4) a surgically removed uterus
(5) one damaged fallopian tube

Items 61 to 63 refer to the following information.

The human body contains many types of tissues and organs. These function as parts of various systems. Five of the systems include the following:

1. Muscular system. Moving skeletal and body parts such as the stomach and heart.
2. Digestive system. Eating, digesting, and absorbing foods; eliminating some wastes.
3. Circulatory system. Transporting foods, waste, oxygen, heat, carbon dioxide, hormones, and other substances.
4. Nervous system. Receiving stimuli from the environment, sending and interpreting data, controlling actions of other body parts.
5. Reproductive system. Producing sex cells for the continuation of the species.

61. The blood, which carries substances around the body, is part of which system?

    (1) muscular system
    (2) digestive system
    (3) circulatory system
    (4) nervous system
    (5) reproductive system

62. With which system are eyes and ears most closely associated?

    (1) muscular system
    (2) digestive system
    (3) circulatory system
    (4) nervous system
    (5) reproductive system

63. Which system is not vital to the survival of an individual human being?

    (1) muscular system
    (2) digestive system
    (3) circulatory system
    (4) nervous system
    (5) reproductive system

Items 64 to 66 refer to the following information and diagrams.

A molecule of water, or H_2O, contains two hydrogen atoms and one oxygen atom. The three atoms are arranged at an angle, not on a straight line.

WATER MOLECULE

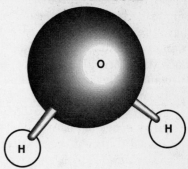

The shape of the water molecule causes it to be polar. This means that one end of the molecule has a positive charge while the other end has a negative charge. As a result, a hydrogen atom in one molecule is attracted to an oxygen atom in another molecule. This attraction is called a hydrogen bond. Groups of water molecules are connected by hydrogen bonds.

HYDROGEN BONDS ◄──►

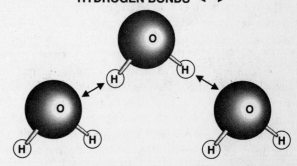

Groups of water molecules look like the diagram at the top of page 261.

WATER

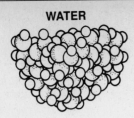

When water freezes, the hydrogen bonds become less flexible. They hold the molecules in a pattern that has more space between the molecules than there is in water in its liquid form.

ICE

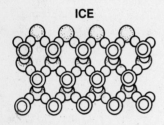

64. A glass jar full of water cracks when the water inside freezes. Why?

(1) Ice takes up more space than an equal mass of water.
(2) Individual water molecules increase in size as the temperature drops.
(3) Glass expands when it becomes cold.
(4) Ice has sharp edges that damage the glass.
(5) Hydrogen gas is liberated, causing the jar to explode.

65. A liquid will boil when heat breaks the bonds between molecules in the liquid. What causes the boiling point of water to be so high compared to substances of similar size and mass?

(1) Water is a pure substance.
(2) Many other substances can dissolve in water.
(3) The groups of molecules in water are tightly bound together and must absorb much heat energy before they break apart.
(4) Water molecules are not attracted to one another.
(5) Oxygen atoms do not absorb heat.

66. Which of the following conclusions can be drawn from the information provided?

(1) Water is very stable; that is, it does not break down easily into hydrogen and oxygen.
(2) There are more molecules in 1 cubic centimeter of water than in 1 cubic centimeter of ice.
(3) Ice is heavier than water.
(4) Water is found in all living things.
(5) Water is transparent and odorless.

Answers and Explanations

1. (Analysis) **(4) in large, flat areas** The area with the most tornadoes—more than 4 per 25,000 square kilometers per year—is the central United States, where the land is flat. The mountainous areas of the United States, near the West and East Coasts, have fewer than 2 tornadoes per 25,000 square kilometers per year; therefore option (1) is incorrect. Option (2) is incorrect because along the seacoasts there are generally fewer than 2 tornadoes, except in the Southeast. Option (3) is incorrect because tornadoes are very rare in most northern areas. Option (5) is incorrect because there are more tornadoes over the land, not over the ocean.

2. (Application) **(5) Alaska** In Alaska during the period covered by this map, there were no tornadoes. Option (1) is incorrect because parts of New England have a relatively high number of tornadoes (for example, 5.1 in Massachusetts). Options (2) and (3) are both incorrect because both states have about 4 tornadoes per 25,000 square kilometers per year. Option (4) is incorrect; the Midwest has more tornadoes than the rest of the country.

3. (Comprehension) **(1) potassium nitrate** If you look at the graph, you can see that potassium nitrate has the steepest line, which indicates the greatest change. At 0°C, 13.3 grams of potassium nitrate dissolve, and at 80°C, 200 grams dissolve. This is the largest change shown, so all the other options are incorrect.

4. (Evaluation) **(4) cerium sulfate** Notice that this line runs downhill. The solubility of cerium sulfate decreases as the temperature increases. Therefore the behavior of cerium sulfate disproves the idea that all solids become more soluble as the water temperature rises. The other options are incorrect because they support the idea.

5. (Comprehension) **(4) releasing laboratory-bred sterile males** This method is described as being used in the Southwest. Option (1) is incorrect because nothing is mentioned about the current use of chemical insecticides. Option (2) is incorrect because it is too expensive. Options (3) and (5) are incorrect because they involve changing the insects in the wild, not in the laboratory.

6. (Evaluation) **(2) Treating cattle individually with chemical insecticides was too expensive.** It was necessary to find a less costly way to control the screwworm. Option (1) is incorrect because the article does not say that chemical insecticides were hard to obtain; they were just too expensive to use. The article does not support option (3). Option (4) is true but was not the reason stated for deciding to try biological control methods. Option (5) is not true. Biological control methods are generally developed by scientists in well-equipped research laboratories, not on farms.

7. (Analysis) **(5) A, B, and C** The author assumes that it is important to protect an agricultural area. Changing the ecology can help the cattle industry by getting rid of a pest. The article does not mention disadvantages because the author assumes they are not very important. The author criticizes chemical insecticides in the case of the screwworm mostly because of the cost. The author does not say that chemical insecticides are unsuitable in general. Therefore, options (1), (2), (3), and (4) are incorrect since they do not include all the assumptions made by the author.

8. (Evaluation) **(2) works best on isolated insect populations** The sterile-male release method described was successful on the island of Curaçao, which is a small, isolated area. When it was used on the mainland, it was less successful. Option (1) is incorrect because it is difficult to apply the method over a wide area like the southwestern United States. Option (3) is incorrect because it is too general a statement to be supported by the article. In some places, chemical insecticides may work better. Option (4) is incorrect because the article holds that the method was successful. Option (5) is incorrect because the article shows that biological control can wipe out an entire population, thereby ending the problem.

9. (Evaluation) **(3) decrease in several generations because many fertilized eggs would be seriously defective and would die** Males would have sperm that could fertilize the eggs, but the fertilized eggs would be defective. This would lower the number of healthy insects

born. Options (1) and (2) are incorrect; eggs would still be fertilized, but they would not develop properly. Option (4) is incorrect because some of the eggs would not produce healthy young; the population would decrease. Option (5) is incorrect because many of the fertilized eggs would be defective, and this would not increase the population.

10. (Application) **(2) spaying animals to prevent their reproduction** Spaying animals and sterilizing male insects reduce the number of offspring the species can produce. Option (1) is incorrect because insecticides directly kill the insects and their eggs. Options (3) and (4) are incorrect because they involve adding a new species, not more of the same species. Option (5) is incorrect because it involves directly killing the pest.

11. (Comprehension) **(1) 120/80** The chart indicates a systolic pressure of 120 and a diastolic pressure of 80, expressed as 120/80. Option (2) is incorrect; it reverses the systolic and diastolic readings. Options (3) and (4) are incorrect; they are not pressures indicated by the chart. Option (5) is incorrect because pressure is expressed as one number over another, not a decimal fraction.

12. (Analysis) **(2) When the heart pumps, blood pressure goes up.** The systolic pressure reading, the higher one, measures the pressure of blood in the arteries right after the heart pumps. The pumping action forces more blood through the arteries and increases the pressure. Option (1) is incorrect because it is the opposite of what happens during pumping action. Option (3) is incorrect because the pumping action of the heart causes changes in blood pressure. Options (4) and (5) are incorrect because when the heart relaxes, blood pressure goes down to the diastolic pressure.

13. (Comprehension) **(2) the amount of energy at the focus of an earthquake** The Richter scale measures the amount of energy where the earthquake starts. Option (1) is incorrect because the focus of an earthquake is deep in the Earth, not on the surface. Option (3) is incorrect because the scale measures energy, not patterns of seismic waves. Option (4) is incorrect because the scale does not measure the effects of an earthquake. Option (5) is incorrect because the scale measures energy, not distance.

14. (Comprehension) **(4) the different effects of an earthquake on various places** The Mercalli scale rates what an earthquake does. Options (1) and (2) are incorrect since the Mercalli scale measures the effects of an earthquake, not its energy. Option (3) is only partly correct. The Mercalli scale is used not only in the area over the focus, but in surrounding areas that feel the earthquake as well. Option (5) is incorrect because the Mercalli scale measures effects, not waves.

15. (Analysis) **(3) had 10 times more energy than they had first estimated** On the Richter scale, an increase of 1 point shows that the earthquake had ten times more energy; the scientists raised the rating when they realized that there was more energy than they had first thought. Option (1) is incorrect because severity is a judgment, which may be based on the Richter scale ratings, but the precise rating is not based on mere judgment. Options (2) and (4) are incorrect because the rating went up 1 point, not down 1 or 2 points. Option (5) is incorrect because the rating went up by 1 on the scale, which means an increase of ten, not of one hundred.

16. (Analysis) **(4) The earthquake caused less damage in Oakland than in San Jose, so the Mercalli rating in Oakland was lower.** The Mercalli scale rates the different effects of an earthquake in various places. Therefore, the same earthquake can have different ratings in the different areas that it affects. Options (1) and (2) are incorrect because the Mercalli scale does not measure energy. Option (3) is incorrect because an earthquake can have more than two Mercalli ratings. Option (5) is incorrect because the earthquake caused less damage in Oakland.

17. (Application) **(5) Call the local poison control center.** When an overdose of medicine has been taken, the correct first-aid action is to call for advice. There are so many types of medicine that treatments for overdoses vary. Option (1) is incorrect because it is not a treatment for swallowed poison. Options (2), (3), and (4) are treatments for swallowed poison, but they are not appropriate when the poison is a medicine.

18. (Application) **(1) Open the doors and windows.** Since carbon monoxide is a deadly gas, the person who finds the victim must first get fresh air into the area. Otherwise the rescuer might be overcome as well. Options (2) and (4) are not suitable treatments for inhaled poisons. Option (3)

should be done after doors and windows are open and after the victim receives artificial respiration, if necessary. Option (5) is part of the first-aid treatment for inhaled poison, but it is the second step after making sure there is fresh air.

19. (Application) **(2) removed her blue jeans** The first emergency action for a poison absorbed through the skin is to remove any clothes that are contaminated. Option (1) is a good idea but not the most important thing to do in the case of a spilled liquid poison. Option (3) is the last step in first aid for poison on the skin. Option (4) is incorrect because it is not a first-aid remedy for any type of poison. Option (5) is incorrect because it is not the first step in treating poison on the skin.

20. (Application) **(4) Call the local poison control center.** First aid may have been successful, but to be sure of complete treatment and recovery, the friend should call the local poison control center or other health professionals. Option (1) is incorrect because the eye was washed for the recommended time. The treatment already seems to have been successful. Option (2) is incorrect because baking soda is not a remedy for poison in the eye. Option (3) is incorrect because the next indicated step is to call for professional help. Option (5) is incorrect because the victim should consult with a health professional.

21. (Analysis) **(2) B only** The chemicals in the dry cell battery may have been used up. The chemical action in a dry cell battery continues even when the switch is off. Therefore, the batteries wear out

even though the item has not been used for a while. Statement A is not a conclusion but a description of what happened when the flashlight was turned on. Therefore, options (1) and (4) are incorrect. Statement C is a conclusion about the relative value of the two types of batteries. There is no information in the article supporting this statement. Therefore, options (3) and (5) are incorrect.

22. (Application) **(3) An outside electric current restored the chemicals in the battery so that they could produce current again.** Option (1) is incorrect because it explains why the battery stopped working but not why it resumed working. Option (2) is not true; chemical action continues at a reduced rate. Option (4) is not true; the CD player is described as being portable. Option (5) may or may not be true, but it does not explain why the CD player started working again.

23. (Analysis) **(4) The life of the battery is shorter when the flashlight is left on.** When the circuit is complete, a dry cell battery works at current-producing levels and the chemicals run out faster. Option (1) is incorrect because batteries do not always produce the same amount of current. Option (2) is incorrect because dry cell batteries cannot be recharged. Option (3) is incorrect because the chemical action is faster when the flashlight is on. Option (5) is incorrect because dry cell batteries continue their chemical action at a reduced rate when the circuit is broken.

24. (Application) **(4) a toy truck** Batteries are most suited for portable items that cannot be

plugged into an outlet. Dry cells are used for smaller toys and machines. Options (1) and (5) are incorrect because they are not portable. Options (2) and (3) also cannot be plugged in while they are operating, so batteries are suitable for them. However, cars and motorized wheelchairs are larger machines that use storage batteries, which can be recharged.

25. (Analysis) **(3) The car's engine generates electricity that recharges the battery as the car runs.** Option (1) is incorrect because the battery is the source of enough current to start the car. Options (2), (4), and (5) are incorrect because car batteries do not need external recharging, so the car would not need to be jump-started.

26. (Evaluation) **(3) On June 21, the Northern Hemisphere is tilted toward the Sun.** The diagram shows the top, or northern half, of the Earth tilting toward the Sun on June 21. Option (1) is incorrect because there is no information about tides in the diagram. Tides are not caused by the nearness of the Earth to the Sun. Option (2) is incorrect; the diagram does not illustrate how tides occur and does not show the Moon. Option (4) is incorrect because the Northern Hemisphere is tilted toward the Sun on that date. Option (5) is not true. It is the tilting of the Earth's hemispheres that causes a longer or shorter day. On March 21 and September 21, days and nights of equal length occur because neither the Northern nor the Southern Hemisphere is tilted toward the Sun.

27. (Application) **(2) The ocean currents in the North**

Atlantic flow from west to east. Because the freighter sails with the current going to England, it makes faster progress in the eastward trip to Liverpool. Option (1) is true but does not explain the difference in the trips. Option (3) is incorrect because the currents flow from west to east in the North Atlantic. Options (4) and (5) do not explain the difference because the freighter does not cross the South Atlantic on either trip.

28. (Analysis) **(3) There must be an equal number of atoms from each element on each side of the equation.** If you examine the equation and the illustration, you will see that before and after the reaction there are four hydrogen atoms, four oxygen atoms, and one carbon atom. They recombine to form different substances, but their number remains the same. Option (1) is incorrect because all reactions do not produce gases. Option (2) is disproved by the information; molecules are said to be made up of atoms. Options (4) and (5) are not supported by the equations given.

29. (Application) **(3) mines** Methane is a hazard in coal mines, where it can cause fire and loss of life. Option (1) is incorrect because people do not come in contact with deposits that are buried deep in the Earth. Option (2) is incorrect because in marshes methane tends to bubble up and disperse into the atmosphere. Options (4) and (5) are incorrect because the use of methane is highly controlled in these environments. Therefore, it is less likely to burn accidently.

30. (Comprehension) **(4) Another person may pick up the viruses by touching the objects.** Since cold viruses survive for a while on objects, washing the objects gets rid of the viruses. Option (1) is not true. Option (2) is also not true; the viruses survive for up to three hours. Option (3) is incorrect because the person with the cold is already infected. It does not matter if he or she picks up a contaminated object. Option (5) is not true.

31. (Analysis) **(5) Avoid going out in the rain.** According to the article, colds are transmitted when rhinoviruses pass from one person to another, either by contact or by air. The weather does not matter. Therefore, the preventive measures related to the spread of viruses by contact or air are based on the facts given in the article. Options (1) to (4) are incorrect.

32. (Analysis) **(2) Rhinoviruses may be transferred from your hands to the mucous membranes in your nose.** Washing your hands gets rid of the viruses. Option (1) may be true, but it is not clear that the hands were actually dirty. Option (3) is not true. Rhinoviruses can survive for up to three hours on the skin. Option (4) is true, but the transmission of the viruses described here is by air, not by contact. Option (5) is not true.

33. (Evaluation) **(1) Warm air rises and cool air sinks.** The diagram shows cool air pushing warm air, which rises. As the warm air becomes cooler, it sinks. Options (2) and (3) are incorrect because they describe the opposite of what actually happens. Option (4) is incorrect because the diagram suggests that the heat of the different types of surfaces influences air movement above them. The truth of option (5) cannot be determined from the diagram. Although the diagram shows a convection cell where land and water meet, there may be other situations in which convection cells occur.

34. (Application) **(2) Face inland.** According to the diagram, the wind is coming off the ocean onto the beach. Facing inland will keep the wind out of your eyes. Option (1) would have the wind coming directly at you. Options (3) and (4) would expose either side of your face to the wind. Option (5) is incorrect because the wind in the diagram is coming from one particular direction.

35. (Application) **(1) a stratovolcano** Mt. St. Helens spewed forth rock, lava, and ashes, making it a stratovolcano. Option (2) is incorrect because cinder cones are formed without lava flows. Option (3) is incorrect because shield volcanoes are formed primarily by lava flows. Option (4) is not a type of volcano; all volcanoes contain magma. Option (5) is incorrect because extinct volcanoes do not erupt.

36. (Evaluation) **(3) Volcanoes are mountains built from deposits of lava, rock, or ashes.** The diagram shows how volcanoes are formed. Option (1) is incorrect because shield volcanoes are not necessarily steep. Option (2) is incorrect because the diagram does not indicate anything about the speed and nature of the eruptions. Option (4) is true, but it is an incorrect answer because the diagram does not show where volcanoes occur. Option (5) is also true, but it is not supported by the diagram.

37. (Application) **(3) ice, water, water vapor** These are three

forms of water: solid, liquid, and gas. Option (1) is incorrect because both lumps and grains of sugar are solids and the option provides no gas. Option (2) is incorrect because rock salt and salt crystals are both solids and the option provides no gas. Option (4) is incorrect because glass and broken glass are both solids, and the option provides no gas. Option (5) is incorrect because a carbonated beverage is not a form of carbon dioxide. It contains bubbles of carbon dioxide gas.

38. (Analysis) **(4) changes in amount of heat** Matter freezes, melts, and boils at different temperatures, changing from one state to another. Option (1) is incorrect because a change in chemical makeup would result in an entirely different kind of matter, not another state of matter. Option (2) is incorrect because a change in position may result from a change of matter but does not cause it. Option (3) is incorrect because it is not a cause. Option (5) may be the result of a change of state, but it is not a cause.

39. (Analysis) **(3) the ciliary muscles** These are in position to pull the lens as needed to alter its shape. Options (1), (2), (4), and (5) are not connected to the lens.

40. (Analysis) **(3) The amount of light entering the eye is changed.** The iris makes the diameter of the pupil larger or smaller, thereby controlling the amount of light that enters the eye. Option (1) is not true. Option (2) is true but is not related to the function of the iris; rather it is related to the position of the optic nerve. Option (4) is true, but the iris does not cause this bulging. Option (5) is not related to the iris.

41. (Application) **(5) the source of activation energy** In this case, the match provides the energy to light the barbecue. Option (1) is incorrect because the match is not a property, it is a form of matter. Option (2) is incorrect since the lit match is an exothermic reaction, releasing energy. Option (3) is incorrect because the lit match is part of a chemical reaction, not a physical change. Option (4) is incorrect; the resulting forms of matter in this reaction are ashes and gases.

42. (Comprehension) **(3) a chemical reaction in which heat energy is absorbed** Option (1) is incorrect because it defines all chemical reactions, not just endothermic reactions. Option (2) is incorrect because it defines an exothermic reaction. Option (4) is incorrect because it defines activation energy. Option (5) is incorrect because all chemical reactions need activation energy to start.

43. (Comprehension) **(3) To the listener, the frequency of the sound waves seems to change.** As the truck is moving closer, the sound waves seem to have a higher frequency because they take increasingly less time to reach the listener. Therefore, the noise of the truck seems greater when it is coming closer. Option (1) is incorrect because the noise coming from the truck's engine is actually the same at all times. It only seems to make more noise when it is near. Option (2) is incorrect; the sound waves do not actually change frequency; they only seem to change frequency. Option (4) is not indicated by any of the information given. Option (5) is not true.

44. (Evaluation) **(2) The Doppler effect can be used to determine whether a source of sound is moving away from or toward someone.** The faster the source of sound moves, the greater the frequency of sound waves. Option (1) is incorrect since the Doppler effect works with other types of waves such as light waves. Nothing in the article suggests that the effect applies only to sound waves. Option (3) is incorrect because the Doppler effect is shown to work with distant sounds. Option (4) is incorrect because the sound waves never actually change frequency, they only appear to change frequency. Option (5) is incorrect because nothing in the diagram gives information about wave interference.

45. (Application) **(3) steak** Steak, an animal product, is high in cholesterol. Options (1), (2), (4), and (5) are not animal products and have little or no cholesterol.

46. (Analysis) **(3) builds up on the artery walls and can block the flow of blood to the heart** Option (1) is incorrect because the cholesterol collects in the arteries, not in the heart. Option (2) is incorrect; cholesterol helps produce cell membranes, not destroy them. Options (4) and (5) are not true and are not supported by the article.

47. (Comprehension) **(2) the arteries become clogged** Although hardening of the arteries in option (1) is a popular term for atherosclerosis, a more accurate description of what happens is that the arteries become narrow and clogged. Options (3), (4), and (5) are not characteristics of atherosclerosis.

48. (Comprehension) **(2) saturated fat** The other four types

of fat listed are not primary sources of cholesterol in the diet. They are not mentioned in the article.

49. (Evaluation) **(3) Most substances expand when heated and contract when cooled.** Option (1) is incorrect because a few substances such as water expand when cooled. Options (2) and (4) are incorrect because most substances do not contract when heated. Option (5) is incorrect because substances do change in volume when heated or cooled.

50. (Comprehension) **(2) Mercury expands when heated.** Option (1) is true but does not explain why mercury rises in a thermometer. Option (3) is not true; mercury expands and does not contract when heated. Option (4) is not possible; heat is not a form of matter and does not take up space. It is a form of energy. Option (5) is also untrue; as with other substances, the glass also expands when heat is added.

51. (Evaluation) **(4) A and B only** When telephone wires expand in the heat, they become longer and sag. When glass is suddenly heated, the quick expansion in one portion causes it to break. Therefore, options (1) and (2) are correct but incomplete. Statement C is not true; soup, made mostly of water, acts like water when frozen and expands. Therefore, options (3) and (5) are incorrect.

52. (Evaluation) **(1) The molecules and atoms of gases are the farthest apart and the most mobile.** It takes progressively more heat to change matter from solid to liquid to gas. Therefore, it follows that gases have atoms and molecules with the most kinetic energy, moving the fastest and farthest apart. Options (2) and (3) are thus incorrect. Option (4) is also incorrect. If the molecules and atoms always acted in the same manner, the states of matter would not change. Option (5) is not true; the molecules and atoms of a solid move slowly.

53. (Analysis) **(1) The Earth's interior is composed of material much more dense than the crust.** Option (2) is incorrect because if the interior were less dense than the crust, the total density of the Earth would have been less than 3 grams per cubic centimeter. Options (3) and (4) are not supported by the diagram. Option (5) is incorrect because a lower density added to the calculation would reduce the final figure and not increase it.

54. (Analysis) **(4) the great pressure on the inner core** The pressure of the upper layers pushes the particles of iron and nickel so close together that they remain solid. Option (1) is incorrect because it suggests that other substances would be in melted or liquid form. The diagram shows the core to be solid. No information is given in the diagram about options (2) or (3). Option (5) is incorrect because the composition of the mantle does not affect the state of matter of the inner core.

55. (Comprehension) **(1) dairy products** Dairy products, like many foods, contain little vitamin C. The foods mentioned in options (2), (3), (4), and (5) are described in the article as high in vitamin C.

56. (Application) **(5) a fresh orange** Fresh, uncooked foods are the best sources of vitamin C. Options (1) and (4) are incorrect because these foods are not high in vitamin C. Options (2) and (3) are incorrect because the drying and cooking of these foods decreases their vitamin C content.

57. (Analysis) **(2) More than one egg will be produced.** Fertility drugs increase egg production. Option (1) is incorrect because the fallopian tubes are blocked; they cannot be opened with fertility drugs. Option (3) is incorrect because the woman's blocked fallopian tubes prevent fertilization from occurring naturally. Option (4) is incorrect because giving fertility drugs to the woman has no effect on the man. Option (5) is incorrect because there is no embryo at this point.

58. (Evaluation) **(3) They want a child who is genetically their own.** Because *in vitro* fertilization uses the couple's eggs and sperm, the resulting embryos have their genes. Many couples feel this result is worth the time, stress, and expense of *in vitro* fertilization. Option (1) is true, but successful *in vitro* fertilization also can take a long time. Option (2) is not true because the health of a child cannot be guaranteed either by adoption or by *in vitro* fertilization. Options (4) and (5) are not true.

59. (Analysis) **(5) The timing of embryo development and uterus development is off.** There are many chemical and biological changes associated with conception, and if they are not timed right, the embryo will not develop properly. When the egg and reproductive system are separated, the chances of bad timing are increased. Option

(1) is incorrect because unfertilized eggs are not transferred. Option (2) is incorrect because defective sperm will not fertilize the egg. Option (3) may occasionally be correct, but the handling of sperm and eggs is strictly regulated to avoid contamination. Option (4) is incorrect because if the sperm are improperly processed, they will not fertilize the egg.

60. (Analysis) **(3) two damaged fallopian tubes** For this woman, there is no way for an egg to get from the ovary to the uterus. Option (1) is incorrect because unhealthy women are not good candidates for *in vitro* fertilization. Option (2) is incorrect because a woman with non-functioning ovaries cannot produce eggs. Option (4) is incorrect because a woman without a uterus cannot carry an embryo. Option (5) is incorrect because a woman with one functioning fallopian tube can become pregnant naturally.

61. (Application) **(3) circulatory system** The blood carries substances to all the tissues in the body through the circulatory system. Although the blood circulates through the other systems, it is part of the circulatory system. None of the other systems listed in options (1), (2), (4), and (5) transport substances.

62. (Application) **(4) nervous system** The eyes and ears receive sensory stimuli and pass it on to the brain, which interprets what we see and hear. Options (1), (2), (3), and (5) are incorrect because these systems do not involve seeing and hearing.

63. (Analysis) **(5) reproductive system** If the reproductive system of an individual is not working properly, the individual will still survive. Options (1), (2), (3), and (4) are incorrect because failure of these systems would cause death.

64. (Application) **(1) Ice takes up more space than an equal mass of water.** The diagram of ice shows more space between molecules than the diagram of water. When water freezes and expands, it exerts more pressure on the glass and breaks it. Option (2) is incorrect because each molecule does not increase in size as the temperature drops. Option (3) is not supported by the information; glass actually contracts when it becomes cold. Options (4) and (5) are not true.

65. (Analysis) **(3) The groups of molecules in water are tightly bound together and must absorb much heat energy before they break apart.** Hydrogen bonds are unusually strong, and they must be broken in order for water vapor to form. Options (1) and (2) are true but are not supported by the information. Option (4) is not true; the polarity of water molecules causes them to be attracted to one another. Option (5) is not true.

66. (Evaluation) **(2) There are more molecules in 1 cubic centimeter of water than in 1 cubic centimeter of ice.** Water molecules are closer together in the liquid form, so there would be more molecules in a given volume. Options (1), (4), and (5) are true, but they are incorrect answers because the information says nothing about these properties of water. Option (3) is not true.

POSTTEST Correlation Chart

Science

The chart below will help you determine your strengths and weaknesses in thinking skills and in the content areas of life science, Earth science, chemistry, and physics.

Directions

Circle the number of each item that you answered correctly on the Posttest. Count the number of items you answered correctly in each column. Write the amount in the Total Correct space for each column. (For example, if you answered 8 comprehension items correctly, place the number 8 in the blank above *out of 12*.) Complete this process for the remaining columns.

Count the number of items you answered correctly in each row. Write that amount in the Total Correct space for each row. (For example, in the Life Science row, write the number correct in the blank before *out of 30*.) Complete this process for the remaining rows.

Cognitive Skills / Content	Comprehension	Application	Analysis	Evaluation	Total Correct
Life Science (pages 38–131)	5, **11**, 30, 47, 48, 55	10, **17**, **18**, **19**, **20**, 45, 56, 61, 62	7, **12**, 31, 32, **39**, **40**, 46, 57, 59, 60, 63	6, 8, 9, 58	_____ out of 30
Earth Science (pages 132–179)	13, 14	**2**, **27**, **34**, **35**	1, 15, 16, **53**, **54**	**26**, **33**, **36**	_____ out of 14
Chemistry (pages 180–205)	3, 42, 50	**29**, 37, 41, **64**	**28**, 38, **65**	4, 49, 51, 52, **66**	_____ out of 15
Physics (pages 206–241)	**43**	22, 24	21, 23, 25	**44**	_____ out of 7
Total Correct	_____ out of 12	_____ out of 19	_____ out of 22	_____ out of 13	Total correct: ____ out of 66 1–54 → Need more review 55–66 → Congratulations! You're ready!

Boldfaced items are based on diagrams, maps, charts, or graphs.

If you answered fewer than 66 questions correctly, determine which areas are hardest for you. Go back to the *Steck-Vaughn GED Science* book and review the content in those areas. In the parentheses under the item type heading, the page numbers tell you where you can find specific instruction about that area of science in the *Steck-Vaughn GED Science* book.

Tests of General Educational Development

TEST _____

TEST TAKEN AT _____

TEST NUMBER

☐

① ② ③ ④ ⑤

TEST ANSWERS   DO NOT MARK IN YOUR TEST BOOKLET

Fill in the circle corresponding to your answer for each question. Erase cleanly.

1 ① ② ③ ④ ⑤	19 ① ② ③ ④ ⑤	36 ① ② ③ ④ ⑤	54 ① ② ③ ④ ⑤
2 ① ② ③ ④ ⑤	20 ① ② ③ ④ ⑤	37 ① ② ③ ④ ⑤	55 ① ② ③ ④ ⑤
3 ① ② ③ ④ ⑤	21 ① ② ③ ④ ⑤	38 ① ② ③ ④ ⑤	56 ① ② ③ ④ ⑤
4 ① ② ③ ④ ⑤	22 ① ② ③ ④ ⑤	39 ① ② ③ ④ ⑤	57 ① ② ③ ④ ⑤
5 ① ② ③ ④ ⑤	23 ① ② ③ ④ ⑤	40 ① ② ③ ④ ⑤	58 ① ② ③ ④ ⑤
6 ① ② ③ ④ ⑤	24 ① ② ③ ④ ⑤	41 ① ② ③ ④ ⑤	59 ① ② ③ ④ ⑤
7 ① ② ③ ④ ⑤	25 ① ② ③ ④ ⑤	42 ① ② ③ ④ ⑤	60 ① ② ③ ④ ⑤
8 ① ② ③ ④ ⑤	26 ① ② ③ ④ ⑤	43 ① ② ③ ④ ⑤	61 ① ② ③ ④ ⑤
9 ① ② ③ ④ ⑤	27 ① ② ③ ④ ⑤	44 ① ② ③ ④ ⑤	62 ① ② ③ ④ ⑤
10 ① ② ③ ④ ⑤	28 ① ② ③ ④ ⑤	45 ① ② ③ ④ ⑤	63 ① ② ③ ④ ⑤
11 ① ② ③ ④ ⑤	29 ① ② ③ ④ ⑤	46 ① ② ③ ④ ⑤	64 ① ② ③ ④ ⑤
12 ① ② ③ ④ ⑤	30 ① ② ③ ④ ⑤	47 ① ② ③ ④ ⑤	65 ① ② ③ ④ ⑤
13 ① ② ③ ④ ⑤	31 ① ② ③ ④ ⑤	48 ① ② ③ ④ ⑤	66 ① ② ③ ④ ⑤
14 ① ② ③ ④ ⑤	32 ① ② ③ ④ ⑤	49 ① ② ③ ④ ⑤	67 ① ② ③ ④ ⑤
15 ① ② ③ ④ ⑤	33 ① ② ③ ④ ⑤	50 ① ② ③ ④ ⑤	68 ① ② ③ ④ ⑤
16 ① ② ③ ④ ⑤	34 ① ② ③ ④ ⑤	51 ① ② ③ ④ ⑤	69 ① ② ③ ④ ⑤
17 ① ② ③ ④ ⑤	35 ① ② ③ ④ ⑤	52 ① ② ③ ④ ⑤	70 ① ② ③ ④ ⑤
18 ① ② ③ ④ ⑤		53 ① ② ③ ④ ⑤	

Tests of General Educational Development

TEST _____

TEST TAKEN AT _____

TEST NUMBER

TEST ANSWERS   **DO NOT MARK IN YOUR TEST BOOKLET**

Fill in the circle corresponding to your answer for each question. Erase cleanly.

① ② ③ ④ ⑤

1 ① ② ③ ④ ⑤   19 ① ② ③ ④ ⑤   36 ① ② ③ ④ ⑤   54 ① ② ③ ④ ⑤
2 ① ② ③ ④ ⑤   20 ① ② ③ ④ ⑤   37 ① ② ③ ④ ⑤   55 ① ② ③ ④ ⑤
3 ① ② ③ ④ ⑤   21 ① ② ③ ④ ⑤   38 ① ② ③ ④ ⑤   56 ① ② ③ ④ ⑤
4 ① ② ③ ④ ⑤   22 ① ② ③ ④ ⑤   39 ① ② ③ ④ ⑤   57 ① ② ③ ④ ⑤
5 ① ② ③ ④ ⑤   23 ① ② ③ ④ ⑤   40 ① ② ③ ④ ⑤   58 ① ② ③ ④ ⑤
6 ① ② ③ ④ ⑤   24 ① ② ③ ④ ⑤   41 ① ② ③ ④ ⑤   59 ① ② ③ ④ ⑤
7 ① ② ③ ④ ⑤   25 ① ② ③ ④ ⑤   42 ① ② ③ ④ ⑤   60 ① ② ③ ④ ⑤
8 ① ② ③ ④ ⑤   26 ① ② ③ ④ ⑤   43 ① ② ③ ④ ⑤   61 ① ② ③ ④ ⑤
9 ① ② ③ ④ ⑤   27 ① ② ③ ④ ⑤   44 ① ② ③ ④ ⑤   62 ① ② ③ ④ ⑤
10 ① ② ③ ④ ⑤   28 ① ② ③ ④ ⑤   45 ① ② ③ ④ ⑤   63 ① ② ③ ④ ⑤
11 ① ② ③ ④ ⑤   29 ① ② ③ ④ ⑤   46 ① ② ③ ④ ⑤   64 ① ② ③ ④ ⑤
12 ① ② ③ ④ ⑤   30 ① ② ③ ④ ⑤   47 ① ② ③ ④ ⑤   65 ① ② ③ ④ ⑤
13 ① ② ③ ④ ⑤   31 ① ② ③ ④ ⑤   48 ① ② ③ ④ ⑤   66 ① ② ③ ④ ⑤
14 ① ② ③ ④ ⑤   32 ① ② ③ ④ ⑤   49 ① ② ③ ④ ⑤   67 ① ② ③ ④ ⑤
15 ① ② ③ ④ ⑤   33 ① ② ③ ④ ⑤   50 ① ② ③ ④ ⑤   68 ① ② ③ ④ ⑤
16 ① ② ③ ④ ⑤   34 ① ② ③ ④ ⑤   51 ① ② ③ ④ ⑤   69 ① ② ③ ④ ⑤
17 ① ② ③ ④ ⑤   35 ① ② ③ ④ ⑤   52 ① ② ③ ④ ⑤   70 ① ② ③ ④ ⑤
18 ① ② ③ ④ ⑤   53 ① ② ③ ④ ⑤

Tests of General Educational Development

TEST _____

TEST TAKEN AT _____

TEST NUMBER

① ② ③ ④ ⑤

TEST ANSWERS   DO NOT MARK IN YOUR TEST BOOKLET

Fill in the circle corresponding to your answer for each question. Erase cleanly.

1 ① ② ③ ④ ⑤   19 ① ② ③ ④ ⑤   36 ① ② ③ ④ ⑤   54 ① ② ③ ④ ⑤

2 ① ② ③ ④ ⑤   20 ① ② ③ ④ ⑤   37 ① ② ③ ④ ⑤   55 ① ② ③ ④ ⑤

3 ① ② ③ ④ ⑤   21 ① ② ③ ④ ⑤   38 ① ② ③ ④ ⑤   56 ① ② ③ ④ ⑤

4 ① ② ③ ④ ⑤   22 ① ② ③ ④ ⑤   39 ① ② ③ ④ ⑤   57 ① ② ③ ④ ⑤

5 ① ② ③ ④ ⑤   23 ① ② ③ ④ ⑤   40 ① ② ③ ④ ⑤   58 ① ② ③ ④ ⑤

6 ① ② ③ ④ ⑤   24 ① ② ③ ④ ⑤   41 ① ② ③ ④ ⑤   59 ① ② ③ ④ ⑤

7 ① ② ③ ④ ⑤   25 ① ② ③ ④ ⑤   42 ① ② ③ ④ ⑤   60 ① ② ③ ④ ⑤

8 ① ② ③ ④ ⑤   26 ① ② ③ ④ ⑤   43 ① ② ③ ④ ⑤   61 ① ② ③ ④ ⑤

9 ① ② ③ ④ ⑤   27 ① ② ③ ④ ⑤   44 ① ② ③ ④ ⑤   62 ① ② ③ ④ ⑤

10 ① ② ③ ④ ⑤   28 ① ② ③ ④ ⑤   45 ① ② ③ ④ ⑤   63 ① ② ③ ④ ⑤

11 ① ② ③ ④ ⑤   29 ① ② ③ ④ ⑤   46 ① ② ③ ④ ⑤   64 ① ② ③ ④ ⑤

12 ① ② ③ ④ ⑤   30 ① ② ③ ④ ⑤   47 ① ② ③ ④ ⑤   65 ① ② ③ ④ ⑤

13 ① ② ③ ④ ⑤   31 ① ② ③ ④ ⑤   48 ① ② ③ ④ ⑤   66 ① ② ③ ④ ⑤

14 ① ② ③ ④ ⑤   32 ① ② ③ ④ ⑤   49 ① ② ③ ④ ⑤   67 ① ② ③ ④ ⑤

15 ① ② ③ ④ ⑤   33 ① ② ③ ④ ⑤   50 ① ② ③ ④ ⑤   68 ① ② ③ ④ ⑤

16 ① ② ③ ④ ⑤   34 ① ② ③ ④ ⑤   51 ① ② ③ ④ ⑤   69 ① ② ③ ④ ⑤

17 ① ② ③ ④ ⑤   35 ① ② ③ ④ ⑤   52 ① ② ③ ④ ⑤   70 ① ② ③ ④ ⑤

18 ① ② ③ ④ ⑤   53 ① ② ③ ④ ⑤

STECK-VAUGHN
GED

The Simulated GED Test begins on page 274.

SIMULATED GED TEST

Science

Directions

The Science Simulated GED Test consists of multiple-choice questions intended to measure your understanding of the general concepts in science. The questions are based on short readings that often include a graph, chart, or figure. Study the information given and then answer the question(s) following it. Refer to the information as often as necessary in answering the questions.

You should spend no more than 95 minutes answering the 66 questions on the Simulated Test. Work carefully, but do not spend too much time on any one question. Be sure you answer every question. You will not be penalized for incorrect answers.

Record your answers to the questions on the answer sheet provided on page 272.

To record your answers, mark one numbered space on the answer sheet beside the number that corresponds to the question on the Simulated Test.

Example

Which of the following is the smallest unit in a living thing?

(1) tissue
(2) organ
(3) cell
(4) muscle
(5) capillary   ① ② ● ④ ⑤

The correct answer is "cell"; therefore, answer space 3 should be marked on the answer sheet.

Do not rest the point of your pencil on the answer sheet while you are considering your answer. Make no stray or unnecessary marks. If you change an answer, erase your first mark completely. Mark only one answer space for each question; multiple answers will be scored as incorrect. Do not fold or crease your answer sheet.

Directions: Choose the <u>best answer</u> to each item.

Items 1 and 2 refer to the following map.

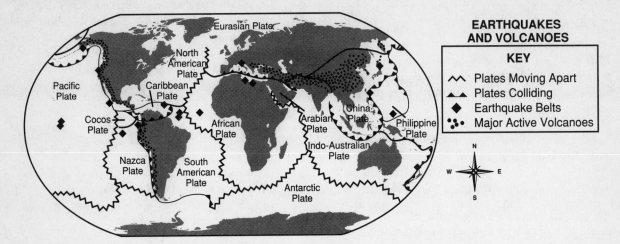

1. Large sections of the Earth's crust that move slowly across the Earth's surface are called plates. When two plates collide, one of the effects is the

   (1) formation of large ocean areas
   (2) creation of islands
   (3) formation of continents
   (4) formation of volcanoes
   (5) creation of rivers

2. Which of the following areas would provide a good opportunity for studying volcanoes and earthquakes?

   (1) northern Europe
   (2) southern Africa
   (3) Australia
   (4) eastern North America
   (5) western South America

Items 3 and 4 refer to the following information.

The density of a substance is its mass per unit of volume. The density of a gas is expressed in grams per liter.

DENSITY OF GASES	
Gas	Density (grams per liter)
Air	1.29
Hydrogen	0.09
Nitrogen	1.25
Oxygen	1.43
Carbon dioxide	1.98
Sulfur dioxide	2.93

3. The specific gravity of a gas is the ratio of its density to the density of air. For example, the specific gravity of oxygen is expressed as:

$$\frac{1.43}{1.29} = 1.11$$

This means that oxygen is 1.11 times as dense as air. Which of the following gases has the highest specific gravity?

   (1) hydrogen
   (2) nitrogen
   (3) oxygen
   (4) carbon dioxide
   (5) sulfur dioxide

4. Nitrogen makes up almost 80 percent of the air. Which of the following is a result of this fact?

   (1) The densities of air and nitrogen are similar.
   (2) Nitrogen is more dense than air.
   (3) Nitrogen is more dense than hydrogen.
   (4) Sulfur dioxide is less dense than nitrogen.
   (5) Nitrogen is more dense than carbon dioxide.

Many kinds of ladybird beetles, or ladybugs, spend their time hunting insects to eat. They favor aphids and scale insects. Ladybugs can clean these harmful insects off plants, leaving the plants with energy to grow. Ladybugs have ample supplies of foul-tasting substances that repel insect-eating birds. Thus, ladybugs can spend the day in plain sight while hunting for insects.

The ladybug's appetite for harmful insects has been put to good use by farmers. An Australian variety of ladybug was imported into California to help control another Australian insect, the cottony cushion scale. This pest was accidentally introduced into America, and it quickly became a major pest in citrus orchards. The Australian ladybugs lay their eggs next to the eggs of the scale insect in citrus trees and other plants. Since a female ladybug lays about a thousand eggs and each hatchling eats about three thousand young scale insects, the benefits to citrus trees are enormous.

Ladybugs hunt until late fall when the cold limits their food supply. Then they take shelter in large groups until the spring. During the sheltering period, ladybugs can be collected in boxes, kept chilled until the spring, sold to gardeners and farmers, and released into the environment.

5. When do ladybugs hunt?

    (1) twenty-four hours a day
    (2) at night
    (3) at dawn and dusk
    (4) from spring to fall
    (5) during the winter

6. An unstated assumption of the author is that

    (1) ladybugs are hunters
    (2) ladybugs are not active all year round
    (3) most ladybugs are harmful to plants
    (4) there are varieties of ladybugs that feed on different types of insects
    (5) ladybugs are a type of beetle

7. People's attitude toward the ladybug has been affected by the insect's

    (1) spots
    (2) ability to repel birds
    (3) habit of sheltering during the winter
    (4) care for its young
    (5) ability to hunt other insects that feed on valuable plants

8. Which of the following conclusions is supported by the information in the article?

    (1) All varieties of ladybugs are hunters.
    (2) Female ladybugs do not bring food to their young.
    (3) Most ladybugs are beneficial to humans.
    (4) Ladybugs have a life expectancy of less than two months.
    (5) Ladybugs mate for life.

9. The winter behavior of ladybugs is most similar to

(1) frogs burying themselves in pond bottoms for the winter
(2) the migration of monarch butterflies south for the winter
(3) hares turning white in winter and brown in summer
(4) trees shedding leaves in the fall
(5) animals growing thicker fur for the winter

10. Suppose your rosebushes are being damaged by aphids. You release some ladybugs in your garden, but two days later, they are gone. Based on the article, which of the following statements could explain what happened?

A. The ladybugs ran out of food and flew away to find more food.
B. Ladybugs eat insects only on citrus trees.
C. The ladybugs were eaten by birds.

(1) A only
(2) B only
(3) C only
(4) A and B
(5) A and C

Items 11 and 12 refer to the following chart.

PULSE RATE AFTER ONE MINUTE OF EXERCISE		
Subject's Physical Condition	Pulse Rate (beats per minute)		
	Light Exercise	Moderate Exercise	Heavy Exercise
Excellent	66	73	82
Very good	78	85	96
Average	90	98	111
Below average	102	107	126
Poor	114	120	142

11. After one minute of moderate exercise, the pulse rate of a person in average physical condition is faster than the pulse rate of a person in

(1) below average condition after one minute of moderate exercise
(2) below average condition after one minute of light exercise
(3) very good condition after one minute of heavy exercise
(4) average condition after one minute of heavy exercise
(5) poor condition after one minute of light exercise

12. Which of the following statements is supported by the information in the chart?

(1) The pulse rate increases with additional minutes of exercise.
(2) People in poor physical condition should not do heavy exercise.
(3) As the pulse rate rises during exercise, so does the number of breaths per minute.
(4) For people in excellent physical condition, the pulse rate increases from light to moderate exercise and decreases from moderate to heavy exercise.
(5) There is a greater difference in the pulse rate between moderate and heavy exercise than between light and moderate exercise.

Tides are the alternate rise and fall of the sea level in oceans. They are caused primarily by the Moon's gravitational pull. The Sun's gravitational pull also contributes to the tides. At any one time, there are two high tides — one on the side of the Earth facing the Moon and one on the opposite side. The average time between high tides at any one location is 12 hours and 25 minutes.

The typical difference in sea level between high and low tides is two feet in the open ocean. Near the coast, the difference is much greater. The greatest difference between high and low tides occurs in the Bay of Fundy in eastern Canada, where the sea level changes by 40 feet.

As the tides change, currents flow to redistribute the ocean's water. Near the coast, the direction of the current changes every 6 1/4 hours. The current first flows toward the shore in what is called a flood current. Then the current flows away from the shore in what is called an ebb current.

13. What is the primary cause of tides on Earth?

   (1)   ocean currents
   (2)   the gravitational pull of the Sun
   (3)   the gravitational pull of the Moon
   (4)   the positions of the planets relative to the Earth
   (5)   the rotation of the Earth

14. In the middle of the ocean, what is the usual change in sea level between high and low tides?

   (1)   2 feet
   (2)   6 feet
   (3)   12 feet
   (4)   25 feet
   (5)   40 feet

15. What is a flood current?

   (1)   an ocean current that carries warm water from the tropics to higher latitudes
   (2)   an undertow
   (3)   a tide that floods low-lying coastal areas
   (4)   a current that carries water away from the shore
   (5)   a current that carries water toward the shore

16. When the Earth, Moon, and Sun are in a straight line, tides are higher than usual. What is the most likely cause of this?

   (1)   The Sun's gravitational pull is added to that of the Moon, which increases the height of the tides.
   (2)   The Moon is in a position in which its gravitational pull is weaker.
   (3)   The positions of the three bodies cause an increase in rainfall, thereby raising the sea level.
   (4)   The positions of the three bodies cause an increase in the amount of polar ice melting, thereby raising the sea level.
   (5)   Tides are higher during the spring.

Items 17 to 20 refer to the following information.

Anyone who recovers from an infectious disease has become immune to that disease. The immunity is caused by the formation of antibodies to fight the disease. Antibodies act to protect the body from disease-causing agents called antigens. Vaccines are specially prepared antigens that stimulate the body's production of antibodies without causing disease. There is a time lag before vaccines take effect, but then the immunity is generally long-term. Antiserums provide immediate short-term protection by giving the patient antibodies produced by another animal.

COMMON IMMUNIZATIONS		
Disease	Type of Immunization	Immunity Period
Diphtheria	Vaccine Antiserum	5 to 10 years 2 to 3 months
Measles	Vaccine Antiserum	Over 10 years A few weeks
Mumps	Vaccine	Probably life
Poliomyelitis	Vaccine	Unknown
Tetanus	Vaccine Antiserum	5 to 10 years A few weeks

17. A thirty-year-old man cut his foot on a rusty nail. He received his last tetanus vaccination when he was twenty-seven years old. What immunization, if any, should he receive?

(1) tetanus vaccine
(2) tetanus antiserum
(3) tetanus vaccine followed by antiserum in a few weeks
(4) tetanus antiserum after he recovers
(5) none

18. A person traveling abroad became ill with diphtheria. He was unsure of the date of his last diphtheria immunization. What immunization, if any, should he receive?

(1) diphtheria vaccine
(2) diphtheria antiserum
(3) diphtheria vaccine followed by antiserum in two months
(4) diphtheria antiserum after he recovers
(5) none

19. There was an epidemic of measles in a town, and a woman was concerned about getting sick. She had had measles as a child many years before. What immunization, if any, should she receive?

(1) measles vaccine
(2) measles antiserum
(3) measles vaccine followed by measles antiserum
(4) measles antiserum after she becomes ill
(5) none

20. Infants are commonly immunized against diphtheria and tetanus at 2, 4, 6, and 18 months. At that point, their bodies are mature enough to maintain long-term immunity. At what approximate age would another immunization against diphtheria and tetanus be necessary?

(1) 2 years
(2) when they enter school
(3) when they graduate from high school
(4) 21 years
(5) Another immunization is not necessary.

Items 21 to 25 refer to the following article.

Sound waves are produced by rapid back-and-forth movements called vibrations. The vibrations cause disturbances in the particles of matter near them. The particles bump into one another and transmit the sound waves over distance. Because sound waves need particles of matter to transmit the disturbance caused by vibration, sound waves cannot travel through outer space.

Two factors that influence the speed of sound waves are the matter through which they travel and the temperature. Matter with particles more tightly packed, such as solids, transmits sounds the fastest. Matter with particles loosely packed, such as gases, transmits sound waves more slowly. In general, the higher the temperature, the faster sound travels. The higher the temperature, the faster the particles of matter move, and the faster they bump into one another, transmitting sound waves.

Light waves, however, are produced by electric and magnetic forces. Light waves differ from sound waves in that they can travel through outer space as well as through matter. Light waves travel faster than sound waves. The speed of sound waves in air at 32°F is 1,085 feet per second. The speed of light is 186,282 miles per second.

21. Why do sound waves travel faster through solids than through gases?

(1) Gases are farther from the source of the vibration.
(2) Particles of solids are tightly packed and transmit disturbances quickly.
(3) Particles of gases are loosely packed and transmit disturbances quickly.
(4) Gases are empty space and do not transmit sound waves.
(5) Solids are warmer than gases.

22. Which of the following situations is an example of the differing speeds of sound and light?

(1) seeing lightning before hearing thunder
(2) watching the light show at a night club
(3) looking at neon signs
(4) the time it takes for sunlight to reach the Earth
(5) flying faster than sound in supersonic jets

23. Which of the following conclusions is supported by the article?

(1) The ground transmits sound faster than the air.
(2) The air transmits sound faster than the ground.
(3) Light waves travel more slowly than sound waves.
(4) Light waves travel only through solids, liquids, and gases.
(5) Light travels only through liquids and gases.

24. Which of the following situations is an example of sound traveling faster through a solid than through a gas?

   (1)   a thunderclap sounding louder overhead than it sounds two miles away

   (2)   the playing of a musical instrument

   (3)   the sound of a siren when an ambulance approaches and then moves away

   (4)   a hunter hearing hoofbeats by putting an ear to the ground

   (5)   a person cupping his or her ear to hear better

25. Which of the following supports the generalization that light waves travel through outer space?

   (1)   the reflection of light off a mirror

   (2)   the passage of light through glass

   (3)   our ability to see the Sun and stars

   (4)   our ability to see through water

   (5)   our ability to see through air

Item 26 refers to the following diagram.

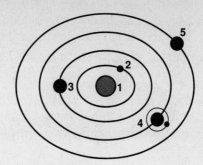

KEY
1 = Sun
2 = Mercury
3 = Venus
4 = Earth and Moon
5 = Mars

26. Which of the following conclusions can be drawn from the diagram?

   (1)   Mars travels around the Sun in less time than the Earth does.

   (2)   The rotation of the planets causes day and night.

   (3)   Mercury is the planet closest to the Sun.

   (4)   Earth is the only planet with significant amounts of water.

   (5)   The Moon is always closer to the Sun than the Earth is.

Item 27 refers to the following diagrams. (The newest rock layer is at the top; the oldest rock layer is at the bottom.)

LIFE CYCLE OF FAULT-BLOCK MOUNTAINS

1. Folding

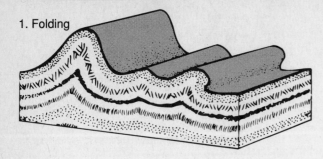

2. Faulting

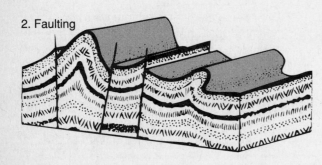

3. Erosion

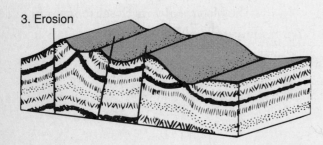

27. Older, buried rock layers become exposed during

   (1) folding
   (2) faulting
   (3) erosion
   (4) folding and faulting
   (5) faulting and erosion

Items 28 and 29 refer to the following chart.

RADIOACTIVE SUBSTANCES AND THEIR USES	
Substance	**Use**
Carbon 14	Estimating age of material that was once living
Arsenic 74	Finding brain tumors
Cobalt	Radiation treatment for cancer Tracing leaks or blockages in pipelines
Iodine 131	Treatment of thyroid gland problems
Radium	Radiation treatment for cancer
Uranium 235	Production of energy in nuclear reactors Atomic weapons

28. If a scientist were interested in determining the dating of a dinosaur bone, what radioactive substance could the scientist use?

   (1) carbon 14
   (2) arsenic 74
   (3) cobalt
   (4) iodine 131
   (5) radium

29. Which of the following conclusions can be drawn from the information in the chart?

   (1) All radioactive substances are expensive.
   (2) Uranium 235 is used only for production of weapons.
   (3) Radioactive substances are always beneficial to humans.
   (4) Radioactive substances have many uses.
   (5) Radioactive substances can be produced by humans.

Items 30 to 32 refer to the following article.

Tumors grow from our own cells. Some tumors are malignant — cancers that are likely to cause death if untreated. Cancer cells are harmful because they grow abnormally and rapidly. They use up nutrients and starve normal cells. Malignant tumors can grow large, causing pressure that interferes with circulation. As the cancer grows, it may spread throughout the body.

What causes cancer? A variety of chemical and physical agents can start cancer. These include chemicals in plastics, cigarette smoke, and asbestos. Radiation can also cause cancer. Certain viruses have been shown to cause cancer in animals. Some researchers think that cancer cells may form but stay inactive for a while. Cancer cells become active only if the body's immune system — which defends against disease — breaks down. Some recognized cancer-causing agents are known to suppress the immune system. People who have had organ transplants take drugs that suppress the immune system, and they show a higher number of cancers.

30. What are malignant tumors?

(1) cancerous growths that may cause death if untreated
(2) cancerous growths that are suppressed by the immune system
(3) harmless growths that can be removed surgically
(4) harmless growths that are a side effect of organ transplants
(5) the major cause of cancer

31. Which of the following practices would be most likely to cause cancer?

(1) eating a vegetarian diet
(2) not exercising
(3) avoiding foods with chemical additives
(4) being exposed to a high level of radiation
(5) drinking water from a well

32. Which of the following statements supports the theory that breakdowns in the body's immune system contribute to the growth of cancers?

(1) Smoking has been linked with lung cancer.
(2) High doses of radiation can cause cancer.
(3) Some cancer-causing agents are chemical agents.
(4) Some cancer-causing agents suppress the immune system.
(5) Cancer cells may suppress the immune system.

Items 33 and 34 refer to the following diagrams.

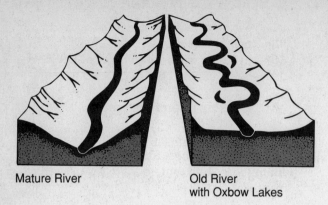

Mature River

Old River
with Oxbow Lakes

33. The diagrams show the river at different stages of its life cycle. During the millions of years between the two stages, the river valley became wider and the river slower. It now winds in S curves called meanders. What best accounts for the widening of the river valley?

(1) annual melting of winter snow
(2) earthquakes
(3) tides
(4) windstorms
(5) erosion by the river

34. Which of the following generalizations is supported by the diagrams?

(1) All rivers drain into an ocean.
(2) Oxbow lakes are found near old rivers.
(3) Flooding is caused by mature rivers.
(4) The valley of an old river is generally steep.
(5) Waterfalls are features of old rivers.

Items 35 and 36 refer to the following diagram.

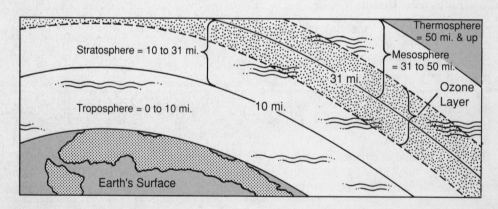

Stratosphere = 10 to 31 mi.

Troposphere = 0 to 10 mi.

Thermosphere = 50 mi. & up

Mesosphere = 31 to 50 mi.

Ozone Layer

31 mi.

10 mi.

Earth's Surface

35. As the distance from the Earth's surface increases, the Earth's gravitational pull decreases. In addition, the air becomes less dense. Gradually the air thins until the atmosphere merges with the near-emptiness of space. In which layer of the atmosphere would a spacecraft operate most efficiently?

(1) troposphere
(2) stratosphere
(3) ozone layer
(4) mesosphere
(5) thermosphere

36. Which of the following generalizations is supported by the diagram?

(1) The ozone layer is part of both the stratosphere and the mesosphere.
(2) The ozone layer is part of both the mesosphere and the thermosphere.
(3) The troposphere extends to a height of 31 miles.
(4) The total height of the atmosphere is 50 miles.
(5) Breathing is difficult in the lower troposphere.

Items 37 and 38 refer to the following information.

All matter is composed of basic substances called elements. An element is a substance that cannot be broken down into simpler substances by chemical reactions. Some common elements are iron, carbon, and oxygen. The smallest particle of an element that still has the properties of that element is called an atom. The smallest particle of the element iron is an iron atom, and a sample of the element iron contains only iron atoms.

When two or more different elements react chemically, a compound is produced. For example, iron oxide, or rust, is a compound made when iron and oxygen react together.

37. Often, when hydrogen and oxygen combine, water is formed. What is water?

   (1)   an atom
   (2)   an element
   (3)   part of an atom
   (4)   energy
   (5)   a compound

38. Which of the following is an unstated assumption by the author?

   (1)   Elements can be broken down into simpler substances by methods other than chemical reactions.
   (2)   All compounds are made of some combination of the elements iron and oxygen.
   (3)   Compounds are usually solids.
   (4)   Elements have at least two different types of atoms.
   (5)   Compounds are more interesting than elements.

Items 39 and 40 refer to the following diagram.

HUMAN DIGESTIVE SYSTEM

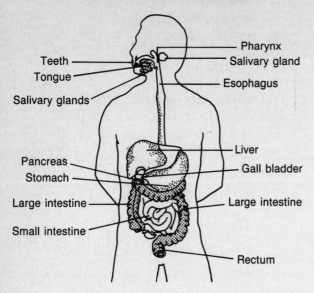

Labels: Pharynx, Salivary gland, Esophagus, Teeth, Tongue, Salivary glands, Liver, Pancreas, Gall bladder, Stomach, Large intestine, Large intestine, Small intestine, Rectum

39. The entire digestive system consists of a long main tube called the alimentary canal. The liver and pancreas are connected to the alimentary canal by small tubes. The alimentary canal ends at the

   (1) liver
   (2) pancreas
   (3) rectum
   (4) large intestine
   (5) small intestine

40. The stomach produces an acid that helps digest food. Heartburn is a painful feeling in the chest caused by stomach acids irritating the

   (1) salivary glands
   (2) esophagus
   (3) large intestine
   (4) small intestine
   (5) rectum

Items 41 and 42 refer to the following information.

Carbon monoxide is a gas produced by coal stoves, furnaces, or gas appliances when they do not get enough air. It is present in the exhaust of internal combustion engines, such as the ones in cars and lawn mowers. Carbon monoxide is a deadly poison. Since it is colorless, odorless, and tasteless, it is very difficult to detect. Victims of carbon-monoxide poisoning become drowsy and then unconscious. Death can occur in minutes.

41. Which of the following actions is likely to be dangerous?

   (1) operating a well-vented coal furnace
   (2) mowing the lawn
   (3) using a properly installed gas stove
   (4) running a car engine in a closed garage
   (5) using a barbecue grill outdoors

42. What is the first symptom of carbon-monoxide poisoning?

   (1) a bad taste
   (2) convulsions
   (3) choking
   (4) unconsciousness
   (5) drowsiness

Items 43 and 44 refer to the following information.

SINGLE PULLEY

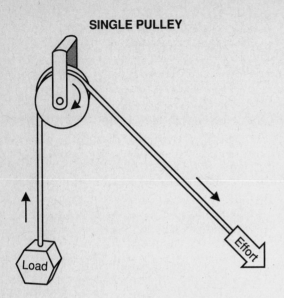

In an ideal pulley, the effort required to pull the rope is equal to the weight of the load.

---

43. When you pull on the rope of a single pulley, what happens?

(1) The load remains stationary.
(2) The load moves downward.
(3) The load moves upward.
(4) The rope moves, but the pulley wheel does not.
(5) The pulley wheel moves, but the rope does not.

44. Which of the following conclusions is supported by the diagram?

(1) A single pulley changes the direction in which force is exerted.
(2) A single pulley does not change the direction in which force is exerted, but it decreases the force required.
(3) A single pulley does not make it easier to lift a very heavy object.
(4) The distance in which the rope is pulled is half as long as the distance in which the load is lifted.
(5) The distance in which the rope is pulled is twice as long as the distance in which the load is lifted.

Items 45 to 48 refer to the following article.

Pathogens are microorganisms that cause disease. To spread disease, a pathogen must leave the body of its host and enter another body. Many pathogens enter the body through a specific route. For instance, pathogens that cause respiratory diseases usually enter through the nose.

The manner in which a pathogen travels from one person to another is related to where the pathogen lives in the host. For example, an intestinal pathogen is likely to be found in the host's feces. Respiratory pathogens can be spread by droplets sneezed or coughed into the air. Such droplets may also be carried on objects. Body fluids, such as saliva, pus, mucus, and urine carry pathogens outside their hosts. Some pathogens that live in the blood are spread by bloodsucking insects such as mosquitoes.

45. A pathogen carried by droplets coughed or sneezed into the air is likely to infect the

(1) respiratory system
(2) intestines
(3) urinary system
(4) blood
(5) respiratory system and urinary system

46. You can prevent the spread of some pathogens by frequently washing your hands. With which pathogens is this effective?

A. respiratory pathogens
B. intestinal pathogens
C. pathogens in the blood

(1) A only
(2) B only
(3) C only
(4) A and B
(5) A and C

47. Malaria is a disease spread from one person to another through the bite of a female Anopheles mosquito. This implies that malaria affects the

(1) blood
(2) respiratory system
(3) skeleton
(4) digestive system
(5) brain

48. The method of spreading a pathogen is related to

(1) its size
(2) where it lives in the body
(3) the temperature
(4) the severity of the disease
(5) the weight of the body

Items 49 to 52 refer to the following article.

Electric currents flowing through wires produce areas of magnetic force called electromagnetic fields. These electromagnetic fields are similar to the magnetic fields produced by magnets. The stronger the electric current, the stronger is the electromagnetic field surrounding it.

Very strong currents produce very strong fields. The force of a strong field can be felt several yards away from the wire. For example, fluorescent lamps held below a 161,000-volt transmission line light up without being plugged into an electric circuit. Many transmission lines are even more powerful.

Wires are not the only producers of electromagnetic fields. All electric appliances produce fields. Appliances with motors, such as hair dryers and washing machines, produce stronger fields than appliances without motors, such as lights and toasters.

Scientists are starting to question the effect of electromagnetic fields on our health. People living near power lines have long complained of headaches, fatigue, memory loss, and more illness than usual. There also seem to be more cases of particular types of cancer among people exposed to strong electromagnetic fields. Although the evidence is inconclusive, it raises some serious public health questions.

---

49. If high-power transmission lines that carry electricity from generating plants to homes and businesses are proved to be health hazards, what would be a likely solution to this problem?

   (1) Move all transmission lines to areas where few people live.
   (2) Move people away from high-power transmission lines.
   (3) Switch to another source of power for everyday use.
   (4) Cover the lines with material that blocks the electromagnetic fields.
   (5) Install small generators in each home and business.

50. A person who wishes to be cautious about exposure to electromagnetic fields should

   (1) use fluorescent rather than regular light bulbs
   (2) stop using magnets
   (3) be careful when plugging in and unplugging appliances
   (4) repair broken appliances immediately
   (5) avoid living near a high-voltage power line

51. Based on the information in the article, which of the following statements is NOT a logical conclusion?

   (1) A vacuum cleaner generates a more powerful electromagnetic field than a 25-watt light bulb.
   (2) People living near high-voltage power lines may be at greater risk for certain types of cancer than people not living near such lines.
   (3) Using an electric blanket causes cancer.
   (4) A 700,000-volt transmission line will light up unplugged fluorescent lamps held below it.
   (5) The stronger the electric current, the stronger is the electromagnetic field.

52. Which of the following statements has not yet been proved?

   (1) Electric currents produce electromagnetic fields.
   (2) The stronger the electric current, the stronger is the electromagnetic field.
   (3) Electromagnetic fields cause health problems.
   (4) Electromagnetic fields are forces similar to magnetic fields produced by magnets.
   (5) The electromagnetic field produced by a high-voltage power line is strong enough to light an unplugged fluorescent lamp.

AIR POLLUTANTS FROM SMOKESTACKS

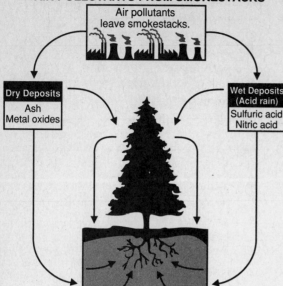

Acidified soil and water leach plant nutrients out of the soil. Heavy metals accumulate in harmful quantities.

53. What causes the air pollutants emitted by smokestacks to become wet deposits?

   (1) industrial processes
   (2) antipollution devices
   (3) petroleum
   (4) oceans
   (5) water vapor in the air

54. What is the eventual effect of dry deposits and acid rain on plant life?

   (1) The plants thrive on the heavy metals accumulating in the soil.
   (2) The root systems are strengthened because of the extra nutrients in the soil.
   (3) The leaves and stems are harmed, but the root systems remain unaffected.
   (4) The plants are weakened by the lack of nutrients and harmed by the accumulation of heavy metals.
   (5) There is no effect on plant life.

About 8 percent of men and 0.5 percent of women are color-blind; that is, they have trouble distinguishing colors. Most color-blind people cannot distinguish red from green. Completely color-blind people see only black, white, and shades of gray; this condition is extremely rare.

55. The fact that many more men than women are color-blind implies that

   (1) color is less important to men than to women
   (2) color blindness is an illness
   (3) color blindness is related to whether a person is male or female
   (4) color blindness is associated with poor vision
   (5) color blindness is a harmless visual defect

56. If a man cannot distinguish red from green, for what occupation would he be unsuitable?

   (1) auto mechanic
   (2) computer operator
   (3) sales person
   (4) railroad engineer
   (5) plumber

Items 57 to 60 refer to the following article.

Acquired immune deficiency syndrome (AIDS) is a disease that destroys the body's ability to protect itself from infection. It is an infection caused by the human immunodeficiency virus (HIV). The AIDS virus is transmitted in semen, vaginal secretions, and blood. The groups most at risk for contracting the disease are homosexual men, intravenous drug users, and their sex partners. Also at risk are people who received transfusions before blood supplies were screened for HIV.

This disease progresses slowly. During the first stage of AIDS, when the body is exposed to HIV, the person may or may not have flu-like symptoms. Within two months, antibodies to protect against HIV are produced. During the second stage, the virus is not active and the person feels fine. The second stage can last for years. In the third stage, swelling of the lymph nodes occurs, but the person is still relatively healthy. During the fourth stage, symptoms gradually appear. AIDS-related infections, tumors, and neurological diseases attack the body, eventually resulting in death.

Presently there is no way to destroy HIV once it is in the body. However, the spread of the virus can be prevented by "safe sex" techniques such as using condoms. The spread of AIDS can also be prevented by not sharing drug needles.

57. What causes AIDS?

(1) taking drugs intravenously
(2) engaging in homosexual activity
(3) receiving a blood transfusion
(4) HIV
(5) antibodies

58. Which of the following generalizations is supported by the information in the article?

(1) The spread of AIDS is limited to the present high-risk groups.
(2) A cure for AIDS will soon be found.
(3) A person with HIV can infect others without being aware of it.
(4) People who recover from AIDS are immune to HIV.
(5) Only men get AIDS.

59. What causes the production of HIV antibodies?

(1) exposure to HIV
(2) blood transfusions
(3) tumors
(4) swelling of the lymph nodes
(5) neurological diseases

60. What effect on the general population will the spread of AIDS have?

(1) Fewer people will use intravenous drugs.
(2) Blood transfusions will no longer be given.
(3) The disease will spread into previously low-risk groups.
(4) The disease will become less serious as it spreads.
(5) There will be no effect on the general population.

FOOD ADDITIVES		
Type	**Examples**	**Possible Effects**
Preservatives	Salt Vinegar Sugar Nitrates	High blood pressure Irritation of the stomach Tooth decay, obesity Cancer

Colorings and
flavorings
	Chlorophyll,
	   Carotene
	
	Synthetic
	   chemicals
	Monosodium
	   glutamate
		Food may appear
		   more nutritional
		   than it actually is.
		Cancer
		
		Allergic reactions

Texturizers	Pectin
	Gelatin
	Glycerol
	   monostear-
	   ate
		Unknown

Supplementary
vitamins and
minerals
	Iron
	Vitamins A,
	   B complex,
	   C, D, etc.
	Iodine
		Both deficiencies
		   and excesses of
		   minerals and vita-
		   mins can cause
		   disease.

61. An example of a food that contains preservatives is

(1)   enriched flour
(2)   homogenized milk
(3)   pickles
(4)   monosodium glutamate
(5)   iodized salt

62. What food additive can give a misleading appearance to food?

(1)   yellow food coloring in commercially prepared baked goods that are made without eggs
(2)   iron in breakfast cereal
(3)   pectin in jelly
(4)   vitamin B in white bread
(5)   vitamin D in milk

63. Why are food additives so important?

A.   They enable food to be transported long distances and be stored for long periods of time without spoiling.
B.   They provide important minerals and vitamins.
C.   Most food additives cause disease.

(1)   A only
(2)   B only
(3)   C only
(4)   A and B
(5)   B and C

Items 64 to 66 refer to the following information.

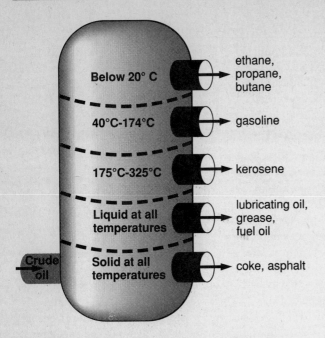

In a process called fractional distillation, crude oil is pumped into a tower where it is heated. Different petroleum products are produced through evaporation and condensation at different temperatures.

---

64. Which of the following statements is supported by the information provided?

(1) Asphalt is a liquid produced during fractional distillation.
(2) Coke is used to make grease.
(3) Kerosene boils at a higher temperature than gasoline.
(4) Burning petroleum fuels creates air pollution.
(5) Crude oil can be used without processing.

65. What causes the substances to separate during fractional distillation?

(1) the force of gravity
(2) the pressure of more crude oil entering the tower
(3) a series of finer and finer filters at each stage of the tower
(4) the pipes carrying the substances away from the tower
(5) the different temperatures at different heights in the tower

66. Which of the following statements is supported by the information in the diagram?

(1) The temperature remains constant throughout the tower.
(2) The higher the location in the tower, the higher is the temperature.
(3) The higher the location in the tower, the lower is the temperature.
(4) All substances produced by fractional distillation boil and then condense in the tower.
(5) All substances produced by fractional distillation condense and then boil in the tower.

Answers and Explanations

Simulated GED Test, pages 275–293

1. (Analysis) **(4) formation of volcanoes** The map shows a relationship between the boundaries of colliding plates and the locations of volcanoes. Options (1), (2), (3), and (5) are not shown to have any relationship to where plates collide.

2. (Application) **(5) western South America** The map shows a high number of earthquakes and volcanoes along the western coast of this continent. The other options have only a few earthquakes and volcanoes and would not be the best areas to study.

3. (Analysis) **(5) sulfur dioxide** Since sulfur dioxide has the highest density of the gases listed, it also has the highest specific gravity:

$$\frac{2.93}{1.29} = 2.26$$

The other gases all have specific gravities of less than 2.

4. (Analysis) **(1) The densities of air and nitrogen are similar.** Since nitrogen makes up almost 80 percent of air, its density has a great influence on the density of air. Option (2) is not true; at 1.25, nitrogen is less dense than air. Option (3) is true, but it is not related to the proportion of nitrogen in the air. Options (4) and (5) are not true.

5. (Comprehension) **(4) from spring to fall** The article states that ladybugs stop hunting when it gets cold and prey is hard to find. Options (1), (2), and (3) are incorrect because the article implies that ladybugs hunt during the day. Option (5) is incorrect because ladybugs are not active during the winter.

6. (Analysis) **(4) there are varieties of ladybugs that feed on different types of insects** This assumption is made when the author explains that an Australian type of ladybug was imported to California to help control the cottony cushion scale. If the American ladybugs had hunted this type of scale insect, the Australian variety would not have been necessary. Options (1), (2), and (5) are stated explicitly in the article; therefore, they are not unstated assumptions. Option (3) is not true.

7. (Evaluation) **(5) ability to hunt other insects that feed on valuable plants** People's attitude toward ladybugs is generally positive, because they are beneficial to humans. Options (1), (2), (3), and (4) are not aspects that affect people, so they do not influence the way people view ladybugs.

8. (Evaluation) **(3) Most ladybugs are beneficial to humans.** The article states that most varieties of ladybugs feed on plant-eating insects. The ladybugs help control the destruction of plants that are valuable to humans. Option (1) is incorrect because the article states that most, not all, varieties of ladybugs hunt insects. Option (2) is incorrect because the article describes how the Australian ladybug lays its eggs near the hatchlings' source of food. Option (4) is incorrect because ladybugs take shelter during the winter and resume hunting the next spring. This implies that they survive more than one season and thus longer than two months. Option (5) is incorrect because the article gives no information about the mating habits of ladybugs. Therefore, no conclusion about this can be made.

9. (Application) **(1) frogs burying themselves in pond bottoms for the winter** Like ladybugs, frogs hibernate during the winter and become active again in the spring. Options (2), (3), (4), and (5) are other forms of adaptation to cold weather, but they differ from the ladybug's adaptation. Option (2) does not occur; ladybugs take shelter where they are. Changes in color, option (3), protect animals that continue to be active during the snowy winter. Ladybugs do not shed anything during the winter, option (4). They do not grow more protection from the cold, option (5).

10. (Evaluation) **(1) A only** The article states that ladybugs hunt for insects, so if the ladybugs ran out of food it is likely that they would fly away in search of more. Options (2) and (4) are incorrect because statement B is not true. The article refers to other plants as well as citrus trees. Options (3) and (5) are incorrect because statement C is not true. Ladybugs produce a foul-tasting substance, so birds do not eat ladybugs.

11. (Comprehension) **(3) very good condition after one minute of heavy exercise** The pulse rate of the person in average condition after one minute of moderate exercise is 98 beats per minute. The chart shows that option (1) is 107 beats, option (2) is 102 beats, option (3) is 96 beats, option (4) is 111 beats, and option (5) is 114 beats. Of these, 98 beats is faster than option (3), or 96 beats.

12. (Evaluation) **(5) There is a greater difference in the pulse rate between moderate and heavy exercise than between light and moderate exercise.** For all types of subjects, the jump in pulse rate between moderate and heavy exercise is greater than the jump between light and moderate exercise. Option (1) is incorrect because the chart does not show what happens after more than one minute of exercise. Options (2) and (3) may be true, but they are not supported by the information given. Option (4) is not true; the pulse rate increases from moderate to heavy exercise.

13. (Comprehension) **(3) the gravitational pull of the Moon** The first paragraph states that the Moon's pull is the primary cause of tides. The Sun's gravitational pull, option (2), has some influence but not much compared to the Moon. Options (1) and (4) do not affect tides. Option (5) contributes to the movement of high tide around the world, but it is not a cause of tides.

14. (Comprehension) **(1) 2 feet** The typical sea-level change in mid-ocean is stated to be 2 feet. The answer is in the second paragraph of the article. Options (2), (3), (4), and (5) are therefore incorrect.

15. (Comprehension) **(5) a current that carries water toward the shore** When the tide is changing from low to high, the flood current redistributes the water in the direction of the coast, as stated in paragraph 3. Options (1) and (2) are not discussed in the article. Option (3) is incorrect because a flood current does not result in flooding. Option (4) is the definition of an ebb current.

16. (Analysis) **(1) The Sun's gravitational pull is added to that of the Moon, which increases the height of the tides.** When the Moon and Sun are in a line on one side of the Earth, their gravitational forces combine and exert more pull on the oceans, increasing the height of the tides. Option (2) is incorrect because if the Moon's gravitational pull were weaker, the tides would be lower. Options (3) and (4) are incorrect because the position of the three bodies does not affect rainfall or ice melting on Earth. Option (5) is not true.

17. (Application) **(5) none** The man was vaccinated only three years ago. According to the chart, his immunity should last for 5 to 10 years. Therefore, he needs no further protection, so options (1), (2), (3), and (4) are incorrect.

18. (Application) **(2) diphtheria antiserum** Because it cannot be determined if he is immune, to be safe, he should be given the antiserum. Options (1) and (3) are incorrect because the vaccine will not give him immediate protection. Option (4) is incorrect because after he recovers, he will have his own antibodies and will not need antiserum. Option (5) is incorrect because he needs immediate protection against diphtheria.

19. (Application) **(5) none** The article states that since the woman had measles as a child, she has antibodies to fight the disease. Therefore, she needs neither the vaccine nor the antiserum of options (1), (2), (3), or (4).

20. (Application) **(2) when they enter school** To ensure that their protection is constant, children are immunized again when they enter school. This occurs about 4 or 5 years after the last vaccination. Since the immunity lasts from 5 to 10 years, immunizing again after 5 years ensures that protection does not lapse. Option (1) is incorrect because the immunity period is at least 5 years. Options (3) and (4) are incorrect because waiting so long would allow the immunity to wear out. The tetanus and diphtheria vaccines last a maximum of 10 years. Option (5) is incorrect because the diphtheria and tetanus vaccines last only 5 to 10 years. To continue the protection, another immunization would be necessary.

21. (Analysis) **(2) Particles of solids are tightly packed and transmit disturbances quickly.** Gases, on the other hand, have particles that are farther apart. Sound takes longer to pass from particle to particle in a gas. Option (1) may or may not be true in a given situation. Option (3) is not true. Sound waves are not transmitted through space but by particles bumping into one another. Option (4) is not true; gases have particles that transmit sound. Option (5) is not necessarily true.

22. (Application) **(1) seeing lightning before hearing thunder** Because light travels faster than sound, lightning is seen before thunder is heard during a thunderstorm. Option (2) is incorrect because the distances are so small that light and sound are perceived at the same time. Option (3) is incorrect because neon signs do not involve sound. Options (4) and (5) are incorrect because they do not compare the speed of light and sound. They involve either light or sound.

23. (Evaluation) **(1) The ground transmits sound faster than the air.** The article indicates that solids (the ground) transmit sound faster than gases (the air). Option (2) is not true. Option (3) is not true; light waves travel more quickly than sound waves. Option (4) is not true because light waves also travel through empty space. Option (5) is untrue because light waves travel through solids and outer space as well.

24. (Application) **(4) a hunter hearing hoofbeats by putting an ear to the ground** Since sound travels faster through a solid, the hunter will hear the hoofbeats sooner through the ground than through the air. Option (1) is incorrect because it involves transmission of sound through air only. Option (2) is incorrect because musical instruments are heard through the air only. Option (3) is incorrect because it involves the transmission of sound through air only. Option (5) is incorrect because the purpose of cupping the ear is to help capture the sound waves traveling through the air.

25. (Analysis) **(3) our ability to see the Sun and stars** Sunlight and starlight reach us from outer space. The other options involve the transmission of light waves through matter, which is not outer space.

26. (Analysis) **(3) Mercury is the planet closest to the Sun.** Option (1) is not true. Mars, being farther from the Sun than Earth, takes longer to travel around the Sun. The diagram does not contain enough information to support options (2) and (4). Option (5) is not true. Since the Moon revolves around the Earth, at times it is farther from the Sun than the Earth is.

27. (Application) **(5) faulting and erosion** Faulting exposes older rock layers when a rock mass slips down past several layers of an adjacent rock mass. Erosion exposes older rock layers as layers above them wear away. Options (1) and (4) are incorrect because folding bends the layers but does not break the newest upper layer. Options (2) and (3) are incorrect because they are only partly correct.

28. (Application) **(1) carbon 14** The chart indicates that carbon 14 is used to estimate the age of material that was once alive, such as a dinosaur bone. Options (2), (3), (4), and (5) all have medical uses.

29. (Evaluation) **(4) Radioactive substances have many uses.** Option (1) is incorrect because the chart does not give information about the value of radioactive substances. Option (2) is incorrect because uranium 235 is also used in nuclear reactors. Option (3) is incorrect because radioactive substances are used in atomic weapons. Option (5), although true, is incorrect because the chart does not give any information about the origin of radioactive substances.

30. (Comprehension) **(1) cancerous growths that may cause death if untreated** This definition appears in the first paragraph of the article. Option (2) is incorrect because if the immune system were successful in suppressing the cancerous growth, the growth would not be malignant. Options (3) and (4) are incorrect because malignant tumors are not harmless. Option (5) is incorrect because malignant tumors are cancers; they are not the cause of cancer.

31. (Analysis) **(4) being exposed to a high level of radiation** Radiation is a known cancer-causing agent. Options (1), (3), and (5) are incorrect because they are practices that lessen the body's intake of artificial substances. Option (2) is incorrect because lack of exercise is not a chemical or physical agent that might cause cancer.

32. (Analysis) **(4) Some cancer-causing agents suppress the immune system.** People who take drugs to suppress the immune system show a higher number of cancers. This statement, found in the second paragraph, supports the theory that the immune system is involved in the development of cancer cells. Options (1), (2), and (3) are true but not related to the role of the immune system. Option (5) is also true, but it deals with a later stage in cancer, not the cause of cancer.

33. (Analysis) **(5) erosion by the river** As the river gets older, it erodes the sides of its channels and starts to loop from side to side. This process widens its valley. The other options are not related to landscape formation in a river valley.

34. (Evaluation) **(2) Oxbow lakes are found near old rivers.** Two oxbow lakes are shown in the diagram of the old river. They are created when a meander of the river is cut off from the main stream. Option (1) is incorrect because the diagram does not show how rivers drain. Option (3) is incorrect because no flooding is indicated in the diagrams. Option (4) is incorrect because the diagram of an old river shows a gently sloping valley. Option (5) is incorrect because the diagram of the old river shows no waterfall.

35. (Application) **(5) thermosphere** The thermosphere is the highest layer shown. It is closest to outer space, and spacecraft would operate very efficiently in its thin air. Options (1), (2), (3), and (4) are all closer to the surface of the Earth, where increasing air density would create more friction for a spacecraft.

36. (Comprehension) **(1) The ozone layer is part of both the stratosphere and the mesosphere.** The diagram shows the ozone layer starting in the upper part of the stratosphere and extending partly into the mesosphere. Option (2) is incorrect because the ozone layer does not reach the thermosphere. Option (3) is incorrect because the troposphere extends to a height of 10 miles. Option (4) is incorrect because the

thermosphere extends beyond the mesosphere, which goes up to 50 miles. Option (5) is incorrect because the troposphere is close to the Earth's surface, where animals live and breathe.

37. (Application) **(5) a compound** Water is a combination of two different elements, which makes it a compound. Since water contains more than one type of element or atom, it cannot be an atom, an element, or part of an atom. Thus, options (1), (2), and (3) are incorrect. Option (4) is incorrect because water is a form of matter, not energy.

38. (Analysis) **(1) Elements can be broken down into simpler substances by methods other than chemical reactions.** In the definition of an element, the author states that an element "cannot be broken down into simpler substances by chemical reactions." By specifically mentioning "chemical," the author implies that there are other methods that can break down elements. In fact, nuclear reactions can break down atoms of an element into smaller atoms. Option (2) is incorrect because the information about iron and oxygen is given as an example of a compound. It does not indicate that iron and oxygen are the only elements that combine into compounds. Option (3) is incorrect because it is untrue and because the article does not discuss the states of matter. Option (4) is untrue. Option (5) is incorrect because nothing in the article indicates a greater interest on the part of the author in atoms, elements, or compounds.

39. (Analysis) **(3) rectum** The diagram shows the digestive system ending at the rectum. Options (1) and (2) are incorrect because the liver and the pancreas are in the middle section of the digestive system. Option (4) empties into the rectum. Option (5) is in the middle part of the alimentary canal.

40. (Application) **(2) esophagus** The esophagus passes downward through the chest. When stomach acids irritate the esophagus, the discomfort is felt in the chest. The other organs listed in options (1), (3), (4), and (5) are not in the chest. Discomfort in any of these organs would not be felt in the chest.

41. (Application) **(4) running a car engine in a closed garage** Since carbon monoxide is part of the exhaust, a car engine should not be run in an enclosed area without proper ventilation. Options (1), (2), (3), and (5) are incorrect because they involve using potentially dangerous items in the proper way, which lessens the danger involved.

42. (Comprehension) **(5) drowsiness** According to the paragraph, drowsiness is the first symptom of carbon-monoxide poisoning. Options (1), (2), (3), and (4) are therefore incorrect.

43. (Comprehension) **(3) The load moves upward.** A downward pull on the rope results in the upward movement of the load. Option (1) is incorrect because the load moves. Option (2) is incorrect because the load moves upward. Option (4) is incorrect because the movement of the wheel helps move the rope. Option (5) is incorrect because the rope does move when you pull on it.

44. (Evaluation) **(1) A single pulley changes the direction in which force is exerted.** Without a pulley, the effort in lifting a load would be upward. With a pulley, the effort exerted is downward. The pulley changes the direction in which effort is exerted. Option (2) is incorrect because the single pulley does change the direction in which effort is exerted. Option (3) is incorrect because it is easier to pull downward on something with your body weight than it is to lift something. Options (4) and (5) are incorrect because the distance in which the rope is pulled is equal to the distance in which the load is lifted.

45. (Comprehension) **(1) respiratory system** The second paragraph states that respiratory infections are spread by the droplets released through sneezing and coughing. Options (2), (3), (4), and (5) are incorrect because they include areas not infected by droplet-carried pathogens.

46. (Analysis) **(4) A and B** Washing your hands is effective against many respiratory and intestinal pathogens because they can be picked up by your hands from sneezes, coughs, or excretions. Pathogens in the blood generally are spread by insects or other animals that bite and draw blood.

47. (Application) **(1) blood** Since pathogens are transmitted by means related to where they reside in the host, a pathogen transmitted by an insect bite would affect the blood. The respiratory system, skeleton, digestive system, and brain mentioned in options (2), (3), (4), and (5) are not accessible to insect bites.

48. (Comprehension) **(2) where it lives in the body** According to the second paragraph of the article, how the pathogen is spread is related to the place in the body where the pathogen lives. Options (1), (3), (4), and (5) are not relevant.

49. (Evaluation) **(4) Cover the lines with material that blocks the electromagnetic fields.** This solution would enable people to continue to use electricity, which is an important source of energy in our society, without moving wires, homes, or businesses. Options (1) and (2) are not practical, since transmission lines are needed where people are located. Option (3) is not a likely solution because it would involve a complete change in the way we live. Option (5) is not only impractical and expensive, but would generate electromagnetic fields anyway.

50. (Application) **(5) avoid living near a high-voltage power line** These lines produce such strong electromagnetic fields, that a person being cautious would try to stay away from them. Option (1) is incorrect; both types of bulbs produce electromagnetic fields. Option (2) is incorrect because magnets produce magnetic fields, not electromagnetic fields. Option (3) is incorrect because the danger involved is electrocution, not exposure to electromagnetic fields. Option (4) is incorrect because a broken appliance is not working and therefore not producing a magnetic field.

51. (Evaluation) **(3) Using an electric blanket causes cancer.** Since the article states clearly that the dangers to health from electromagnetic fields have not yet been proved, it cannot be stated as fact that an electric blanket causes cancer. Options (1), (2), (4), and (5) are all true statements relating to information provided in the article. Note that the key words making option (2) true are "may be." In other words, option (2) states a possibility, not a fact.

52. (Evaluation) **(3) Electromagnetic fields cause health problems.** Although links between certain health conditions and electromagnetic fields have been suggested, so far this relationship has not been proved. Options (1), (2), (4), and (5) are established and proven principles of physics.

53. (Analysis) **(5) water vapor in the air** The water vapor in the air combines with the solid pollutants to form sulfuric acid and nitric acid, which are components of acid rain. Options (1), (2), and (3) are involved in the production or control of air pollutants; they cannot affect pollutants that have been released. Option (4) is incorrect because although the oceans are polluted by dry deposits and acid rain, they are not involved in the production of wet deposits.

54. (Analysis) **(4) The plants are weakened by the lack of nutrients and harmed by the accumulation of heavy metals.** The diagram shows how the tree is a victim of the action of acidified soil as nutrients are removed and harmful heavy metals are concentrated. Option (1) is incorrect because the diagram indicates that heavy metals are harmful in quantity. Option (2) is incorrect because there are no extra nutrients in the soil. Option (3) is incorrect because the entire plant is weakened, not just the upper part. Option (5) is incorrect because the acidified soil interferes with the normal interaction of plants and soil.

55. (Analysis) **(3) color blindness is related to whether a person is male or female** Color blindness is an inherited characteristic that is linked with sex. Option (1) is not true. Options (2) and (4) are untrue, and they are also unrelated to the distribution of color blindness between the sexes. Option (5) is true under most circumstances, but it is unrelated to the greater incidence of color blindness among men.

56. (Application) **(4) railroad engineer** Since railroad engineers must be able to distinguish red from green on railroad signals, a color-blind person would not be considered for such a position. Options (1), (2), (3), and (5) involve occupations in which the ability to distinguish red from green is not critical.

57. (Analysis) **(4) HIV** The virus HIV is the cause of AIDS. Options (1), (2), and (3) are not causes of AIDS but activities associated with the spread of the AIDS virus. Option (5) involves substances that are produced by the body to fight HIV.

58. (Evaluation) **(3) A person with HIV can infect others without being aware of it.** Since the infection is not active

during the second stage, which can last for years, many people who assume that they are healthy can unknowingly spread the virus. Option (1) is incorrect because AIDS is spreading into other parts of the population through the sex partners of people in the high-risk groups. Option (2) is incorrect because nothing in the article implies that researchers are near discovery of a cure. Option (4) is incorrect because people do not recover from AIDS. Option (5) is incorrect because intravenous drug users and the sex partners of people in the high-risk groups are also women.

59. (Analysis) **(1) exposure to HIV** HIV antibodies are produced only in response to infection by HIV. Option (2) is incorrect because most transfused blood does not contain HIV. Options (3), (4), and (5) are symptoms of the later stages of AIDS.

60. (Analysis) **(3) The disease will spread into previously low-risk groups.** Because the sexual contacts of people in the high-risk groups are not limited to other members of the high-risk groups, AIDS is increasingly found in people thought to be at little or no risk. Option (1) may or may not be true. Option (2) is not true; blood is now screened for HIV to make it safe for transfusion. Option (4) has not been true up to now, and nothing indicates that AIDS will become less serious. Option (5) is not true.

61. (Application) **(3) pickles** Vinegar and salt are used to preserve cucumbers, a process that results in pickles. Option (1) is flour containing supple-

mentary minerals and vitamins. Option (2) is milk processed so that the cream is distributed throughout rather than just remaining on the top; the process does not preserve the milk. Option (4) is added to change the taste of food; it is not a food. Option (5) is a product with a supplementary mineral.

62. (Application) **(1) yellow food coloring in commercially prepared baked goods that are made without eggs** The rich yellow color makes the product look as if it contains eggs. Options (2), (3), (4), and (5) do not affect the appearance of food.

63. (Analysis) **(4) A and B** Since most people no longer grow their own food, food additives are important in preventing the spoiling of food that is transported long distances and stored for long periods. In addition, some food additives provide minerals and vitamins that may otherwise be lacking in the diet. Most food additives do not cause disease.

64. (Evaluation) **(3) Kerosene boils at a higher temperature than gasoline.** The diagram shows that kerosene boils between 175°C and 325°C, and gasoline boils between 40°C and 174°C. Option (1) is incorrect because the diagram shows that asphalt remains solid in the tower. Option (2) is incorrect because grease is shown as a separate substance made from petroleum. Option (4) is true, but it is not supported by the information provided. Option (5) may be true, but it is not supported by the information provided.

65. (Analysis) **(5) the different temperatures at different heights in the tower** Variations in temperature mean that different substances will boil and condense at different levels, thus separating out. Option (1) is contradicted by the diagram. Option (2) simply causes the petroleum to move, not separate. Option (3) is not true; the process has nothing to do with filters. Option (4) is not correct; the pipes simply transport the substances.

66. (Evaluation) **(3) The higher the location in the tower, the lower is the temperature.** The diagram shows the temperature decreasing as the height increases. Options (1) and (2) are contradicted by the diagram. Options (4) and (5) are not true because some substances remain liquid or solid at all temperatures in the process.

SIMULATED GED TEST Correlation Chart

Science

The chart below will help you determine your strengths and weaknesses in thinking skills and in the content areas of life science, Earth science, chemistry, and physics.

Directions

Circle the number of each item that you answered correctly on the Simulated Test. Count the number of items you answered correctly in each column. Write the amount in the Total Correct space for each column. (For example, if you answered 8 comprehension items correctly, place the number 8 in the blank above *out of 11*.) Complete this process for the remaining columns.

Count the number of items you answered correctly in each row. Write that amount in the Total Correct space for each row. (For example, in the Life Science row, write the number correct in the blank before *out of 30*.) Complete this process for the remaining rows.

Cognitive Skills / Content	Comprehension	Application	Analysis	Evaluation	Total Correct
Life Science (pages 38–131)	5, **11**, 30, 45, 48	9, **17**, **18**, **19**, **20**, **40**, 47, 56, **61**, **62**	6, 31, 32, **39**, 46, 55, 57, 59, 60, **63**	7, 8, 10, **12**, 58	_____ out of 30
Earth Science (pages 132–179)	13, 14, 15, **36**	**2**, 27, 35	1, 16, **26**, **33**, 53, 54	**34**	_____ out of 14
Chemistry (pages 180–205)	42	**28**, 37, 41	3, 4, 38, **65**	**29**, **64**, **66**	_____ out of 11
Physics (pages 206–241)	**43**	22, 24, 50	21, 25	**23**, **44**, 49, 51, 52	_____ out of 11
Total Correct	_____ out of 11	_____ out of 19	_____ out of 22	_____ out of 14	Total correct: ____ out of 66 1–54 → Need more review 55–66 → Congratulations! You're ready!

Boldfaced items are based on diagrams, maps, charts, or graphs.

If you answered fewer than 55 questions correctly, determine which areas are hardest for you. Go back to the *Steck-Vaughn GED Science* book and review the content in those areas. In the parentheses under the item type heading, the page numbers tell you where you can find specific instruction about that area of science in the *Steck-Vaughn GED Science* book.

GLOSSARY

acceleration an increase in speed

acid a substance that releases hydrogen ions in a water solution; (example: vinegar)

activation energy the energy that must be added to start a chemical reaction

adaptation a process that helps an organism survive

adaptive radiation the process by which many species evolve from one species

adrenal gland an endocrine gland that secretes the hormones adrenaline and cortisone

adrenaline a hormone that prepares the body to meet emergencies

air a colorless, odorless, and tasteless mixture of gases; the lower atmosphere

air mass a large body of air that has the same temperature and moisture throughout

alimentary canal the tube that makes up the digestive system

alveoli small air sacs in the lungs

amplitude the distance between the rest position and crest of a wave

anaphase the fourth stage of mitosis, or cell division

angiosperms flowering seed plants; (examples: fruit trees and lilies)

antibodies substances produced by the body that fight disease

aorta a large artery that carries oxygen-rich blood from the heart to the body

arterioles small blood vessels that branch from arteries

artery a blood vessel that carries blood away from the heart

atmosphere the layers of gases that surround the Earth

atom the smallest particle of an element that has all the properties of that element

atomic number the total number of protons in the nucleus

atria the upper chambers of the heart

bacilli rod-shaped bacteria

bacteria small one-celled organisms

base a substance that releases hydroxide ions in a water solution; (example: milk of magnesia)

biome an ecosystem that covers a large area of the world; (example: a desert)

biosphere the thin layer of the Earth where life exists

boiling point the temperature at which a liquid changes into a gas

bronchi the two tubes that run from the trachea into each of the lungs

bronchioles small tubes that branch off from the bronchi in the lungs

byproduct a waste material produced during a process; (example: oxygen is a byproduct of photosynthesis)

cambium a one-cell-thick layer of growth cells in stems, branches, and roots of woody plants

capillary a tiny blood vessel with walls one cell thick

capsid the outer protein layer of a virus

capsule the thick outer slime layer of a bacterium

carbon monoxide a gas produced when fossil fuels do not burn completely

cell the microscopic unit of matter that makes up all living organisms

cell membrane a thin layer of matter enclosing a cell

cell wall a layer of matter outside the cell membrane that supports and protects plant cells

chemical reaction a process in which the properties of one or more substances change, resulting in a new substance

chemistry the study of matter and changes in matter

chlorophyll the green coloring of plants that is needed for photosynthesis

chloroplast the structure in a plant cell that contains chlorophyll

chromatin the part of a cell's nucleus that contains hereditary information

chromosome a part of a cell that contains hereditary information and is visible during mitosis, or cell division

cinder cone volcano a type of volcano made of cinders and rock particles that forms from explosive eruptions

circuit a complete path in which an electric current flows

cocci bacteria shaped like spheres

composite volcano a type of volcano made of alternating layers of hardened lava and cinders with alternating quiet and violent eruptions

compound the chemical combination of two or more elements; (example: water is a compound of hydrogen and oxygen)

compression the pushing together of molecules in a longitudinal wave

condensation the changing of matter from a gas to a liquid

continental polar air mass a large body of air originating over land near the poles; (example: over northern Canada)

continental tropical air mass a large body of air originating over land near the equator; (example: over Mexico)

cortisone a hormone that maintains salt balance

covalent bond a chemical combination of two or more atoms in which electrons are shared

crest the high point of a wave

crust the outer layer of the Earth, including the Earth's surface

culture in laboratory work, a sterile, pure food medium

cytoplasm the parts of the cell (other than the nucleus) that carry out the cell's activities

daughter cells the cells produced as a result of mitosis

decomposers in a food chain, organisms that break down dead plants and animals; (examples: bacteria and fungi)

denitrification the process by which certain bacteria change nitrates into gaseous nitrogen

density a measure of mass in relation to volume

dicotyledon a flowering plant whose seedling has two leaves; (example: a tomato)

direct respiration the process by which single-celled organisms exchange gases directly with the environment

distillation the process of removing dissolved material from a liquid

DNA the molecule that contains hereditary information and controls the activities of each cell

dominant trait a trait that will appear in an offspring if a parent contributes it; dominant traits will hide recessive traits

drag the friction created when objects move through air

earthflow the slow, downhill movement of soil and plants

earthquake the shaking and trembling that results from sudden movements of rock deep within the Earth

Earth science the study of the Earth and the space around the Earth

echolocation a process in which an animal locates objects and food by producing a sound that is reflected to the animal

ecosystem a selected area where living and nonliving things interact; (example: a swamp)

electric charge a basic property of protons and electrons

electric circuit a continuous, unbroken pathway over which electric current can flow

electric current the flow of electrons through a wire

electric field the area of force that surrounds a charged particle

electromagnetic wave a wave that does not need a medium through which to travel; (example: a light wave)

electromagnetism the relationship between electricity and magnetism

electron a tiny, negatively charged particle that revolves around the nucleus of an atom

element a substance that cannot be broken down into simpler substances by chemical means; (examples: hydrogen and oxygen)

embryo the early stage of development of a fertilized egg

endocrine system the glands that secrete substances that affect the functions of other parts of the body

endoplasmic reticulum a system of tubes that carries materials throughout a cell

endothermic reaction a chemical reaction that absorbs energy

energy the ability to move matter from one place to another or to change matter from one substance to another

environment the surroundings in which an organism lives

epicenter the point on the Earth's surface directly above the focus of an earthquake

era the largest period of geological time

erosion the gradual wearing away and moving of rock, soil, and sand along the Earth's surface

esophagus the tube that connects the mouth and stomach

estrogen a hormone that controls the development of secondary sex characteristics in females

evaporation the process of changing a liquid into a gas; (example: water to vapor)

evolution an orderly change; usually refers to the development of a species over time

exothermic reaction a chemical reaction that releases energy

external respiration the exchange of gases between the environment and the blood

extinction the disappearance of a species

fault a break in the Earth's crust

first law of motion a law stating that objects at rest tend to stay at rest, and objects in motion tend to stay in motion, until they are acted upon by outside forces

flagella whiplike structures that move bacteria through water and other fluids

focus the point under the Earth's surface where rocks break and move during an earthquake

food chain the movement of food through an ecosystem

force a push or pull acting on an object; (example: friction)

fossil the preserved remains of a once-living thing

fossil fuel a source of energy formed over long periods of time from the remains of plants and animals; (example: petroleum)

freezing point the temperature at which a liquid changes to a solid

frequency the number of waves that pass a given point in a given amount of time

friction a force that slows a moving object

galaxy a group of millions or billions of stars; (example: the Milky Way)

gamete the sexual reproductive cell formed during meiosis that contains half the chromosomes of the parent cell

gas a form of matter that has no definite shape or volume; (example: oxygen)

gene a part of the genetic molecule that determines a particular trait

gene frequency how often a particular gene occurs in a population

genetic code the particular structure of DNA that determines the traits of an organism

genetic drift the chance increase or decrease in the frequency of a particular gene

genetics the study of inherited characteristics

genotype the genes that are inherited

germination the process by which seeds sprout

glacier a large mass of moving ice

gonad a reproductive organ

gravity a force of attraction that exists between any two objects in the universe

gymnosperms nonflowering seed plants; (example: pine trees)

heart the major organ of the circulatory system that pumps blood through the system

herbicide a chemical that kills plants

hereditary capable of being passed from a parent to an offspring

homologous structures body parts from different organisms that have similar structure but perform different functions; (example: a bird's wing and a human's arm)

hormone a substance secreted by the endocrine system that affects the function of other parts of the body; (example: adrenaline)

host cell a living cell that provides food and energy to another cell

hybrid an organism that has a mixture of dominant and recessive traits

hypothesis a theory considered to be true for the purpose of investigation

ice age a period of thousands of years during which the Earth's climate grows colder and the polar ice caps spread north and south

igneous rock a type of rock formed when hot liquid rock cools into crystals; (example: granite)

immune system the body's main defense against disease

indirect respiration the process by which many-celled organisms exchange gases with the environment

inertia the tendency of an object to keep moving or remain at rest

infectious capable of being spread from one organism to another; (example: the cold virus is infectious)

inner core the solid iron and nickel center of the Earth

insulin a hormone that decreases the level of sugar in the blood

internal respiration the exchange of gases between the blood and the body's cells

interphase the first stage of mitosis, or cell division

ion a charged particle formed from an atom

ionic bond a chemical combination of two or more atoms in which electrons are transferred from one atom to another

joule a measurement of work

kinetic energy the energy of an object in motion

kingdom one of the four main groups of living things; (example: Kingdom Plantae)

landslide a form of rapid mass wasting in which rocks and soil fall quickly down a mountain

lava molten rock that breaks through the surface of the Earth in a volcanic eruption

leucoplast the structure in a plant cell that helps change glucose into starch and stores the starch

life science the study of living things and how they interact with each other and their environment

liquid a form of matter that has definite volume but no definite shape; (example: water)

longitudinal wave a wave that pushes and pulls molecules back and forth parallel to its direction

lymphocyte a type of white blood cell that produces antibodies

machine a device that helps people do work by changing the force or distance (or both) involved in a task; (example: a lever)

magma hot liquid rock beneath the Earth's surface

magnetic field the area of force that surrounds a magnet

malnutrition a poor state of health that results from an unbalanced diet

mantle a layer of the Earth below the crust

maritime polar air mass a large body of air originating over the ocean near the poles

maritime tropical air mass a large body of air originating over the ocean near the equator

mass amount of matter

mass wasting the downhill movement of rocks and soil caused by gravity

matter anything that has mass and takes up space

mechanical advantage the amount of help a particular machine provides

medium the matter through which a wave travels

meiosis the process in which sexual reproductive cells are formed

melting point the temperature at which a solid changes into a liquid

meristems areas in seed plants that contain growth cells

mesosphere the layer of the atmosphere above the stratosphere

metamorphic rock a type of rock formed when other rocks are subjected to extreme heat or pressure; (example: gneiss)

metamorphosis the process by which the immature form of an organism turns into a very different adult form; (example: caterpillars metamorphose into moths or butterflies)

metaphase the third stage of mitosis, or cell division

mitochondria in a cell, rod-shaped structures that produce most of the cell's energy

mitosis the process by which a cell divides into two daughter cells identical to the parent cell

mixture a combination of elements in which each element keeps its own properties; (example: air)

molecule the smallest particle of a compound that has all the properties of that compound; (example: a water molecule contains two atoms of hydrogen and one atom of oxygen)

monocotyledon a flowering plant whose seedling has one leaf; (example: grass)

mudflow a form of rapid mass wasting in which mud slides down a slope

natural selection the idea that individuals having characteristics that help them adapt to their environment are more likely to survive and reproduce

neutron a particle without a charge in the nucleus of an atom

newton a measure of force

nicotine a poisonous chemical found in tobacco smoke

nitric oxide a gas released by burning gasoline

nitrification a process by which certain soil bacteria convert the nitrogen from decomposing organisms into forms that plants can use

nitrogen fixation a process by which certain bacteria take nitrogen from the atmosphere and combine it with other substances into a form plants can use

nonrenewable resource a resource that cannot be replaced once it is used up; (example: coal)

nuclear fission the splitting of an atomic nucleus into two smaller nuclei

nuclear membrane in a cell, the layer of matter that separates the nucleus from the cytoplasm

nucleolus the part of a cell's nucleus that helps make protein

nucleus in life science, the part of the cell that controls cell activities; in chemistry, the small dense core of an atom that consists of protons and neutrons

offspring the direct descendants of an animal or plant

ohm a measure of resistance to the flow of electrons

ore a rock that contains valuable minerals

organ a group of different tissues working together

organelle a structure in a cell; (example: mitochondria)

organic molecules molecules that contain carbon combined with nitrogen, hydrogen, or oxygen, which are the building blocks of all living things

organ system several organs working together to perform a specific function; (example: the digestive system)

outer core the layer of the Earth between the mantle and the inner core

ovaries the female reproductive endocrine glands

oxytocin a hormone that causes uterine contractions during labor

ozone a form of oxygen in the atmosphere that absorbs ultraviolet sunlight

pancreas an endocrine gland that secretes the hormone insulin

parallel circuit an electric circuit in which the current divides and flows in two or more separate paths

parathormone a hormone that regulates the body's use of calcium and phosphorus

parathyroid gland an endocrine gland that secretes a hormone called parathormone

parent cell the cell undergoing mitosis, or cell division

particle accelerator a long, narrow tunnel, charged with electric and magnetic fields, used to accelerate and collide particles in order to release energy and create new particles

pathogen a microorganism that causes disease; (example: the AIDS virus)

periodic table an arrangement of the elements according to their properties

petroleum oil

pH scale a measurement from 0 to 14 of the strength of an acid or a base

phagocyte a type of white blood cell that fights disease-causing organisms by engulfing them

phenotype in genetics, the appearance of an individual

phloem a type of plant tissue that carries food from the leaves to other parts of the plant

photosynthesis the process by which plants use light energy to make food

physics the study of matter and energy and how they are related

pituitary an endocrine gland that produces growth hormone, oxytocin, and other hormones

plasmids small structures in bacteria that may contain genetic material

population density the number of people who live in a specific area

potential energy energy stored in the position of an object at rest

primary consumers in a food chain, animals that eat plants; (example: cows)

primary waves the fastest type of seismic waves

producers the lowest organisms in a food chain, usually green plants

product a substance formed as a result of a chemical reaction

prophase the second stage of mitosis, or cell division

proton a particle with a positive charge in the nucleus of an atom

pulmonary veins blood vessels that carry oxygen-rich blood from the lungs to the heart

Punnett square a chart used by scientists to predict which traits will be inherited

purebred in genetics, an organism that when bred always produces the same trait

radioactivity the release of energy and matter that results from changes in the nucleus of an atom

rarefaction the spreading apart of molecules in a longitudinal wave

ray a narrow beam

reactant a substance that is an ingredient of a chemical reaction

receptors nerve endings that receive stimuli

recessive trait a trait that does not appear when combined with a dominant trait, and that must be contributed by both parents in order to appear in an offspring

reflection the bouncing of light rays off a surface

reflex an automatic response to a condition in an organism's environment; (example: squinting in strong light)

refraction the bending of light rays as they pass from one medium to another

renewable resource a resource that can be replaced

resistance the opposition a material offers to the flow of electrons; in physics, a force that opposes or slows motion; (example: the weight of an object)

resources the things organisms need to live, such as water, food, and energy

respiration the process by which living things take in oxygen and release carbon dioxide to obtain energy

rhinoviruses a group of viruses that cause most colds

ribosome a part of a cell that has a role in making proteins

Richter scale a measure from 1 to 10 of the amount of energy released by an earthquake

rock cycle the process by which rocks continue to change from one type to another

salinity the amount of salt in ocean water

salt a neutral compound that results from the chemical combination of an acid and a base; (example: sodium chloride, or table salt)

secondary consumers in a food chain, animals that eat primary consumers; (example: lions)

secondary waves seismic waves that can travel through solids but not through liquids or gases

second law of motion a law stating that objects will accelerate in the direction of the force that acts upon them; mass and force will affect the rate of acceleration

sedimentary rock a type of rock formed by the hardening of particles of sand, mud, clay, or other sediments; (example: sandstone)

seismic waves vibrations caused by movement of rock during an earthquake

semicircular canals structures in the ears that help maintain balance

septum the wall of tissue that separates the right and left sides of the heart

series circuit an electric circuit in which there is only one path for the electric current

shield volcano a type of volcano with quiet lava flows that form a gently sloping mountain made of hardened lava

slime layer the outer covering of a bacterium

soil creep a form of mass wasting in which soil particles gradually move downhill

solid a form of matter that has definite shape and volume; (example: brick)

solubility the greatest amount of a solute that will dissolve in a given amount of solvent at a certain temperature

solute the substance being dissolved in a solution

solution a mixture in which two or more substances are dissolved in one another; (example: salt water)

solvent the substance doing the dissolving in a solution

speciation the development of a new species from an old one

species a group of genetically similar organisms that can mate and produce fertile offspring; (example: dogs)

spindle protein fibers formed during cell division, or mitosis

spirilla corkscrew-shaped bacteria

stomach the part of the alimentary canal where fat begins to be broken down, minerals are dissolved, and bacteria in food are killed by stomach acids

stratosphere the layer of the atmosphere above the troposphere

sulfur dioxide a gas released by the burning of fossil fuels

sunspots dark spots that appear on the Sun's surface

surface waves the slowest type of seismic waves that travel from the focus up to the surface of the Earth

synthetic not found in nature

talus slope the rocks and soil that come to rest at the bottom of a mountain or hill

taxonomy the classification of organisms into groups based on evolutionary relationships

telophase the fifth and final stage of mitosis, or cell division

temperature a measure of heat content

tertiary consumers in a food chain, animals that eat secondary consumers

testes the male reproductive endocrine glands

testosterone a hormone that controls the development of secondary sex characteristics in males

thermal energy conversion a process of producing energy, in which the difference in temperature between surface and deep seawater is used to generate electricity

thermodynamics the study of the relationship between heat and the energy of motion

thermosphere the uppermost layer of the atmosphere

thymine a nitrogen base that is paired with adenine in DNA

thymosin a hormone that may affect the formation of antibodies in children

thymus an endocrine gland that produces the hormone thymosin

thyroid gland an endocrine gland that produces the hormone thyroxin

thyroxin a hormone that controls how quickly food is converted to energy in cells

trace fossil the evidence or mark of a living thing; (example: footprints in rock)

trachea the tube through which air passes from the back of the mouth to the lungs

trait an inherited characteristic of an organism

troposphere the lowest layer of the atmosphere

trough the low point of a wave

vacuole storage spaces in a cell that contain food or water, or that collect and excrete waste

vaporization the process of adding heat to a liquid in order to change it into a gas

vein a blood vessel that carries blood back to the heart

velocity the distance covered in a specific unit of time

vena cava a large vein that carries oxygen-poor blood into the heart

vent an opening in a volcano

ventricles the lower chambers of the heart

venules small blood vessels that carry blood into the veins

vestigial structure a body structure that is poorly developed or not functioning but that is similar to a well-developed, functioning structure in another organism; (example: the vestigial leg bones of some snakes)

virions new virus particles produced by a virus that takes over a host cell

virus a single molecule of genetic material surrounded by a coat of protein

volcano the place where lava breaks through the Earth's surface

volt a measure of the strength of the source of energy producing an electric current

voltage the source of energy for an electric current

water cycle continuous movement of water from the Earth's surface to the air, then back to the surface again

wave a disturbance that travels through space or matter; (example: sound waves)

wavelength the distance between the crests of two consecutive waves

weathering the process by which large rocks are broken down into smaller rocks

wind moving air

work the result of a force moving an object

xylem a type of plant tissue that carries water upward from the roots to the stems and leaves

zygote the cell that is produced by the combining of two gametes during sexual reproduction

INDEX

rhinoviruses, 96
ribosomes, 42
Richter scale, 168
rock, 140–41, 145, 176
rock cycle, 176

S

salinity, 151
salt, 181, 197
    sodium chloride, 151
sand bars, 164
secondary cell wall, 46
secondary consumer, 112
secondary seismic wave, 168
second law of motion, 210
sediment, 138, 166, 176
sedimentary rock, 140, 176
seed plant, 72, 76–77
seismic wave, 136, 168
semicircular canal, 87
septum, 84
series circuit, 239
shape, of wave, 230
shield volcano, 170, 252
shock, 61
skin, 41
slime layer, 94
small intestine, 82
smelting, 141
sodium bicarbonate, 198
sodium carbonate, 198
sodium chloride, 151
soil bacteria, 56–57
soil creep, 166
solar system, 134–35
solid, 184–185, 202–3
solubility, 196
solute, 181, 196
solution, 181, 196
solvent, 181, 196
sound waves, 207, 230, 232
speciation, 66
species, 39, 66, 111
spindle, 44
spirilla, 94
spits, 164

starch, in photosynthesis, 46
static electricity, 218
stomach, 82
stratosphere, 146, 174
sugar
    in DNA, 106
    in photosynthesis, 46, 52, 57
    in respiration, 54
sulfur, in atmosphere, 159
sulfur dioxide, 117
Sun, 112, 149, 155
sunspots, 155
supporting detail, 40
supporting statement, 110
surface seismic waves, 168

T

talus slope, 166
taxonomy, 74
telophase, 44
temperature, 144
tertiary consumer, 112
testes, 86
testosterone, 86
theory of evolution, 62
theory of origin of life, 64–65
thermal energy conversion, 177
thermodynamics, 207, 212–13
thermosphere, 146, 174
thymine, 106
thymosin, 86
thymus, 86
thyroid gland, 70–71, 86
thyroxin, 86
topic sentence, 40
trace fossil, 138
trachea, 55
trait, 64, 67, 102
trees, 47
tree surgery, 47
trichina worm, 81
trichinosis, 81
troposphere, 146, 174
trough, of wave, 230
TSH hormone, 86

U–V

undernutrition, 51
unstated assumption, 80
vacuole, 42
valley, erosion of, 169
vaporization, 184
vein, 84–85
velocity, of wave, 231
vena cava, 84
ventricle, 84
venule, 85
vestigial structure, 125
virion, 95
virus, 90, 92, 94–95, 96–97
vitamin C, 97
volcano, 170–71
voltage, 222–23
volts, 222–23

W–Z

water
    in the Earth's atmosphere, 146
    in photosynthesis, 46, 52, 57, 75
    in respiration, 54
    pollutants in, 110
    in seed germination, 77
water cycle, 123, 133, 150
Watson, James D., 106
wave, 207
    characteristics of, 230–33
wavelength, 231
weathering, 145
weight gain during pregnancy, 50–51
white light, 235
wind, 158
work, measuring, 209
xylem, 75
zygote, 45